AF615814

# The genus *Littoraria*
# (Mollusca, Gastropoda, Littorinidae)

# The littorinid molluscs of mangrove forests in the Indo-Pacific region

## The genus Littoraria

David G. Reid

British Museum (Natural History)

First published in 1986 by
British Museum (Natural History)
Cromwell Road, London SW7 5BD



Publication No. 978

British Library Cataloguing in Publication Data

Reid, David G.
The littorinid molluscs of mangrove
forests in the Indo-Pacific Region
The genus *Littoraria*
1. Gasteropoda——Classification
I. Title    II. British Museum (Natural History)
594'.3    QL430.4

ISBN 0-565-00978-5

Printed and bound in Great Britain by
Butler & Tanner Ltd, Frome and London

**Frontispiece** Shell colour polymorphism in *Littoraria* species (x1.8). From top:
Row 1: *Littoraria* (*Littorinopsis*) *filosa* (Sowerby), Magnetic I., Queensland (DGR).
Row 2: *Littoraria* (*Littorinopsis*) *philippiana* (Reeve), Magnetic I., Queensland (DGR).
Row 3: *Littoraria* (*Littorinopsis*) *pallescens* (Philippi), Ao Nam-Bor, Phuket I., Thailand (DGR).
Row 4: *Littoraria* (*Littorinopsis*) *luteola* (Quoy & Gaimard), Kurnell, New South Wales (DGR).
Row 5: *Littoraria* (*Lamellilitorina*) *albicans* (Metcalfe), Santubong, Sarawak (DGR).

# CONTENTS

# SYNOPSIS

The taxonomy of the *'Littorina scabra'* species complex is revised. Twenty species are recognized from mangrove forests in the Indo-Pacific region and are assigned to the genus *Littoraria* Griffith & Pidgeon. One new subgenus, two new species and one new subspecies are described. Complete synonymies, descriptions of shell, radula and reproductive anatomy, accounts of habitat and of geographical distribution are provided for each species. The most useful taxonomic characters of the shell are sculpture, microsculpture and columellar form. Amongst the anatomical characters, reproductive anatomy is emphasized and features of the penis, sperm nurse cells and pallial oviduct are used to define species and to differentiate four subgenera. In the genus as a whole, thirty-six Recent species are recognized worldwide. A cladistic analysis of the phylogeny of the genus is presented, and possible relationships with other genera in the family Littorinidae are discussed.

# ACKNOWLEDGEMENTS

I wish to thank all those who by their help have made the present study more complete. In particular, my grateful thanks to W.F. Ponder (AMS) for discussion, advice and encouragement, and also to J.D. Taylor (BMNH) and J. Rosewater (USNM). By his numerous publications on the family Littorinidae, J. Rosewater has laid all the foundations for the present work and has generously made available his own unpublished material and specimens. For permission to study the collections in their care, I thank: the curators and staff of BMNH, AMS, USNM, K. Boss (MCZ), C. Christensen (BPBM), G.M. Davis (ANSP), C.M. Yang (NUS), the Linnean Society of London; and for the loan of specimens: P. Bouchet (MNHNP), S. Boyd (NMV), E. Gittenberger (RNHL), R.N. Kilburn (NM), T. Okutani (NSMT), G. Oliver (NMW), T. Schiøtte (Zoologisk Museum, København), J. Stanisic (QM), N.V. Subba Rao (ZSI), C. Vaucher (MHNG) and F.E. Wells (WAM). Specimens were received from: D.R. Bellwood, K. Fujiwhara, S.T. Garnett (all James Cook University of North Queensland), B.S. Morton (University of Hong Kong), M. Nishihira (Kyoto University) and Z. Wang (Institute of Oceanology, Academia Sinica). This work was carried out in the Department of Zoology, James Cook University of North Queensland, and I thank the Head of the Department, C. Burdon-Jones. Elsewhere, laboratory facilities were provided by: J. Hylleberg (Phuket Marine Biological Center, Thailand), E.A. Kay (University of Hawaii), J.E. Ong (Universiti Sains Malaysia, Penang) and A. Sasekumar (Universiti Malaya, Kuala Lumpur). For help in the field I am grateful to M. Gilham (Darwin, N.T.), S. Pripanapong (Kanchanadit, Thailand) and N. Sarti (Dept. of Fisheries and Wildlife, Boome, W.A.). For assistance with techniques of electron microscopy and histology I thank J. Darley and L. Winsor respectively, and for German translation A. von Wallenstern (all James Cook University of North Queensland). In addition to many of the above, I am indebted to R. Cleevely (BMNH), V. Fretter, A. Graham (both Univeristy of Reading), R.S. Houbrick (USNM), N.J. Morris (BMNH), W.B. Rudman (AMS) and R.D. Turner (MCZ) for useful discussion.

This study was made possible by the award of a scholarship for postgraduate research from the Drapers' Company of London, to which charitable institution my deepest gratitude. I thank the trustees of the Australian Museum for the Keith Sutherland Award which financed travel around Australia. Financial support was also received as a grant from the short-term visitor programme of the Smithsonian Institution, Washington.

My coverage of the literature was made more complete by the use of the bibliography of the family Littorinidae by C.W. Pettitt (1974*a,b*, 1979), and by access to the looseleaf system of the Department of Malacology, AMS, compiled by W.F. Ponder and the late C. Hedley.

# INTRODUCTION

It has long been recognized that within the family Littorinidae a group of predominantly tropical species can be distinguished by the possession of large, thin shells, which are often brightly and variably coloured. In addition, many of these species are remarkable for their occurrence in mangrove forests and for their supposed adaptations to a largely terrestrial life. The most well known member has hitherto usually been referred to as '*Littorina scabra*', but in the present work the group is recognized as the genus *Littoraria*. Although formally based upon several uniquely derived anatomical features, this classification also reflects the distinctive shells and specialized ecology of the component species.

In the present treatment of the genus, anatomical features have been examined in detail for the first time. This has led to the recognition of many more species than have been admitted by previous authors. While all Recent species of *Littoraria* have been revised, the major change to accepted classification has been the recognition of twenty species occurring in mangrove forests of the Indo-Pacific province, where only three were commonly recognized previously. Only these species, here collectively referred to as the *scabra* group, will be described in detail in the taxonomic section. Other species of *Littoraria* (as listed p. 72) are more well known; only minor changes to their classification are suggested and they are discussed here mainly for purposes of comparison.

While most members of the family occur on intertidal rocks, the majority of *Littoraria* species are found upon mangrove trees, salt marsh vegetation, driftwood and wooden pilings. Several of these species can also occasionally be found on intertidal rocks in sheltered situations. A few species of the genus typically occupy the high intertidal zone of more exposed rocky shores (p. 73), but these will not be considered in detail here. It is in mangrove forests that the group attains its greatest diversity and abundance, and very few other littorinids are found there. Of all marine molluscs associated with the mangrove habitat, *Littoraria* species are amongst the most ubiquitous, being found even upon isolated trees on rocky and coral shores.

Although most littorinids inhabit the intertidal zone and are wetted at least by high spring tides, certain *Littoraria* species show amongst the most nearly terrestrial habits in the family. These snails may be found up to 5 m above the ground in the branches of trees, and a few occur even at the landward edges of swamps. All will climb vertically to avoid submersion by a rising tide, so that respiration is at all times in air. Perhaps in consequence the gill leaflets are somewhat smaller in size than those of intertidal *Littorina* species. Another unusual adaptation, thought by some authors to be associated with semi-terrestrial life, is the ovoviviparity of some *Littoraria* species. In these forms the embryos are retained between the gill lamellae in the mantle cavity until the early veliger stage. However, even at the highest tidal levels, some species continue to produce pelagic eggs.

A number of littorinids, in particular members of the genus *Littorina* in Europe, have been noted for their extreme variability in shell colour and sculpture. The *Littoraria scabra* group has been considered equally polymorphic, with yellow, pink or brown, and variously patterned shells, ranging in sculpture from smooth to strongly carinate. There has, however, been a recent trend to re-evaluate some of the supposed polymorphic species of the family.

The history of the *Littorina saxatilis* complex may be taken as an instructive example. During the late eighteenth and early nineteenth centuries nineteen species were described (Fischer-Piette & Gaillard, 1971), but this diversity was reduced, on the basis of shell characters, to seven subspecies and twelve varieties by Dautzenberg & Fischer (1912). As interest in polymorphism and variation increased, the distribution of the colour and form varieties was studied, notably in the long series of papers by Fischer-Piette & Gaillard (e.g. 1971). James (1968) used anatomical characters, including the penis, radula and pigmentation pattern, as well as shell shape and colour, to define five subspecies, which were said to be distinct in Britain but to interbreed elsewhere. In subsequent investigations greater emphasis was placed on penial anatomy and the method of development, whether oviparous or ovoviviparous, leading to the recognition of three species within the complex (Sacchi, 1975; Heller, 1975*a*; Raffaelli, 1979). Genetic analysis based upon isoenzyme patterns has confirmed the interpretation of the morphological evidence, by demonstrating that the species are reproductively isolated (Wilkins & O'Regan, 1980). Most recently, Hannaford Ellis (1979) separated a new species from the well known *Littorina rudis*, primarily on the basis of female reproductive anatomy and the method of development, even though the shells of the two were sometimes indistinguishable. The validity of a species definition based entirely upon reproductive anatomy and method of development has been questioned (Caugant & Bergerard, 1980; S. M. Smith, 1982), but in this case is supported by isoenzyme analysis (Ward & Warwick, 1980) and differences in breeding seasons (Hannaford Ellis, 1983). The status of *Littorina saxatilis* (Olivi) itself, described from Venice, remains in doubt, but it will probably prove to be a senior synonym of *Littorina rudis* (Maton) (S. M. Smith, 1982; Raffaelli, 1982; Hannaford Ellis, 1983; but see J. E. Smith, 1981) and will be used as such here. The disjunct distribution is explained by the probable introduction of the species to Venice (S. M. Smith, 1982) as has also occurred in South Africa (Hughes, 1979*b*). Once discrete species were recognized within the *Littorina saxatilis* complex, it became possible to investigate ecological segregation of species and to consider the adaptive significance and maintenance of the shell polymorphisms (Heller, 1975*b*, 1976).

Other variable taxa which have recently been shown, on the basis of anatomical characters, to comprise several species, include *Littorina obtusata* (L.) (Sacchi & Rastelli, 1967), *Nodilittorina ziczac* (Gmelin) (Borkowski & Borkowski, 1969; Bandel & Kadolsky, 1982) and *Littorina scutulata* Gould (Murray, 1979). These case histories set precedents with important implications for any taxonomic work on the family Littorinidae. While certain shell characters are highly variable and subject to local adaptation, the range of variation within species may not be as great as has been supposed. Nevertheless, certain species cannot be separated using shell characters alone, and reproductive anatomy appears to be of primary taxonomic significance. Specimens must be examined from throughout the geographical range to determine the status of distant populations. The history of the *L. scabra* complex shows certain parallels with the example given.

The specific name *scabra* was published by Linnaeus in 1758, but as early as 1705 Rumphius had described and illustrated *'Buccinum foliorum'*, noting its habitat on mangrove trees. Between 1830 and 1857 thirty-one specific and varietal names were introduced for Indo-Pacific members of the *scabra* group. A further ten names appeared from 1871 until 1900, but previous to the three new taxa described herein only one new variety has been described this century. The first and most discriminating monographic treatment was that of Philippi (1847–1848), who recognized thirteen Indo-Pacific species in the *scabra* complex, of which seven are retained unchanged here. Although the first anatomical drawings were made by Quoy & Gaimard in 1833, Philippi described only shells. He correctly recognized the extreme colour variation of *L. angulifera* from the tropical Atlantic and described seven colour varieties of *Litorina 'scabra'* and three of *Litorina 'intermedia'*. These latter two species are now known to be rather uniform, and three of Philippi's varietal names are here raised to specific rank. The monograph of *Littorina* by Reeve (1857) increased the number of recognized species to eighteen, but since his species concepts were narrow, six of these fall into synonymy. Weinkauff (1878, 1882) made some attempt to synonymize, broadening the concept of *Litorina 'scabra'*, but essentially compiled the work of previous authors. The work of Nevill (1885) was not illustrated, his system of varieties and subvarieties is confusing, and his descriptions often inadequate. Nevertheless, with many specimens before him Nevill was able to make some sensible suggestions concerning variation, sexual dimorphism and synonymy.

The concept of *Littorina 'scabra'* as a single, widely variable, pantropical species was established by Tryon (1887). In addition to the nominate form (which covered five of the species here recognized), Tryon admitted a variety *lineata* (the tropical Atlantic species *L. angulifera*), a variety *intermedia* (comprising three of the smaller species) and a variety *filosa* (including all the colourful ribbed and carinate forms, amongst which seven species can be distinguished). Remaining names in the *scabra* group were distributed between ten other species. Working at about the same time, von Martens was not influenced by Tryon, but followed Philippi, so that his concepts of species were essentially correct, as shown by his list from the East Indies (von Martens, 1897).

Subsequent work, until 1965, consisted mainly of faunistic lists and, latterly, figures of shells in popular texts. Authors often followed Philippi and von Martens in recognizing several species (e.g. Casto de Elera, 1896; Hidalgo, 1904–1905; Annandale & Prashad, 1919; Prashad, 1921; Oostingh, 1927; Dautzenberg, 1929; Yen, 1942; Kuroda & Habe, 1952), but in doing so the earlier errors were perpetuated, particularly in regard to the several species confused under the name *Littorina 'intermedia'*, and few new contributions were made. Some authors followed Tryon's broad species concept (e.g. Fischer, 1891; Melvill & Standen, 1901; Dautzenberg & Fischer, 1905; Schepman, 1909). The first use of radular characters in the taxonomy of species of *Littoraria* was by Adam & Leloup (1938) who, on the basis of supposed similarity in radular teeth, reduced *Littorina 'filosa'* (=*L. pallescens*) to a variety of *L. scabra s. s.* A new standard in littorinid taxonomy was set by Abbott (1954) and Whipple (1965), who described not only the characters of the shell, but also of the radula, male reproductive anatomy and egg capsules. Whipple (1965) gave a description of *L. intermedia* (as *Littorina scabra*).

The comprehensive monograph of the Littorinidae of the Indo-Pacific by Rosewater (1970, 1972) has provided an invaluable source of reference for all subsequent studies of the family. This work established generic and subgeneric groupings based on penial and radular characters, but returned to Tryon's concept of *Littorina 'scabra'* as a widely variable species. Three species were admitted in the *scabra* complex, and placed together in the subgenus *Littorinopsis*. These were *Littorina 'scabra'* (here divided into seventeen species), *Littorina 'carinifera'* (here divided into two species) and *Littorina melanostoma*. Rosewater (1963, 1970, 1980*b*, 1981) regarded *Littorina 'scabra'* as a pantropical species with subspecies *Littorina 'scabra scabra'* in the Indo-Pacific, *Littorina scabra angulifera* in the Atlantic and *Littorina scabra aberrans* in the Eastern Pacific. The status of *L. angulifera*, described by Lamarck (1822) from the Caribbean, has been the subject of debate since 1833 when Quoy & Gaimard applied the name to the Indo-Pacific *L. scabra s. s.*, while in 1842 d'Orbigny determined the Caribbean shells as *Littorina scabra*. Subsequent authors, with such exceptions as Tryon (1887) and Bequaert (1943), have mostly recognized that *L. angulifera* is a distinct species, probably basing this decision largely upon the fact of its geographical isolation from the Indo-Pacific *L. scabra* group, since similarities of the shells have usually been stressed. More recently, the two have been separated on the basis of supposed radular differences (Marcus & Marcus, 1963; Bandel, 1974; but see Rosewater, 1980*b*), although no other anatomical comparisons have been made. Now that species of the *L. scabra* group are more clearly defined, it is evident that *L. angulifera* and *L. scabra s. s.* can easily be separated by shell characters alone, and that the penes of the two species are entirely different (p. 103). The anatomy of *L. aberrans* is as yet unknown, but the shell, and particularly the protoconch (p. 14), is sufficiently distinct that it must be given full specific status also.

Most subsequent authors have accepted Rosewater's classification, with the exceptions of Fischer (1970), Higo (1973) and Brandt (1974). Accounts of reproduction and ecology of some Indo-Pacific *Littoraria* species have been given by Abe (1942), Kojima (1958*c*), Struhsaker (1966), Berry & Chew (1973) and Muggeridge (1979), but in each case only one species was involved so that no taxonomic problems were raised. However, Nielsen (1976), describing zonation of littorinids in a mangrove forest in Thailand, observed that a large form with a white, wide columella (=*L. scabra s. s.*) occurred only at the seaward edge, while smaller shells (i.e. *L. pallescens* and *L. intermedia*) were found throughout the forest. Cook (1983) has examined proportions of colour morphs of *Littoraria* in a mangrove forest in New Guinea and separated three (unnamed) species using shell and penial characters (*L. pallescens, L. scabra, L. intermedia*).

The present work has developed from an ecological study of mangrove littorinids on Magnetic Island, Queensland, of which one of the aims was to investigate the extreme shell variation of *Littorina 'scabra'* as described by Rosewater (1970). At this locality it was found that shells could be classified into a number of discrete types, occupying different habitats or zones within the mangrove forest. Various anatomical characters were found to be correlated with the shell types, suggesting that these were distinct species. Differences were not merely direct effects of the environment on a plastic phenotype, for while habitats overlapped, phenotypic differences remained and no intermediate forms were encountered. Furthermore, transfer of young snails between habitats during manipulative ecological experiments failed to produce any change in morphology. The existence of a degree of reproductive isolation was suggested by differences in penial shapes and was confirmed by the observation of only a very low frequency of mis-pairing during copulation in the field. Extension of the morphological study to embrace the entire Indo-Pacific province showed that the species as defined in North Queensland remained distinct throughout their ranges of distribution. Although certain characters showed regional or clinal changes, penial shapes remained constant. Using those characters found to vary least within species, as discussed in the following sections, further species were then recognized in other geographical areas. In only a few cases were problems encountered in the classification of allopatric populations, probably because species are widely dispersed as pelagic larvae.

# MATERIALS AND METHODS

## Material, Types and Synonymies

This monograph is based largely upon material collected by the author throughout Australia, in South East Asia and other parts of the Indo-Pacific. In addition, all the collections of the following institutions have been examined: British Museum (Natural History); National Museum of Wales; Australian Museum; Queensland Museum; National Museum of Victoria; Western Australian Museum; National Museum of Natural History, Smithsonian Institution; Academy of Natural Sciences of Philadelphia; Museum of Comparative Zoology, Harvard University; Natal Museum; Rijksmuseum van Natuurlijke Historie; Zoological Research Collection, National University of Singapore; Sarawak Museum; and Phuket Marine Biological Center, Thailand.

All type specimens referred to in the synonymies have been examined unless otherwise noted. Holotypes of new species described herein, and also neotypes, are deposited in the British Museum (Natural History), with the exception of the new subspecies *L. cingulata pristissini*, located in the Australian Museum. Paratypes are deposited in both institutions, and in the National Museum of Natural History, Smithsonian Institution.

The status of type material of the many species and varieties described by Philippi deserves special comment. In 1846 Philippi published descriptions of new species of *Littorina* in the collection of H. Cuming, and these were subsequently figured in his monograph of the genus (1847–1848). Lectotypes of most of these species were designated by Rosewater (1970) from the Cuming Collection in the British Museum. Often Cuming's original label accompanies the specimens and in all such cases the specific name is inscribed in a different hand and followed, as in Philippi's monograph, by the abbreviation 'Ph.'. The handwriting has been authenticated as that of Philippi himself by comparison with labels in the collection of the Senckenberg Museum, Frankfurt (R. Janssen, pers. comm.). Usually Philippi acknowledged Cuming when he illustrated specimens from his collection, but did not always do so. Several additional species and varieties were described by Philippi in his monograph and in those cases in which material from Cuming was acknowledged, specimens in the British Museum can be identified as types with some confidence. Types of the remaining taxa are not present amongst the collections of Philippi in either the Museum für Naturkunde, East Berlin (R. Kilias, pers. comm.) or in the Museo Nacional de Historia Natural, Santiago, Chile (N. Bahamonde, pers. comm.). Philippi described two species from material received from Largilliert, and in these cases lectotypes (now housed in the MNHNP) have been designated from the Largilliert Collection in the Natural History Museum of Rouen. In a few cases specimens originating from localities mentioned by Philippi and closely resembling his figures, have been discovered in the British Museum, with Cuming's labels and named by Philippi as discussed above. On the basis of this evidence these specimens can be accepted as lectotypes. Types of the remaining species must be presumed lost, and Philippi's figures are designated as lectotypes. Lectotype designation is especially important for the varieties of *Litorina 'scabra'* and *Lito-*

*rina 'intermedia'* illustrated in Philippi's plate 5, of which figures 6 to 11 were incorrectly cited in the text, as has also been noted by Weinkauff (1878) and Nevill (1885).

Lectotypes of the taxa described by Nevill (1885), housed in the Zoological Survey of India and designated herein, were selected by N. V. Subba Rao on the basis of measurements given by Nevill.

Problems have been encountered in the compilation of synonymies, owing to the frequent confusion in the literature of species of similar appearance. Shell characters are often diagnostic, so that published figures can usually be correctly determined. In the absence of adequate descriptions or figures, references by authors to original descriptions and to earlier figures have aided interpretation of species concepts, and in a few cases the known geographical distributions have also been of use. All new names are included in synonymies, but where an original figure could not be identified with certainty or a description of a new species was inadequate, the entry is preceded by a query. Other doubtful references have been omitted. In each synonymy an attempt has been made to compile the major taxonomic works, significant contributions to the classification of the species and references which contain lengthy synonymies. Certain major faunal lists have been included, while for rare or unusual species the mere recognition of the species as distinct has been a sufficient criterion. Obvious spelling errors in specific names have not been listed separately, but are included under the corrected spelling. Of generic names, *Melarhaphe* and *Littorina* have been emended or misspelt by several authors (p. 72) and are entered separately in the synonymies.

## Methods

### *Shell characters*

Most of the recent taxonomic studies of littorinids have employed indices of shell shape derived as simple ratios of length, width and aperture size (e.g. James, 1968; Heller, 1976; Hannaford Ellis, 1979), but the precise measurements taken have varied. The parameters defined by Raup (1966) from a geometrical analysis of shell coiling have been used to examine shell variation in *Littorina 'saxatilis'* by Newkirk & Doyle (1975), but are not readily visualized and so are unsuitable for the purposes of descriptive taxonomy.

The measurements taken here are illustrated in Fig. 1. Shell height (H) is the maximum linear dimension of the shell from the apex to the anterior edge of the lip. Erosion of the shell seldom occurs in the mangrove habitat and apices are usually intact, so that shell height is a suitable measure of shell size. The species investigated here possess a prominent peripheral keel or a system of spiral ribs; shell breadth (B) was measured from the junction of the peripheral rib with the outer apertural lip to the corresponding point half a revolution earlier. In species with a flared lip, B was measured just behind the apertural expansion. These measurements of shell height and breadth were preferred to those parallel and perpendicular to the axis of coiling, for they are more accurately reproducible. Apertural length (LA) was measured as the maximum external dimension, and apertural width (WA) as the maximum external dimension perpendicular to LA. These four measurements were combined as follows to give indices of shell shape:

Shell proportion P = height H/breadth B

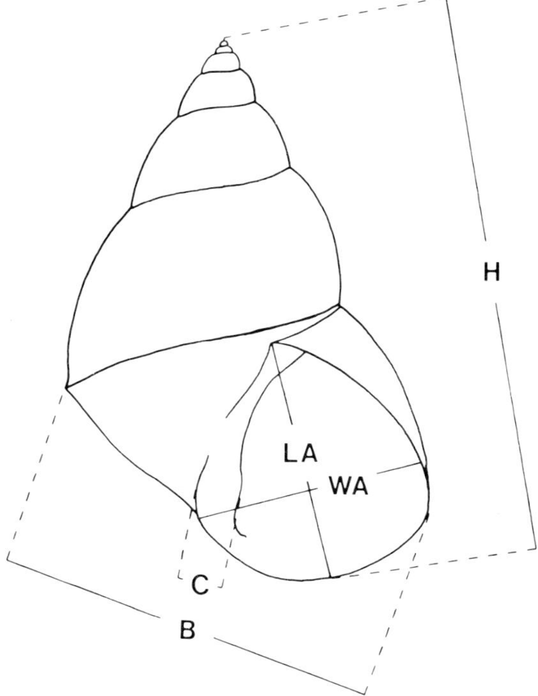


**Fig. 1** Shell dimensions: **B**, breadth; **C**, columellar width; **H**, height; **LA**, apertural length; **WA**, apertural width.

Apertural shape (circularity) S
= apertural width WA/apertural length LA
Spire height SH = height H/apertural length LA

Dimensions of type specimens and of lectotype figures are given in the species descriptions, together with measurements of a number of shells to indicate the range of size and shape encountered. Approximate columellar width (C) is also given; using a micrometer eyepiece the measurement was made perpendicular to the axis of the columellar pillar, from the mid point of its height to the furthest point of the inner apertural lip. All measurements were made on 'adult' shells, as defined by a thickening or flaring of the apertural lip and a laying down of opaque callus within. This development merely indicates a slowing or cessation of shell growth and does not necessarily correspond to the point at which sexual maturity is attained. In young, actively growing shells the peristome is thin and brittle.

Sexual dimorphism of adult shells was demonstrated by comparison of mean shell heights and of the three shape indices of a random sample of 10 male and 10 female shells from a representative locality. Ratios were compared using the non-parametric Mann-Whitney U-test, but to compare mean heights the more powerful parametric t-test was used (Zar, 1974). Means and standard errors are quoted for each variable. The sample size was small, but the object was to quantify dimorphisms which are often strikingly apparent, rather than to detect minute statistical differences.

In the species descriptions the spire is simply described as relatively tall or low (SH approximately greater than 2.0 or less than 1.7 respectively), depending upon both apical angle and aperture size. Spire outlines are described as convex if the apical angle decreases with whorl number, or as concave if the angle shows an increase. The ratio P provides an indication of the acuteness of the apical angle at the last whorl. The number of whorls of the teleoconch can be counted from the sinusigera ridge terminating the protoconch (e.g. Fig. 40b).

In the descriptions of shell sculpture, spaces between the grooves are referred to as ribs, whether or not they are prominent or rounded. Primary ribs and grooves are defined as those present up to and including the fourth whorl of the teleoconch. Secondary and higher orders of sculpture appear subsequently, formed by division or intercalation of ribs. A similar system of description of orders of sculpture has been used by Rosewater (1982) and Bandel & Kadolsky (1982). The majority of species possess a peripheral keel or an enlarged peripheral rib, at least until the last whorl, where the outline may become more rounded. As the shell grows, the suture with the succeeding whorl overlies the peripheral rib. The number of primary grooves quoted in the species descriptions is that present above the peripheral rib. Usually this will be the number visible on the early spire whorls, but sometimes the most anterior groove is hidden in the suture. It is then necessary to trace the primary grooves to the last whorl in order to be sure of the identification of the peripheral rib. In the few species with a rounded periphery and uniform ribs, the peripheral rib is defined as that to which the suture is attached on the penultimate whorl. Groove width is quoted as a fraction of the average rib width and, since relative groove width increases with whorl number, the figure given is the maximum value. On the last whorl ribs are more prominent than grooves and sculpture is described as the total number of ribs, including those on the base below the periphery. Other descriptions of sculpture on the last whorl refer only to the area between the suture and the periphery.

Details of shell microsculpture and of the protoconch are visible at low magnification, but photographs were taken with the scanning electron microscope after coating of specimens with gold and palladium. The terminology of Thiriot-Quiévreux (1972) and Robertson (1974) is adopted in designating as the protoconch the entire larval shell formed prior to metamorphosis. The embryonic shell, formed by the shell gland, is termed 'protoconch I', and the remainder of the larval shell, deposited by the mantle edge, is termed 'protoconch II' (see review by Jablonski & Lutz, 1983). The protoconch is terminated by a strong axial rib, the sinusigera ridge, marking the point at which metamorphosis occurred. The post-larval shell or teleoconch is formed by the mantle edge after metamorphosis.

Where ranges of figures are given, values in parentheses are extremes of the range which are rarely encountered.

### *Anatomical characters*

Colours of animals were described from living specimens where possible. Penes were drawn by camera lucida, in most cases using living animals relaxed in an approximately 1% solution of propylene phenoxetol in sea water, or otherwise fixed in formalin. It was found that boiling the animals produced the same degree of penis extension as relaxation. Penis length was measured from filament tip to attachment of the base to the head-foot.

Spermatozoa were removed from the vas deferens of living animals and fixed in a 1% solution of glutaraldehyde in sea water before examination with a light microscope and drawing by camera lucida. Each group of nurse cells illustrated was taken from a single individual. Where living specimens were not available, material fixed in formalin was used; although eupyrene sperm were then agglutinated, nurse cells were often well preserved and comparison with fresh material showed that their shape and structure were normal. Dimensions of nurse cells are maximum lengths including projecting rods, but excluding flagella.

The pallial oviducts were drawn from material fixed in formalin. Their complex structure was investigated by cutting gross serial transverse sections under a dissecting microscope. The sections drawn in the systematic account are those passing through the apex of the spiral of the oviduct. Shading of the several glandular elements of the oviduct follows that used in Figs 6 and 7. Dimensions of the largest oviduct seen are recorded in each description; the seminal receptacle is not included in the measurement of overall length; the diameter of the spiral section is the maximum in any direction; the length of the straight section of the pallial oviduct extends from the most anterior whorl of the spiral section to the terminal papilla or pore. Egg capsules of *L. articulata* were released by snails kept in containers half filled with sea water, and spawning occurred on the day after collection from the field.

The following histological techniques were used to investigate penial and oviducal structure: staining in haematoxylin and eosin; the Mallory-Heidenhain rapid one-step trichrome (Cason, 1950); and the alcian blue-periodic acid-Schiff technique for the histochemical differentiation of mucins (Mowry, 1956).

In order to assess the variability of the anatomical features described, from six to ten specimens of each species, from a wide geographical range, were dissected in detail. Penes were examined in many more animals. For the species *L. scabra*, *L. intermedia*, *L. philippiana*, *L. filosa* and *L. articulata*, the sperm, penes and oviducts of ten of each sex were examined each month during the course of a twelve month study of reproductive condition at Magnetic Island, North Queensland, Australia (Reid, in prep.). As discussed in the section on taxonomic characters, the form of the reproductive structures of these five species were not found to vary greatly during this time.

Radulae from at least four specimens of each species were dissected from material fixed in formalin. The radulae were soaked in 10% potassium hydroxide solution for 2 hours, cleaned by hand, stored in 70% ethanol and cleaned ultrasonically for 15 seconds before examination with the scanning electron microscope. All radulae were mounted flat and uncoated and were viewed from above. Total radular length was measured and the range of the ratio of radular length to shell height was recorded.

Of the species described in the present monograph, anatomical data was obtained for all but *L. delicatula* and *L. flammea*. In addition, all other species of the genus *Littoraria* (list p. 72) were dissected, with the exception of *L. aberrans*. For purposes of comparison and discussion of phylogenetic relationships, the following members of other littorinid genera were dissected:

| | |
|---|---|
| *Bembicium:* | *auratum* (Quoy & Gaimard); *nanum* (Lamarck) |
| *Cenchritis:* | *muricatus* (L.) |
| *Echininus:* | *antoni* (Philippi) (=*nodulosus* auctt.); *cumingi* (Philippi) |
| *Fossarilittorina:* | *meleagris* (Potiez & Michaud); *mespillum* (Mühlfeld) |
| *Littorina:* | *keenae* (Rosewater) (=*planaxis* Philippi); *littorea* (L.); *obtusata* (L.); *scutulata* (Gould) |

| | |
|---|---|
| *Melarhaphe:* | *neritoides* (L.) |
| *Nodilittorina:* | *acutispira* (Smith); *angustior* (Mörch) (=*lineata* (Orbigny)); *australis* (Gray); *aspera* (Philippi); *dilatata* (Orbigny); *hawaiiensis* Rosewater & Kadolsky (=*picta* Philippi); *knysnaensis* (Philippi); *millegrana* (Philippi); *modesta* (Philippi); *praetermissa* (May); *pyramidalis* (Quoy & Gaimard); *sundaica* (Altena); *unifasciata* (Gray); *ziczac* (Gmelin) |
| *Peasiella:* | *roepstorffiana* (Nevill) |
| *Tectarius:* | *grandinatus* (Gmelin); *pagodus* (L.) |

Generalizations concerning the characters of littorinid genera are based upon these species and upon published accounts by other authors as quoted.

## Zonation and Distribution

Except where otherwise acknowledged, notes on habitat and zonation are based upon personal observations at the localities in the list of records which are followed by the abbreviation DGR. Vertical distribution of snails on the trees was measured at low tide. For the identification of mangrove trees in the field, works by Jones (1971), Percival & Womersley (1975), Lear & Turner (1975) and Semeniuk *et al.* (1978) were used.

Species distribution maps were compiled from the localities of the museum specimens seen. The locality records listed are those marked on the distribution maps and are not a complete list of all collections examined. So far as possible only reliable modern records have been used; in those few cases in which doubtful records are listed the locality is preceded by a query and plotted as an open circle. Literature records have only been included if they extend the known distribution significantly; they are noted as such and plotted as open circles on the maps.

## Abbreviations

The following abbreviations are used in the text, synonymies and tables:

| | |
|---|---|
| A | Identification confirmed by anatomical data from preserved specimens (specified for *L. articulata* and *L. strigata* only). |
| AMS | Australian Museum, Sydney |
| ANSP | Academy of Natural Sciences of Philadelphia |
| B | shell breadth |
| BMNH | British Museum (Natural History), London |
| BPBM | Bernice P. Bishop Museum, Honolulu |
| C | columellar width |
| DGR | collection by the author; majority of material, including all figured specimens, now in BMNH |
| H | shell height |
| L. | genus *Littoraria* |
| LA | apertural length |
| MCZ | Museum of Comparative Zoology, Harvard University, Cambridge, Massachusetts |
| MHNG | Muséum d'Histoire Naturelle, Geneva |
| MNHNP | Muséum National d'Histoire Naturelle, Paris |
| NM | Natal Museum, South Africa |

| | |
|---|---|
| NMV | National Museum of Victoria, Melbourne |
| NMW | National Museum of Wales, Cardiff |
| NSMT | National Science Museum, Tokyo |
| N.S.W. | New South Wales, Australia |
| N.T. | Northern Territory, Australia |
| NUS | National University of Singapore |
| P | shell proportion = H/B |
| Qld. | Queensland, Australia |
| QM | Queensland Museum, Brisbane |
| RNHL | Rijksmuseum van Natuurlijke Historie, Leiden |
| S | apertural shape (circularity) = WA/LA |
| SH | spire height = H/LA |
| SM | Sarawak Museum, Kuching |
| USNM | National Museum of Natural History, Smithsonian Institution, Washington, D.C. |
| W.A. | Western Australia |
| WA | apertural width |
| WAM | Western Australian Museum, Perth |
| ZSI | Zoological Survey of India, Calcutta |

In particular it should be noted that the abbreviation *L.* refers only to the genus *Littoraria*. The appearance of a specific epithet in quotation marks indicates that the authority quoted may have confused several species under the one name, but that insufficient information was provided for their subsequent determination.

# MORPHOLOGICAL CHARACTERS

In the following section the morphology of shell and animal in the genus *Littoraria* is described, and features are evaluated as taxonomic characters. Comparisons are drawn with other genera in the family and, where possible, character states are assessed as ancestral or derived, as a basis for decisions concerning generic classification and for a discussion of phylogenetic relationships (p. 71). For taxonomic purposes, shell characters are the most convenient to use. Although colour and size are highly variable within species, shape and sculpture are relatively constant, so that shell characters alone are adequate for the identification of the majority of specimens of *Littoraria*. Amongst the anatomical characters, the form of the penis is diagnostic of most species.

## Shell Characters

*Shape, size and thickness*

Statistical analyses of shell shape and size have sometimes been used to distinguish between closely related species of littorinids (Borkowski & Borkowski, 1969; Smith, 1981). However, most studies have emphasized the variability of these characters within species particularly in relation to the exposure to wave action (James, 1968; Struhsaker, 1968; Newkirk & Doyle, 1975; Heller, 1976; Raffaelli, 1979; Janson, 1982*b*) or selection by crab predators (Heller, 1976; Elner & Raffaelli, 1980) experienced by individual populations. In the species of *Littoraria* associated with mangroves shell shape is rather constant within species, even in those with the widest geographical ranges, and is therefore a reliable taxonomic character. This constancy is perhaps a consequence of the planktonic dispersal of the species (p. 65), combined with the uniformly sheltered conditions prevailing in the mangrove environment. In contrast, the accounts of variability have mostly referred to species in which genetic exchange between populations is limited by ovoviviparous or benthic development, favouring adaptation of local populations to the wide range of exposures experienced on rocky shores.

Although shape may be relatively uniform, adult shell size shows a two- to three-fold variation in most of the species of *Littoraria* considered herein. In general, individuals from unusual or extreme habitats, such as high level salt marshes, stunted mangrove bushes in full sunlight, or sheltered rocky shores, tend to be of the smallest size. Gallagher & Reid (1974) observed that *L. angulifera* and *L. irrorata* in Florida attain larger size in an apparently more favourable habitat. Trematode parasites are thought to produce gigantism in certain molluscan hosts (Wright, 1966). However, this effect can be ruled out in *Littoraria* species from mangroves, for of the several thousand specimens examined during the present study only five contained parasites, and these shells were not of especially large size. Muggeridge (1979) found no parasitic infection in *L. luteola* (as *Littorina scabra*) at Patonga, New South Wales.

The shape of the aperture does not vary greatly between species of *Littoraria*. In most species the peristome is coplanar and the angle between this plane and the coiling axis of the shell (the angle of elevation of the coiling axis, Vermeij, 1971) is such that the peristome lies flush with a flat surface when the shell is placed upon it. This form may be adaptive for snails living on the predominantly flat surfaces of leaves and trunks. In *L. scabra* the apertural plane is generally hollowed anteriorly to fit the narrow aerial roots of the mangrove trees (*Rhizophora*) upon which it is often found; this is probably a direct mechanical effect of the substrate, as observed in *L. irrorata* by Bingham (1972*a*).

In young, actively growing shells the apertural lip is thin, sharp and brittle. Certain species of *Littoraria* exhibit a conspicuous flaring and thickening of the outer lip of the aperture when growth slows or ceases, while in other species the lip is merely somewhat thickened from within. If growth is later resumed, a flared lip remains as a prominent varix interrupting the body whorl of the shell. Flaring of the lip is particularly common amongst those species of the subgenera *Littorinopsis* and *Lamellilitorina* with thin shells, which occupy higher tidal levels. Regular measurement of marked individuals in the field at Magnetic Island, Queensland, has shown that in *L. filosa* and *L. philippiana* growth ceases and the lip is flared and thickened during a period coinciding with the annual breeding season, which is also the time of year with highest rainfall and temperature. Sewell (1924) suggested that such varices were indicative of adverse conditions. However, it seems likely that the interruption of growth is directly correlated with reproductive condition, since duration of the interruption differs in the two species and matches their respective peak breeding periods; furthermore, young prereproductive snails are not affected. Cessation of growth during the breeding season has been observed in other littorinids (Moore, 1937; Williams, 1964; Borkowski, 1974). In *L. filosa* and *L. philippiana* not every mature individual forms a flared lip when growth ceases, although the majority do so. The tendency to develop a flared lip and varices is more marked in males than in females in all species with this habit, despite the observation of Sewell (1924) to the contrary in a small sample of '*Littorina scabra*'. The distribution of varices in *L. luteola*, *L. delicatula* and *L. ardouiniana* suggests that in these species also they may be formed annually. However, *L. albicans* may possess as many as twenty varices and an annual breeding season is perhaps unlikely to account for their formation, unless the species is unusually long lived.

A feature of the aperture which has not hitherto been stressed as an important taxonomic character amongst littorinids is the form, and sometimes also the colour, of the columella. The range of variation of columellar form amongst species of *Littoraria* is illustrated in Fig. 2. Of especial significance are the relative width of the columella and its shape, whether excavated or rounded. Columellar form is constant within species and can be a useful diagnostic character, as in the separation of shells of *L. filosa* and *L. philippiana*. A narrow and rounded columella is usually, but not invariably, associated with a thin shell.

Shell thickness is only described qualitatively in the taxonomic section. There is considerable variation within the genus, ranging from the thin and fragile *L. delicatula* which is easily crushed between the fingers, to the solid *L. sulculosa*, up to 1.5 mm in thickness at the outer apertural lip, which can only be crushed by several blows of a heavy hammer.

Species of *Littoraria* show marked correlations of shell thickness, size and shape with their patterns of zonation on the trees (whether measured above the ground or relative to tidal datum) (Reid, in prep.). These trends are briefly described here, and their adaptive significance considered.

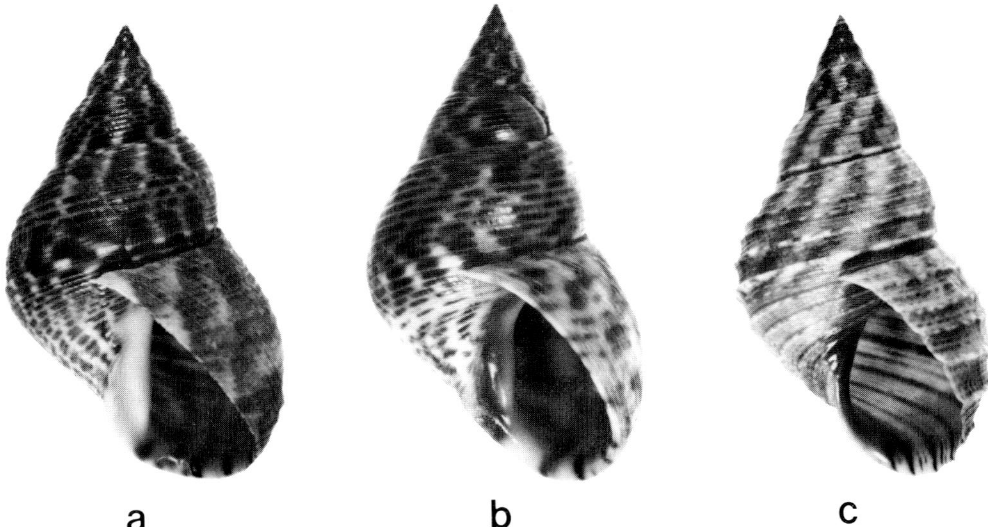


**Fig. 2** Examples of columellar types in *Littoraria*: **(a)** *L. scabra*: wide, excavated, with straight pillar; **(b)** *L. pallescens:* excavated, with convex pillar, pinched at base, and rounded inner apertural lip; **(c)** *L. filosa*: narrow, rounded, with concave pillar.

Species from high levels, found mainly on the leaves of mangrove trees, typically have thin shells, for example *L. filosa*, *L. luteola* and *L. albicans*, whilst those zoned at the lowest levels have thick shells, as in *L. sulculosa*, *L. intermedia* and *L. articulata*. Various influences of the biotic and physical environment select for and constrain shell thickness or calcification, as reviewed by Vermeij (1978). Of the environmental factors which may have a direct influence on shell calcification, low salinity and high acidity of ground water in mangrove swamps would not appear to affect *Littoraria* species, since all remain above the water level. Although a few species are somewhat variable in thickness, there is at present no evidence that shells from more estuarine areas are any thinner than those in fully marine situations, as has also been noted by Marcus & Marcus (1963) in *L. angulifera*. The high level species may be analogous to terrestrial and freshwater pulmonates, in which low availability of calcium carbonate may limit shell thickness (Graus, 1974; Vermeij & Covich, 1978). However, calcium carbonate is unlikely to be in short supply in a marine environment, even at supralittoral levels. *L. carinifera*, occurring at the landward edge of the forest, is reached only by the highest spring tides, and yet develops a thick shell. It has also been suggested (Vermeij, 1973*a*) that in high intertidal gastropods calcification may be constrained by the increase in tissue acidity caused by anaerobic respiration during periods when the shell is tightly sealed to reduce evaporation. However, respiration rates of inactive littorinids are likely to be very low (Vermeij, 1973*a*). Calcification is less efficient at low temperatures (Graus, 1974), but all species of *Littoraria* have tropical or subtropical distributions and there is no obvious trend of increasing thickness at lower latitudes, either within or between species.

Several selective factors may influence shell thickness. It might be proposed that species living at high levels on leaves possess thin, light shells in order to minimize the danger of dislodgement while they are attached only by mucus. An additional or alternative hypothesis is that species occupying lower levels are more susceptible to predation by crabs, which cause selection for thicker shells. Crab predation is intense in mangrove habitats and while certain crabs, especially species of *Metopograpsus*, are arboreal, it appears that the overall intensity of crab predation decreases with distance above the ground (Reid, in prep.). A similar interspecific gradient in shell thickness has been found in the *Littorina* species of rocky shores in Britain, and explained in terms of adaptation to crab predation (Heller, 1976; Elner & Raffaelli, 1980; Reimchen, 1982).

At many localities it can be observed that the *Littoraria* species from higher tidal levels are of larger size than those found at lower levels on the trees. Small species from low levels include *L. articulata*, *L. strigata*, *L. sulculosa* and *L. intermedia*. This trend cannot be accounted for by growth to larger size in more favourable habitats, where there may be a higher productivity (Vermeij, 1980; McQuaid, 1981) or a longer time available for feeding (Berry, 1961; Janson, 1982*a*), for then the larger littorinids would be expected at lower levels. Amongst littorinids from rocky shores, intraspecific size gradients have sometimes been explained in terms of resistance of small individuals to dislodgement by waves (North, 1954; Struhsaker, 1968; but see McQuaid, 1981) or their ability to shelter in crevices (Emson & Faller-Fritsch, 1976; Raffaelli & Hughes, 1978). Since *Littoraria* species remain above the water level and because mangrove environments are always sheltered, wave action and crevice availability should not influence shell size. A pattern of upshore size increase is found within many species of gastropods from the high intertidal zone on rocky shores, and is believed to reflect the greater tolerance of larger individuals to rigorous physical conditions (Vermeij, 1972*b*). The same explanation may account for the larger adult size of mangrove littorinids from higher tidal levels, as also suggested by Vermeij (1973*b*). In contrast, on tropical rocky shores supralittoral littorinids show no consistent trends in shell size with zonation level. Although the advantage of larger size must still apply, it may be counteracted by the fact that temperature regulation in the full sun is more effective in smaller gastropods (Vermeij, 1973*a*). In the shaded habitat of mangrove forests, dissipation of heat may be less important.

Large shell size may also be of adaptive significance in connection with immunity from crushing by crab predators (Zipser & Vermeij, 1978; Elner & Raffaelli, 1980; Reimchen, 1982). However, as mentioned above, the intensity of crab predation is greatest at lower levels on the trees, where species are of smaller size. Reimchen (1982) has made the interesting suggestion that under certain circumstances small shell size may confer resistance to crushing. This may be so because a thick shell is an alternative method of defence. When adult size is reached, littorinids thicken the shell internally and marginally, and smaller species may reach adult size, and hence immunity to predation, at an earlier age.

Shell shape is also correlated with zonation level, those species from higher levels being generally narrower and more high spired (e.g. *L. filosa*, *L. philippiana*, *L. luteola*). Apertural shape shows no consistent trends. A decrease in the apical angle of the shell during ontogeny is common in gastropods (Vermeij, 1980), so that the trends in shell shape may in part represent an allometric consequence of the corresponding gradient in shell size. On tropical rocky shores a similar trend of relatively taller spires at higher levels has been interpreted as an adaptive response, reducing the relative area of the foot for conduction of heat from the substrate and for evaporative water loss (Vermeij, 1973*a*). A similar explanation could apply to the mangrove littorinids (Vermeij, 1973*b*). In temperate *Littorina* species, shore level gradients in shell shape are less distinct, perhaps because exposure to wave action is of more significance (Vermeij, 1973*a*). For example, populations of the ovoviviparous *Littorina saxatilis* from more exposed localities show lower spires and larger apertures (Newkirk & Doyle, 1975; Raffaelli, 1979; Smith, 1981; Janson, 1982*b*).

### *Sexual dimorphism*

In the family Littorinidae larger mean size of female shells, or a preponderance of females in the larger size classes, has been reported in many species (Pelseneer, 1926; Struhsaker, 1966; Sacchi, 1968; Daguzan, 1977) and usually explained by greater growth rates of females (Sewell, 1924; Moore, 1937; Lenderking, 1952, 1954; Borkowski, 1974; Underwood & McFadyen, 1983). A contributory factor may also be the greater longevity of females, as proposed by Sewell (1924), Pelseneer (1926) and Daguzan (1977). In *Littoraria* species larger females have been recorded in *L. 'scabra'* (Sewell, 1924), *L. pallescens* (Abe, 1942; as *Melaraphe scabra*), *L. luteola* (Muggeridge, 1979; as *Littorina scabra*), *L. intermedia* (Struhsaker, 1966, as *Littorina scabra*), *L. angulifera* (Lenderking, 1952; Gallagher & Reid, 1974) and *L. irrorata* (Bingham, 1972*a*; Gallagher & Reid, 1974; Hamilton, 1978).

In the present study dimorphism was examined in each species by comparison of ten adult shells of each sex. Despite small sample size, significant size dimorphism was found in fourteen of the nineteen species and subspecies for which data were available (Table 1). In each case male shells were of smaller size.

Dimorphism of shell shape in other littorinids has been noted by several authors (e.g. Sacchi, 1968; Daguzan, 1977). Within the genus *Littoraria*, the type and extent of shape dimorphism is variable, as shown by the shape indices summarized in Table 1. Where

**Table 1** Summary of sexual dimorphism in the mangrove-associated *Littoraria* species of the Indo-Pacific.

| Species | Locality | H | P | S | SH |
|---|---|---|---|---|---|
| *L.* (*Littoraria*) *vespacea* | Santubong, Sarawak | .001 – ** | .883 – | .970 – | .218 – |
| *L.* (*Lamellilitorina*) *albicans* | Santubong, Sarawak | .028 – * | .280 – | .352 – | .012 – * |
| *L.* (*Littorinopsis*) *scabra* | Moa I., Qld. | <.001 – ** | .352 – | .630 – | <.001 – ** |
| *L.* (*Littorinopsis*) *lutea* | Ubin I., Singapore | .001 – ** | .740 + | .106 – | .002 – ** |
| *L.* (*Littorinopsis*) *pallescens* | Ubin I., Singapore | .067 – | .740 – | <.001 – ** | <.001 – ** |
| *L.* (*Littorinopsis*) *philippiana* | Magnetic I., Qld. | .348 – | 1.0 + | .657 – | .014 – * |
| *L.* (*Littorinopsis*) *intermedia* | Magnetic I., Qld. | <.001 – ** | .630 – | .264 – | .004 – ** |
| *L.* (*Littorinopsis*) *subvittata* | Aldabra | .001 – ** | .002 – ** | .684 – | <.001 – ** |
| *L.* (*Littorinopsis*) *filosa* | Darwin, N. T. | .259 – | .166 + | .796 + | .297 – |
| *L.* (*Littorinopsis*) *c. cingulata* | Broome, W. A. | .029 – * | .076 – | .854 = | .012 – * |
| *L.* (*Littorinopsis*) *c. pristissini* | Denham, W. A. | .027 – * | .106 – | .825 + | <.001 – ** |
| *L.* (*Littorinopsis*) *luteola* | Kurnell, N. S. W. | .031 – * | .018 – * | .218 + | <.001 – ** |
| *L.* (*Littorinopsis*) *ardouiniana* | Hong Kong | .021 – * | .004 – ** | .218 + | <.001 – ** |
| *L.* (*Littorinopsis*) *delicatula*[1] | Port Canning, Bengal, India | | | | – |
| *L.* (*Palustorina*) *melanostoma* | Kanchanadit, Thailand | .252 + | .106 + | .280 – | .394 + |
| *L.* (*Palustorina*) *conica* | Santubong, Sarawak | .008 – ** | .712 – | <.001 – ** | .028 – * |
| *L.* (*Palustorina*) *carinifera* | Sungei Merbok, Malaysia | .852 – | .314 + | .166 – | .166 – |
| *L.* (*Palustorina*) *sulculosa* | Broome, W. A. | <.001 – ** | .064 – | .218 – | .090 – |
| *L.* (*Palustorina*) *articulata* | Broome, W. A. | .001 – ** | .394 + | .415 – | .106 – |
| *L.* (*Palustorina*) *strigata* | Penang, Malaysia | .032 – * | .970 + | .012 – * | .436 – |

H = shell height; P = shell proportion; S = aperture shape; SH = spire height. Figures are probability levels for comparison of mean values of H (t-test) and shape indices (Mann-Whitney U test) between samples of 10 males and 10 females. +, – or = indicates direction of difference when value for males is compared with that for females. * indicates a difference significant at 0.05 level, ** at 0.01 level. [1] from Nevill (1885).

dimorphism of relative spire height (SH) is significant, males are characterized by a lower spire, as measured by the ratio of shell height to aperture length. The apparently lower spire of males is not the result of a larger apical angle during the course of growth. Rather, it is caused by the distension of the last whorl and or flaring of the lip, which enlarge the aperture relative to the shell height. Only if expansion of the last whorl is considerable does this produce a dimorphism of shell proportion (P). Aperture shape (S) is sometimes more elongate in males, which contributes to the apparent lowering of spire height. It must be emphasized that such dimorphism is only achieved at the last whorl of the shell and is not apparent in younger specimens. In species with significant dimorphism, shells with an adult aperture can, with experience, be immediately identified as male or female on the basis of their shape. Nevill (1885) observed the lower spire and patulous aperture of male shells of *L. delicatula* and *L. scabra*, while Abe (1942) noted the more elongate shell of the male in *L. pallescens* (the elongate aperture of the male of this species does in fact cause the whole shell to appear elongate, although there is no significant difference in shell proportion). However, Struhsaker (1966) and Muggeridge (1979) have reported an absence of shape dimorphism in *L. intermedia* and *L. luteola* respectively, both species for which highly significant dimorphism has been found in the present study. Probable explanations for this discrepancy are that both authors used the ratio of shell height to breadth, which does not measure the most obvious aspect of dimorphism, and furthermore that shape dimorphism is only evident in adult shells, which were not specifically selected by these authors.

### *Protoconch*

The protoconch is of similar form in most of the species of *Littoraria* described in detail in the present work (e.g. Figs 28b, 40b, 87b, 91b). The first whorl (protoconch I) is smooth, but thereafter (protoconch II) is sculptured by strong spiral ribs of which five are usually visible, and by oblique axial ridges which are most obvious in the middle of the whorls. After a further three whorls the termination of the protoconch is marked by a prominent sinusigera ridge, by a discontinuation of sculpture and often also by a change from the horn colour of the larval shell to a colour closer to that of the adult shell. Only in *L. albicans* is the protoconch distinctive (Fig. 24b), being low spired, sculptured by low spiral ribs only, and of unusually large size. In the majority of species the length of the protoconch is 320–415 μm, but in *L. albicans* 610–660 μm. The apical angle of the protoconch is usually similar to or greater than that of the first teleoconch whorl; only in *L. conica* is it smaller, producing a papillose apex (Fig. 79a). Species of *Littoraria* from mangrove forests are unusual amongst littorinids in that the protoconch is not infrequently intact even in adult shells, presumably as a consequence of the sheltered nature of the environment.

Characters of the protoconch would not appear to be useful for classification at the generic level in this family, for amongst species with planktotrophic development the protoconch is rather uniform, varying only in the degree of sculpture of the protoconch II. In species of *Littorina*, *Melarhaphe* and *Nodilittorina* in which the adult shell is rather smooth, the protoconch II is sculptured by spiral rows of minute tubercles (Pilkington, 1971; Thiriot-Quiévreux & Babio, 1975; Fretter & Manly, 1977; Fish & Fish, 1977). In strongly sculptured *Nodilittorina* species, as in many species of *Littoraria*, the tubercles tend to fuse, forming spiral ribs which are often discontinuous or undulating (Rosewater, 1981; Bandel & Kadolsky, 1982). Several authors have reported variability within species in the degree of fusion of tubercles and development of spiral ribbing (Struhsaker & Costlow, 1968; Bandel & Kadolsky, 1982). Within the genus *Littoraria* most species show strong sculpture of the protoconch, but in the western Atlantic *L. irrorata* the tubercles are not fused into spiral ridges (Thiriot-Quiévreux, 1980). A pattern of more or less fused tubercles on the protoconch II appears also in some rissoacean genera (Thiriot-Quiévreux & Babio, 1975) and in certain other mesogastropods and neogastropods (e.g. Bandel, 1975).

The length of the protoconch formed by planktotrophic larvae ranges from 250 to 450 μm in the accounts quoted above, so that the 610 to 660 μm protoconch of *L. albicans* would seem to be exceptional for the family. Although the shape of the protoconch in this species is unusually broad, the small size of the protoconch I and the strong sinusigera ridge terminating the protoconch II suggest that the larva does undergo planktotrophic development (Shuto, 1974). Protoconchs of species of the subgenus *Littorinopsis*, which are ovoviviparous, are not distinguishable from those of oviparous species of *Littoraria* which spawn pelagic egg capsules, for the larvae of the former are retained only until the early veliger stage and a long phase of planktotrophic development ensues (p. 66). In littorinids which undergo nonplanktotrophic, so-called 'direct', development (in which larvae are brooded or encapsulated until hatching as benthic juveniles; see Jablonski & Lutz, 1983) the protoconch is of relatively large size, almost smooth, composed of two whorls or less, and lacks a sinusigera ridge (e.g. Rosewater, 1982). Amongst the species of *Littoraria* such a protoconch is seen only in the Eastern Pacific *L. aberrans*. Although the anatomy and method of development of this species are unknown, a nonplanktotrophic larval form can be predicted.

### *Shell sculpture*

The sculpture of the teleoconch is the most important of the taxonomic characters of the shell of *Littoraria* and is diagnostic for the majority of the species. In previous taxonomic studies shell sculpture has usually been described only in general terms, although in the *Nodilittorina ziczac* species complex of the western Atlantic the number of spiral grooves has been used to distinguish several similar species (Borkowski & Borkowski, 1969; Bandel, 1974; Bandel & Kadolsky, 1982). No detailed comparisons, of the type presented here for *Littoraria*, have hitherto been made for any other group of littorinids, perhaps because in other species the character is often too variable to be diagnostic. Many studies have emphasized variability of sculpture in the family (e.g. Fischer-Piette & Gaillard, 1971; Borkowski, 1975), especially in relation to exposure of the habitat (James, 1968; Struhsaker, 1968; Heller, 1975*a*; Smith, 1981). These accounts have concerned species from exposed rocky shores and it appears that *Littoraria* from sheltered mangrove habitats are less variable in this respect. Even in this genus there is considerable variation within some species, often between geographically distant populations, as in *L. carinifera* and *L. filosa*, but occasionally also within local populations, as in *L. cingulata pristissini* and *L. pallescens*.

From the faint axial growth lines visible on the apical whorls and from newly settled shells collected in the field, it is evident that in *Littoraria* growth of the postlarval shell begins by the infilling of the areas on each side of the beak of the larval aperture, thus forming a planar aperture. In many species the first one to three whorls of the postlarval shell are smooth and the number of such whorls before the appearance of spiral sculpture is recorded in the species descriptions. In several species spiral sculpture begins immediately after the sinusigera ridge formed at metamorphosis, but the sculpture produced is clearly different from that of the larval shell. Wherever spiral sculpture develops, it consists of from six to twenty-six narrow grooves visible above the suture with the following whorl. Often the more posterior of these grooves appear first and may be deeper and more closely spaced than the rest, but all are developed in the space of one to two revolutions. These grooves are here termed primary grooves, and the spaces between, the primary ribs. The point of appearance of the primary grooves, their number and spacing, are characteristic of each species.

On subsequent whorls the number of ribs is increased by either or a combination of two processes: primary ribs may become divided by secondary grooves (each division forming a continuing primary rib and an anterior secondary rib), or secondary ribs may appear in the primary grooves by intercalation of small riblets which may expand into ribs. Occasionally one or more ribs may be added by intercalation or division on the early whorls; the convention is adopted that all ribs present on the fourth whorl of the postlarval shell are regarded as primary. Secondary sculpture does not occur in all species, but if and when it is formed it usually develops over most of the surface of the whorl in the space of one revolution. Division of primary ribs by secondary grooves occurs either centrally or towards the anterior face of the rib. Division and intercalation may occur side by side on the same whorl, at different stages on the same shell, or one process may occur exclusively. Where tertiary and higher orders of sculpture appear the scale is so small and the ribs so numerous that the distinction between formation by intercalation or division can be difficult to draw. Furthermore, towards the end of the last whorl spiral sculpture may become indistinct as axial growth lines become stronger.

On the last whorl of the adult shell grooves often become wider and ribs prominently rounded. The relative width of ribs and grooves is an important character. In species with the strongest sculpture the primary ribs may become especially prominent on the last whorl, developing into raised carinae (e.g. *L. carinifera*, *L. filosa*, *L. pallescens*). In many species the peripheral rib is the largest and most prominent, and enhances the natural angulation of the shell, thus imparting a strongly keeled appearance. The basal part of the shell below the keel is sculptured in rather similar fashion to the area above the periphery, although the ribs become smaller and more closely spaced anteriorly. Since only the last whorl of the base is visible, primary and secondary sculpture cannot always be distinguished; also, differences in rib width are less pronounced on the base. For these reasons basal sculpture is of lesser importance as a taxonomic character and is not considered in detail in the species descriptions.

Nodulose sculpture is entirely lacking in all known species of *Littoraria*, although it is conspicuous in the littorinid genera *Tectarius*, *Echininus*, *Cenchritis* and *Bembicium*, in many species of *Nodilittorina* and a few of *Littorina* and *Melarhaphe*. The distribution of nodulose sculpture in the family has been reviewed by Bandel & Kadolsky (1982) and, as pointed out by these authors, the character cannot be used as evidence of close phylogenetic relationship. In the genus *Nodilittorina* sculpture is especially variable between species, ranging from regular spiral ribbing to strong nodulation. Spiral sculpture is present in most littorinid genera, although it is weak or absent in *Melarhaphe* (Rosewater, 1981).

The significance of shell sculpture has been related by various authors to temperature control, hydrodynamic properties and defence against predation, as reviewed by Vermeij (1978). In an examination of interspecific variation in shell sculpture of littorinids with shore level, Vermeij (1973*a*) found examples of both stronger and weaker sculpture at higher levels. Stronger sculpture at high levels was explained in terms of temperature regulation, the relatively greater surface area increasing heat loss by convection and reradiation. In mangrove environments, littorinids occurring within the forest, at whatever level, are shaded by foliage, so that they do not experience temperature stress. Only those few species inhabiting foliage at the seaward edges of forests will normally experience direct sunlight. It is therefore probably significant that of four species known to occur on leaves in the outer *Avicennia* fringe in sunny positions, two are the most strongly sculptured in the genus. These are *L. filosa*, with strong spiral carinae, and *L. albicans*, with strong axial varices. It is interesting that the sculpture is produced in different ways in the two species and that both are in contrast to the nodulose sculpture of high level *Nodilittorina* species on rocky shores. A third species, *L. pallescens*, is not restricted to the edges of swamps, but occurs also on *Rhizophora* trees and within forests; here shell sculpture is variable, but specimens with strong spiral ribs are common. The fourth species from *Avicennia* foliage is *L. luteola*, in which the sculpture is not especially strong, but since this is a subtropical species it may not experience such temperature stress as the others. The typical habitats of *L. delicatula* and *L. ardouiniana* are not known, but both have relatively thin shells and are colour polymorphic, so that they are likely to be found on foliage (pp. 10, 18). Although *L. ardouiniana* often forms a few varices, both species show only weak spiral sculpture. Since the *Littoraria* species discussed here occur above the water level in sheltered environments, shell sculpture will not have any significance in relation to wave action.

Vermeij (1978) accounted for increasingly strong sculpture in gastropods occupying lower levels on the shore in terms of resistance to more intense crab predation. As discussed above (p. 10), mangrove littorinids at low levels have thick shells, probably as a defence against predators. In addition the thickest species, *L. sulculosa*, also develops strong spiral ribs. These may further strengthen the shell, but since the species has been found in open, sunny situations a role in temperature regulation cannot be ruled out. *L. carinifera* occurs near ground level at the landward side of dense, shady forests and possesses a thick shell with a strong peripheral keel and often spiral ribs also. In this case the sculpture is probably related to predation, as supported by the fact that *L. conica*, from the same habitat, but found higher above the ground, has a considerably thinner and smoother shell.

In addition to the conspicuous spiral sculpture described, microsculpture is visible under low magnification, especially on the last two whorls where the scale is larger. In the subgenus *Littorinopsis* regular spiral striae are visible in the grooves, but are faint or absent on the ribs. If the grooves are narrow, microsculpture may be visible only in the wider posterior grooves. In the subgenus *Palustorina* spiral striae are restricted to the ribs, while the grooves contain strong, regular axial lines. The axial sculpture is most clearly visible in species with wider grooves; when narrow the grooves appear pitted. In *L.* (*Lamellilitorina*) *albicans* spiral microsculpture is absent and axial striae are prominent in the grooves. Other species of the genus show spiral striae developed over the whole surface of the shell. The form of the microsculpture can be a useful diagnostic feature, as for example in the differentiation of shells of *L. cingulata* and *L. sulculosa*.

In most species the surface appears glossy under low magnification; the periostracum is not evident and is presumably thin and closely adherent. In *L. melanostoma, L. carinifera* and *L. flammea* the layer is thicker and occasionally flakes off, showing that in these three species the spiral microsculpture on the ribs is largely produced by ridges on the periostracum. Reimchen (1981) has described spiral ridges on the periostracum of *Littorina mariae*. In *L. vespacea* the periostracum is occasionally produced into short bristles.

## *Shell colour*

Most of the authors writing on the *Littoraria scabra* complex since Philippi (1847–1848) have not failed to comment on the wide range of shell colour forms. Despite the recognition of many species within the complex in the present revision, the majority still show considerable colour variation. In all *Littoraria* species shell colour is best described as a ground colour, either a shade of white to yellow, or of orange pink, with a superimposed pattern of dark pigment, usually brown or black. Dark pigment is deposited in the form of spiral dashes, usually confined to the shell ribs. The dashes are often discrete, but if very dense, they may run together or appear smudged. In most species the dark dashes are aligned to some degree, especially at the suture and periphery of the whorls, to form axial flames. This alignment is often most apparent on the spire whorls, and some shells show complete alignment from suture to base, which produces oblique axial stripes. Within most species the development of dark pigmentation is variable, ranging from complete absence, through faint mottling, to dark dashes and stripes which sometimes cover the surface. Continuous spiral colour bands are unusual.

Although variation in these species is striking, not all can be described as polymorphic, since strictly the term can only be used of discrete variation (Ford, 1945). In *Littoraria* species the variation of the dark pattern is apparently continuous, but, at least in paler shells, the ground colour is either pale yellow or orange pink, with no intermediate shades. In shells with the darkest patterning, the ground colour is obscured and appears brown. For convenience of description, those species in which shells may be either predominantly yellow, pink or brown are termed 'polymorphic' in the species descriptions, as opposed to 'variable' species in which only the degree of patterning changes. Amongst the polymorphic species the same range of colour forms is encountered in each and the polymorphisms may be homologous (see Frontispiece).

Colour polymorphism is most striking in members of the subgenus *Littorinopsis*, being shown in all or part of their range by ten of the twelve species (the exceptions are *L. scabra* and *L. subvittata*). In the other subgenera of *Littoraria*, the only strikingly polymorphic species is *L.* (*Lamellilitorina*) *albicans*. Pink shells are very occasionally found in *L.* (*Palustorina*) *melanostoma* and *L.* (*P.*) *articulata*, but these species are not described as polymorphic. As discussed below, the occurrence of polymorphism is correlated with habitat rather than probable phylogenetic grouping.

Several of the European *Littorina* species are well known to show great variation in colour, which is in some species continuous (e.g. Heller, 1975*a,b*) whilst in others discrete morphs can be recognized (e.g. Smith, 1976; Naylor & Begon, 1982; Raffaelli, 1982). In *Littorina mariae* Reimchen (1981) has demonstrated that the polymorphism has a genetic basis. In *Littoraria* nothing is known of the inheritance of shell colour, but in the most highly polymorphic species (e.g. *L. filosa, L. albicans, L. cingulata pristissini*), where all colour forms can be found together on the same trees, colour would seem likely to be determined genetically. However, in some of the other polymorphic and variable species (e.g. *L. pallescens, L. philippiana, L. luteola, L. scabra*) the frequency of colour forms varies greatly between habitats. In particular, brown shells (with dense dark pigment) predominate on *Rhizophora* trees, while yellow and pink shells (with little or no dark pigment) are most numerous on *Avicennia* and *Sonneratia* trees. In the European species of *Littorina* differences in morph frequency have been explained by selection for crypsis by visual predators (Heller, 1975*b*; Smith, 1976; Reimchen, 1979). There is some evidence for visual selection in the highly polymorphic species of *Littoraria*, but in the less polymorphic species preliminary transfer experiments suggest a more direct influence of substrate upon colour, perhaps mediated by an effect of diet (Reid, in prep.).

It is striking that the most highly polymorphic species of *Littoraria* (*L. filosa, L. luteola, L. albicans, L. pallescens*) are those found on the foliage of mangrove trees several metres from the ground. Of other polymorphic species, *L. cingulata pristissini* occurs in a variety of habitats including foliage of samphires and mangroves, whilst although details of the habitats of *L. delicatula* and *L. ardouiniana* are not known, their thin shells suggest that they may live at high levels above the ground, perhaps on leaves. The remaining species occur at lower levels, on trunks rather than leaves, and are mostly more uniform, with dark coloration predominating. An association between the tree climbing habit and colour polymorphism is evident in several groups of pulmonate land snails, including helicids (Jones *et al.*, 1977).

At present it is only possible to speculate upon the significance of shell colour and colour polymorphism in *Littoraria*. One attractive hypothesis is that birds prey upon the snails living at upper levels in the trees and since they hunt visually may cause selection for a genetic polymorphism. Work on polymorphic land snails, especially *Cepaea* in Europe (reviewed by Jones *et al.*, 1977), has suggested several mechanisms by which visual predation could maintain polymorphism. Of these, apostatic selection (e.g. Smith, 1975) and frequency dependent selection for crypsis against the heterogeneous background of leaves and branches (Reimchen, 1979; Cook, 1983) may be mentioned. At lower levels on the trees dark coloured shells are cryptic against the uniformly brown trunks, perhaps further contributing to avoidance of attack by crabs. Vermeij (1973*a*) gave several examples of littorinids from the highest levels of rocky shores with light coloured shells, and suggested that pale colours minimize absorption of visible radiation. If this effect were significant in *Littoraria*, uniformly pale, rather than polymorphic, shells would be expected at high levels. However, it is noteworthy that in *L. albicans* the young shells are polymorphic while in the adults, which inhabit sunny foliage at higher levels, the shells fade to white. Determination of shell colour in *Littoraria* and the possible selective forces maintaining polymorphism will be considered in detail in future work (Reid, in prep.).

In general, shell colour is so variable in *Littoraria* species as to be a poor guide to identification. Nevertheless, the relative abundance of colour forms, the size of the pigment dashes, their degree of alignment and the number of axial stripes per whorl, may serve to characterize species. A few species show a more or less constant shell colour and pattern (e.g. *L. carinifera, L. vespacea, L. cingulata cingulata*).

The columellar pillar is pale or white in most specimens, but the colour of the excavated area, and sometimes also of the parietal callus, can be useful in distinguishing species. Columellar colour was used as a taxonomic character in the *Nodilittorina ziczac* species complex (Borkowski & Borkowski, 1969). In polymorphic species columellar colour is often correlated with external shell colour, being white in unpigmented yellow or pink shells, and purple in darkly patterned shells. Internal shell colour usually reflects the external pattern.

Since shell coloration is so variable in many members of the family, the character cannot be regarded as of great significance in a consideration of relationships at the generic level. Nevertheless, it is noteworthy that a pattern of spiral dashes aligned to form axial markings is common to species of *Littoraria* and to those of *Nodilittorina* which lack nodulose sculpture (e.g. Bandel & Kadolsky, 1982). In the genus *Littorina*, however, spiral bands of colour predominate. Of possible phylogenetic significance is the presence or absence of an unpigmented spiral band in the anterior part of an otherwise dark brown aperture. This pattern is typical of the genera *Nodilittorina* and *Melarhaphe*, and is seen also in *Laevilitorina* and *Rissolittorina* (Ponder, 1966; Ponder & Rosewater, 1979) and in some species of Lacunidae. In other groups this pale apertural stripe is generally absent, although *Littorina keenae* appears to be an exception. In the genus *Littoraria* this pattern is often visible in specimens of *L. pintado* and *L. mauritiana*, both of which show several ancestral character states (Fig. 18), and is also seen in *L. carinifera*. Conceivably, the white apertural stripe is an ancestral character in the genus.

### *Operculum*

Several authors have used characters of the operculum in the taxonomy of littorinids (Abbott, 1954; Rosewater, 1972, 1981; Bandel & Kadolsky, 1982). Members of the genera *Tectarius* and *Cenchritis* show a mesospiral operculum, and *Echininus* and *Peasiella* a multispiral one. In *Nodilittorina* both paucispiral and mesospiral opercula are found, and Bandel & Kadolsky (1982) have argued that tighter coiling of the operculum is an adaptation to fit a more circular aperture and to thicken the operculum, in order to reduce water loss. Other littorinids, as reviewed by these authors, show a paucispiral operculum, which appears to be the ancestral condition in the family. In the genus *Littoraria* the operculum is thin and paucispiral (as illustrated by Rosewater, 1981) and no differences could be detected between the opercula of different species.

## Anatomical Characters

### *Coloration of head-foot*

The coloration of the animal has sometimes been used as a taxonomic character in littorinids, for example in the descriptions of supposed subspecies of *Littorina 'saxatilis'* by James (1968). Amongst the Indo-Pacific species of *Littoraria* treated here, pigmentation of the animal is seldom of use in distinguishing species since in the majority the pattern is the same. The sole the foot is pale, usually whitish or cream, and the sides mottled with grey or black pigment. The head is grey to black, paler at the tip of the snout and often with an unpigmented, short, longitudinal streak between the tentacles, which is most conspicuous in the largest individuals. The red buccal mass is visible within the head. Irregular bands of pigment reach almost to the tips of the tentacles; the bases are darkly pigmented but for a prominent white stripe on inner and outer sides. This pattern is illustrated by *L. scabra* (Fig. 3) and is characteristic of most members of the genus. The pattern may be contrasted with that typical of *Nodilittorina Echininus*, *Melarhaphe* and *Fossarilittorina*, with an unpigmented patch or band over the eye and sometimes an absence of pigment on the distal parts of the tentacles.

In species of *Littoraria* with colour polymorphic shells the pigmentation of the animal is correlated with the colour of the shell, as has been observed in *Littorina obtusata* in Europe (Barkman, 1955; Bakker, 1959). For example, in *Littoraria filosa* the animals with pure yellow shells are entirely unpigmented, in darker shells animal pigmentation becomes more pronounced, especially on the head, until animals are dark grey in brown shells. Only in *L. albicans*, with red tentacles lacking basal stripes, is the coloration of the animal diagnostic.

### *Male reproductive tract*

The anatomy of the male reproductive system of *Littoraria* (e.g. *L. scabra*, Fig. 3) is similar to that of the European *Littorina* species described in detail by Linke (1933) and Fretter & Graham (1962). Lobules of the testis, orange or red brown in colour, ramify in the digestive gland and join to form a duct running close to the surface of the visceral mass against the columella of the shell. The more distal, convoluted portion of the testicular duct is distended with stored sperm during the breeding season, functioning as a seminal vesicle, and leads to the pallial vas deferens via a short renal section. In species of *Littoraria* the pallial section is a closed tube with a central slit-like lumen, surrounded by a swollen, glandular prostate. This condition has been previously reported in *L. angulifera* and *L. flava* by Marcus & Marcus (1963) and in *L. melanostoma* by Berry & Chew (1973) and contrasts with the open prostate found in most other genera of littorinids, as discussed below. In the majority of *Littoraria* species the prostate opens to a ciliated groove which carries sperm forward over the lateral surface of the head-foot to the conspicuous penis, situated on the right side of the head behind the eye. In a few species the entire vas deferens is a closed tube, the groove over the head-foot and penial groove being closed as a shallow duct, leaving only a minute pore communicating with the mantle cavity adjacent to the distal end of the prostate.

The form of the penis shows considerable variation within the Littorinidae and is perhaps the single most important taxonomic character of the anatomy of the family. It has been used to define the generic groups of the family by Rosewater (1970, 1972, 1981) and in the revision of his classification by Bandel & Kadolsky (1982), besides being employed as a specific character in many other recent taxonomic studies (e.g. Whipple, 1965; Heller, 1975*a*; Ponder & Rosewater, 1979). In two cases dimorphism of penial shape has provided the first indication of the existence of a pair of sibling species (Sacchi & Rastelli, 1967; Murray, 1979).

In general the littorinid penis is differentiated into a thick, wrinkled and muscular basal region, often with glandular appendages, and a narrow, smooth, distal filament, along which sperm passes in a deep, but usually open, groove at the posterior (dorsal) edge. The conspicuous glandular elements are of two types. In all species of *Littorina*, *Nodilittorina*, *Tectarius*, *Echininus* and *Peasiella* penial glands are found. Each gland is visible externally as a papilla, within which the duct may appear hyaline and is referred to as an 'accessory flagellum' by Abbott (1954) and Rosewater (1970, 1981). The internal structure of penial glands has been described by Linke (1933) and by Marcus & Marcus (1963). The small

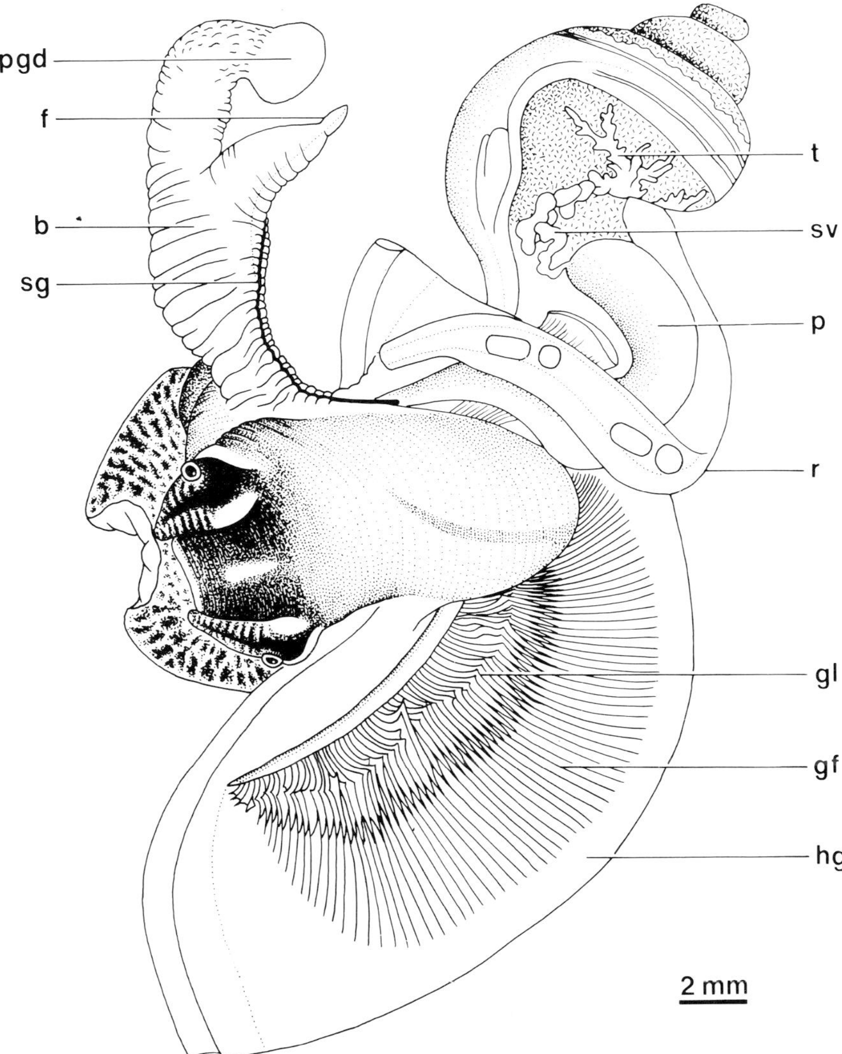


**Fig. 3** *Littoraria* (*Littorinopsis*) *scabra:* male reproductive tract; mantle cavity opened between rectum and hypobranchial gland; **b**, base of penis; **f**, filament of penis; **gf**, gill fold; **gl**, gill leaflet; **hg**, hypobranchial gland; **p**, prostate gland; **pgd**, penial glandular disc; **r**, rectum; **sg**, penial sperm groove; **sv**, seminal vesicle; **t**, testis.

penial papillae of *Tectarius*, *Echininus* and *Nodilittorina pyramidalis*, as well as the very large basal glands of *Echininus antoni*, show the same structure as penial glands. The second glandular element occurs in the genera *Littoraria* and *Nodilittorina* and has been variously described as a 'disk or sucker' (Leidy, 1845), 'clasper' (Lenderking, 1954), 'haftlappen' (Marcus & Marcus, 1963), 'penial gland' (Whipple, 1965), 'lateral thickened appendage' or 'basal flap' (Rosewater, 1970, 1981), 'attachment gland' (Bingham, 1972*b*) and as an 'adhesive flagellum' (Bandel & Kadolsky, 1982). It is suggested that all these terms refer to a similar, probably homologous, structure, for which the name penial glandular disc seems appropriate. Only in species of *Nodilittorina* do both glandular types occur on the penis together.

The structure of the penial glandular disc has not hitherto been described in detail. When relaxed, the penis is folded back into the mantle cavity, lying against the head-foot with the filament nearest the mid-line. In species of *Littoraria* the glandular disc appears bulbous or resembles a sucker, and in the resting position the secretory surface is lowermost. During the present study, sections of four species were examined: *L. scabra*, *L. philippiana*, *L. melanostoma* and *L. articulata*. In each case the epithelium of the penis is columnar and, except in the sperm groove, unciliated. The penial disc is composed of glandular tissue lying below the epithelium, staining dark pink in haematoxylin and eosin, blue in Mallory-Heidenhain trichrome and magenta (indicating neutral mucins) by means of the alcian blue-periodic acid-Schiff (ABPAS) technique. The granular secretion is discharged through numerous fine cytoplasmic extensions passing between the epithelial cells. Over most of the base the epithelium stains magenta in ABPAS and is not secretory, but the epithelium overlying the secretory surface of the disc is taller and contains numerous goblet cells staining bright blue in ABPAS (indicating acidic mucins). Elsewhere these cells are abundant on the penial filament, where they extend also beneath the epithelium, and presumably their secretion serves for lubrication. Similar goblet cells are found in the female oviduct, especially in the sperm groove. The subgenera *Littorinopsis* and *Palustorina*, as represented by the four species examined, differ slightly in that the disc of *Littoraria* is composed almost solely of glandular cells whilst in *Palustorina* the glandular tissue is interspersed amongst the muscle fibres and blood spaces within the penial base.

The penis of *Littoraria* species is differentiated into filament and base, penial glands are absent, the penial glandular disc is well developed and the sperm groove usually, but not always, open. Within the genus penial shape varies widely, but is in the majority of cases diagnostic of the species (Fig. 4, and in systematic account). Specific differences can be seen in the relative lengths of filament and base, position of the glandular disc, coloration of both disc and base, and in the closure of the sperm groove. There are few clear trends in penial form above the species level. In the subgenus *Littorinopsis* (as in the type species, *L. angulifera*, Fig. 4o) the base is bifurcate, the glandular disc being carried on a limb well separated from the penial filament; the junction of filament and base is marked by a constriction, and although relative proportions of base and filament vary between species, the entire penis is, even in the relaxed state, of large size relative to the length of the shell. In three species of this subgenus (*L. intermedia*, *L. philippiana*, *L. subvittata*) the sperm groove is closed as a duct. Penes of four Indo-Pacific species of *Littorinopsis* (*L. scabra*, *L. intermedia*, *L. luteola*, *L. pallescens*) have been illustrated by previous authors, as indicated in the synonymies, and that of the Atlantic *L. angulifera* by Leidy (1845), Marcus & Marcus (1963) and Rosewater (1981).

Members of the subgenus *Palustorina* (as represented by the type species, *L. melanostoma*, Fig. 75a–g) show a rather characteristic penial shape. In this group the penis is not bifurcate, the glandular disc being closely incorporated into the base; the penis is of relatively small size and the sperm groove is always open. Penes of species placed in the subgenus *Palustorina* have not hitherto been figured.

Amongst species of the subgenus *Littoraria*, penial form is rather diverse (Figs 4, 21a–f). The base is often clearly bifurcate, although not so in *L. pintado*, *L. flava*, *L. fasciata*, *L. irrorata* and *L. vespacea*, in which species the penial glandular disc is closely incorporated

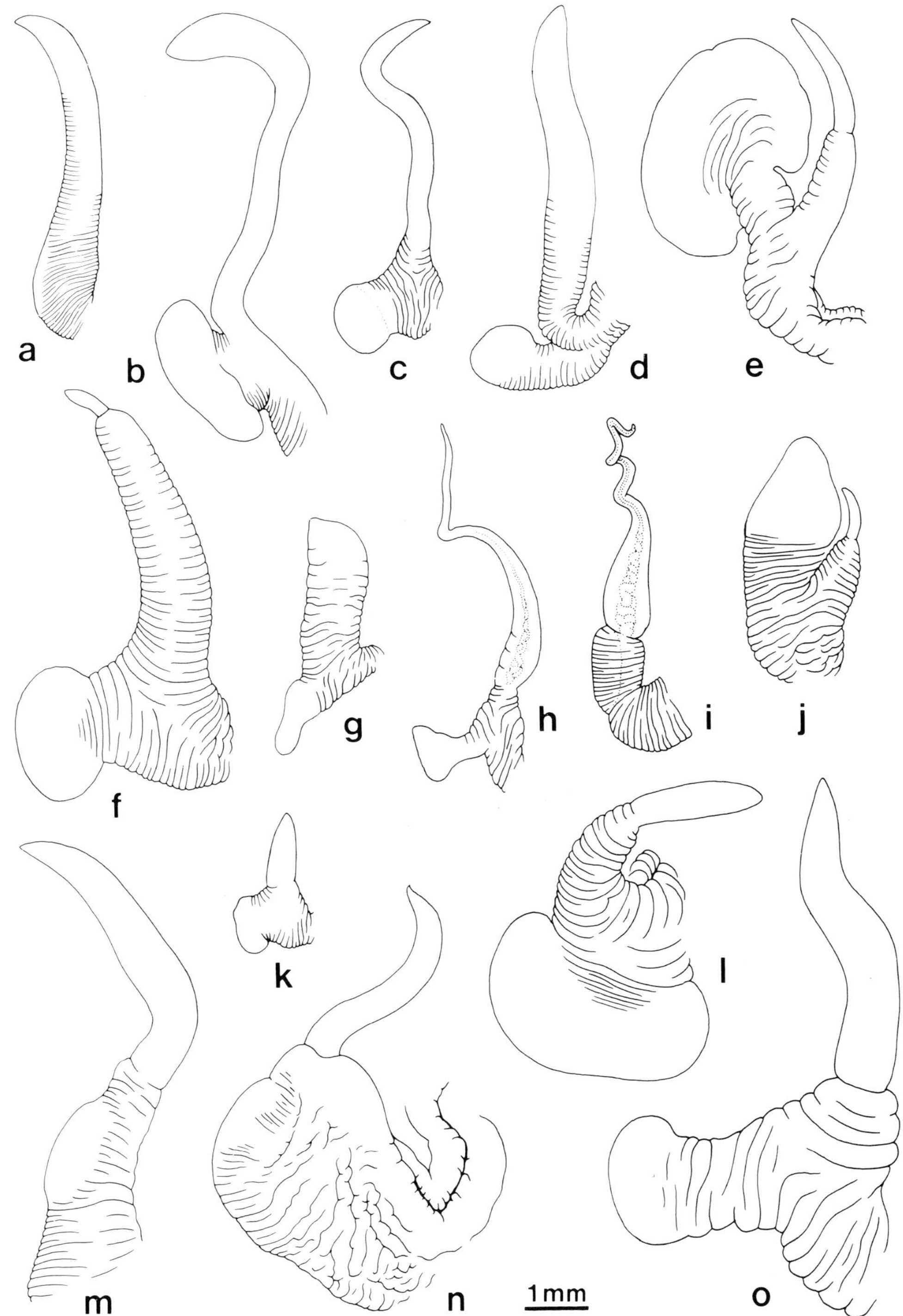


**Fig. 4** Penes of *Littoraria* species other than those described in the systematic section: **(a)** *L. pintado*; **(b)** *L. coccinea*; **(c)** *L. glabrata*; **(d)** *L. mauritiana*; **(e)** *L. undulata*; **(f)** *L. nebulosa*; **(g)** *L. cingulifera*; **(h)** *L. tessellata*; **(i)** *L. flava*; **(j)** *L. varia*; **(k)** undescribed species from tropical Eastern Pacific; **(1)** *L. zebra*; **(m)** *L. irrorata*; **(n)** *L. fasciata*; **(o)** *L. angulifera*.

into the base. In this subgenus the sperm duct is internal in *L. flava*, *L. tessellata* and in an undescribed species from the Eastern Pacific.

Despite the undoubted taxonomic importance of penial characters, there are several difficulties to be considered. Most serious is that although shapes are diagnostic to an experienced eye, it is usually necessary to examine a number of specimens in order to gain the necessary appreciation of the range of variation and of the aspects of shape which are most significant. A morphometric approach (as used by Hannaford Ellis, 1979) might be of use in the initial separation of certain difficult species, but is hardly appropriate in a broad taxonomic treatment, since not only are penes irregular and hard to measure precisely, but their shapes are influenced by the degree of relaxation and method of fixation of the specimens. In this monograph most penes have been drawn from living specimens removed from the shell and relaxed in a 1% solution of propylene phenoxetol, of which the shapes are the same as those seen when the shells of crawling animals are lifted. If animals are dropped into boiling water the penes are also fixed in a similarly extended state. The penes of animals fixed in alcohol or formalin without prior relaxation are often somewhat contracted and usually contorted. Therefore, to illustrate penial shapes several examples have been drawn for each species.

Once artificial differences in shape due to contraction or contortion are recognized, penial form is found to be rather constant. Although immature males lack a fully developed penis, once maturity is reached there is little change in shape of the organ with increasing size of the animal. Penes of smaller animals are sometimes paler in colour. In certain temperate littorinids the penis is reduced in size, or even shed, outside the breeding season (Linke, 1933; Berry, 1961; Palant & Fischelson, 1968; Grahame, 1969; Le Breton, 1970). However, in a study of reproductive condition of five species of *Littoraria* at Magnetic Island, Queensland, over one year, males were found to be mature virtually throughout the year in *L. scabra*, *L. intermedia* and *L. articulata*, while in *L. filosa* and *L. philippiana* with more restricted breeding seasons the change in size and shape of the penis with reproductive condition was slight (Reid, in prep.). Even in *L. luteola*, with one of the most temperate distributions of the Indo-Pacific species, no significant reduction in size of the penis occurs (Muggeridge, 1979). It is possible that trematode parasitism may reduce penis size (Lysaght, 1941), but this condition is rare in *Littoraria*.

Although the species-specific form of the penis is so striking in many littorinid groups, its functional significance is not clear. Linke (1933) suggested that the secretion of the penial glands secures the base of the penis in position in the mantle cavity of the female, while only the elongated filamentous tip enters the bursa copulatrix. Bingham (1972*b*) suggested a similar function for the penial glandular disc of *L. irrorata*. This possibility is strengthened by the discoidal or sucker-like shape of the appendage and by personal observation of the application of its secretory surface to the roof of the mantle cavity (or occasionally to the inside of the shell or side of the head-foot) of the female during copulation in the species of the subgenus *Littorinopsis* in Queensland.

Heller (1975*a*) was so impressed by the differences in penial morphology of four sympatric European *Littorina* species that he suggested that penial shape is of importance in species recognition and perhaps as an isolating mechanism preventing interspecific mating. In several littorinids males are rather indiscriminate in their choice of mate (p. 59) and the success of copulation may depend upon the recognition response of the female, which could be based on penial shape. However, as pointed out by Raffaelli (1979) there is much variation in the number and arrangement of the penial glands in the species considered by Heller, which would make specific recognition by this means difficult. In other *Littorina* species Sacchi & Rastelli (1967) found a conspicuous difference in penial shape between the two species in the *obtusata* group, with very similar conchological characters, although Hannaford Ellis (1979) found penial shapes to be characteristic but not entirely diagnostic of two species in the *saxatilis* group. There is stronger evidence for penial morphology as an isolating mechanism in certain other gastropod groups, for example in endodontid land snails Solem (1976) reported character displacement of penial shapes in localities where congeneric species are sympatric. In the genus *Littoraria* penial shape is more consistent than in *Littorina* and might therefore indeed be a species recognition character.

Fretter & Graham (1962, p. 352) have described examples of correspondence in structure of penis and pallial oviduct in several prosobranchs. In *Littoraria* there is no very obvious correspondence, at least when the penis is relaxed. For example, the bursa copulatrix of *L. scabra* is large, despite the small size of the penial filament. In the conchologically similar pair *L. philippiana* and *L. pallescens*, as in the pair *L. articulata* and *L. strigata*, relative length of the filament is the most important diagnostic character, yet in neither case do the oviducts show corresponding differences. The more posterior position of the bursa copulatrix in many *Littoraria* species is not associated with any particular penial shape. However, the shape of the penis during copulation may be very different from that at rest, as illustrated by Bingham (1972*b*) in *L. irrorata*.

Although characters of the male reproductive anatomy are of importance in classification at both generic and specific levels, they do not permit more than speculation upon phylogenetic relationships, at the present state of knowledge. The pallial genital ducts of prosobranchs have evolved from open grooves (Fretter & Graham, 1962) and the retention of an open prostate in at least seven littorinid genera (Fig. 17) is probably primitive. In all species of *Littoraria*, and in *Melarhaphe*, the prostate is closed, a condition probably derived independently in the two groups, which show few other derived characters in common (Fig. 17). Similarly, the closed penial sperm duct found in a few species of the subgenera *Littoraria* and *Littorinopsis*, and elsewhere in the family in *Melarhaphe*, *Fossarilittorina*, *Bembicium* and also in *Rufolacuna* (Ponder, 1976), has probably been independently derived in each case.

Of the glandular elements of the penis, the histology of the penial glands described by Linke (1933) in *Littorina* species and by Marcus & Marcus (1963) in *Nodilittorina lineolata* (Orbigny) (as *Littorina ziczac*, but see Bandel & Kadolsky, 1982) is so similar that the glands are almost certainly homologous. In *Littorina* the number of penial glands ranges from zero (in *Littorina scutulata* Gould *s. s.*; Murray, 1979) to fifty-nine (in *Littorina obtusata* (L.); Linke, 1933; Sacchi & Rastelli, 1967) and both within species (Linke, 1933; Raffaelli, 1979) and to a certain extent also between species, the number is correlated with shell size. Almost all species of *Nodilittorina* show a single penial gland (Bandel & Kadolsky, 1982), although in *N. lineolata* the number varies from zero to two (Marcus & Marcus, 1963) and in *N. pyramidalis* from zero to one (pers. obs.). There are neither penial glands nor glandular disc in *Nodilittorina striata* (King & Broderip) (Rosewater, 1981). Penial glands of identical external appearance occur in *Echininus*, numbering two to twelve (Rosewater, 1972), in *Tectarius*, numbering approximately one hundred, and occur singly in *Peasiella*. The variation in number of penial glands between genera is correlated with shell size, and is possibly connected with the allometric relationship of gland number and size seen in *Littorina*. The distribution of penial glands in the family (Fig. 17) suggests that they are probably an ancestral feature of the genera considered here, with the exception of *Bembicium*. The variation in number of glands within genera, and even within species, suggests that they might readily be lost, to derive the condition found in the genera *Littoraria*, *Cenchritis*, *Melarhaphe* and *Fossarilittorina*.

Another penial feature of note is the presence of numerous small papillae on the filament of species of *Tectarius* and *Echininus* (Rosewater, 1972; pers. obs.) and in *Nodilittorina pyramidalis* (Rosewater, 1970; pers. obs.). The presence of papillae seems correlated with large size of the shell, for they are found in all *Tectarius* species, absent in the small *Echininus viviparus* (Rosewater, 1982), and present in only one relatively large *Nodilittorina* species.

The homology of the penial glandular disc is more difficult to determine, since the gross appearance is less characteristic. Amongst *Littoraria* species the disc varies in appearance from a prominent sucker to a thin flap, or even a gland so closely incorporated into the base as to be scarcely visible from the surface. However, judging by the four species examined histologically, the structure is similar in each case. The secretion of both glandular types consists of acidophil granules (Linke, 1933; pers. obs.), although in the glandular disc these are secreted through cytoplasmic extensions between epithelial cells, while in penial glands the secretion passes into intercellular canals. Amongst other littorinid genera, glandular structures of similar external appearance to the glandular disc occur only in the genus *Nodilittorina*, although their histology has not been examined. Penes of *Nodilittorina* species have been illustrated by Abbott (1954), Marcus & Marcus (1963), Rosewater (1970) and Ponder & Rosewater (1979), and show a range of differentiation of the glandular disc, from a swelling adjacent to the penial gland, to a well defined disc very similar to that in *Littoraria* species (e.g. *N. tuberculata* (Menke); Abbott, 1954). If the penial glandular discs are homologous in *Littoraria* and *Nodilittorina*, a close relationship is indicated between these genera.

In the littorinid genus *Bembicium* the penis bears distal, presumably glandular swellings, and in *Rissolittorina* (Ponder, 1966) and *Rufolacuna* (Ponder, 1976) long penial appendages have been described. In *Laevilitorina* and *Macquariella* the penis is simple (Ponder, 1976; Ponder & Rosewater, 1979) and no glandular appendages have been noted in the European Lacunidae (Fretter & Graham, 1962). Consideration of the relationships of these groups must await further information on their anatomy.

The functional significance of the evolutionary modifications of the penis remains obscure. The presence of glandular elements is not correlated with habitat, being absent in *Cenchritis muricatus* and *Melarhaphe neritoides*, both from the supralittoral zone, but present in *Nodilittorina* from similar habitats. As discussed above, the presence of glandular elements is to some extent correlated with size, and possibly tied to it by an allometric relationship in the case of penial glands, but again *Cenchritis muricatus* is an exception. This correlation with size suggests that the main function of glandular elements may be support of the organ by adhesion, as concluded by various authors. The penial glandular disc might fulfill this role more effectively when carried on a projection of the base; if so, this adaptation appears to have arisen repeatedly in *Littoraria* species (compare Figs 4 and 18). If, on the other hand, the principal function of the glands is concerned with species recognition, then penial shapes cannot be interpreted as mechanical adaptations.

*Sperm cells*

In many littorinids, as in certain other prosobranchs (reviewed by Fretter & Graham, 1962), the testis produces not only the typical (or eupyrene) sperm responsible for fertilization, but also atypical sperm. The typical sperm of *Littoraria* appear thread-like, 100 to 350 μm in length, with further details only visible at the highest magnifications of the light microscope. Typical sperm of *L. angulifera* and *L. nebulosa* were described by Reinke (1912), while Buckland-Nicks (1973) gave an account of the ultrastructure of the spermatozoa of *Littorina scutulata*.

The atypical sperm of littorinids are known as nurse cells and are rounded, 8 to 55 μm in diameter, and packed with yolk granules. The development of the nurse cells in the testis has been described in several littorinid species (Reinke, 1912; Ankel, 1930; Linke, 1933; Woodard, 1942*a*; Battaglia, 1952; Buckland-Nicks & Chia, 1977). Briefly, they are derived from the germinal epithelium in the tubules of the testis by an unusual reductional division. Each nurse cell is initially attached to the tubule wall by a stalk, and by pseudopodia to developing spermatids. As the cell detaches, the nucleus degenerates and yolk accumulates in the cytoplasm. In certain littorinids, including the majority of species of *Littoraria*, the nurse cells contain conspicuous rod-shaped inclusions, which according to Woodard (1942*a*, in *L. irrorata*) are formed from eupyrene sperm heads, which enter the nurse cells in a process analogous to fertilization. Numerous typical sperm become attached by their tips to the nurse cells (Fig. 5) and the composite bodies, called spermatozeugmata, are transferred to the female in a mucous spermatophore (Woodard, 1942*b*; Buckland-Nicks, 1973). Within the bursa copulatrix the nurse cells degenerate and the sperm reattach to the epithelium of the bursa (Fretter & Graham, 1962, p. 342) before eventual storage in the seminal receptacle.

Most authors have stressed the probable role of the nurse cells in the nutrition of the eupyrene sperm, while Grahame (1973) suggested that the yolk of the nurse cells may contribute to the energetic cost of egg production in the female. In addition, the coordinated beating of the flagella of the attached sperm may propel the unit more effectively (Borkowski, 1971; Buckland-Nicks, 1973). Woodard (1942*b*) observed that nurse cells may also control sperm agglutination and ingest excess sperm in the oviduct.

The possible taxonomic significance of nurse cells was first pointed out by Borkowski (1971), who noted that the cytoplasmic rods were present in some littorinids and absent from others. Further, Borkowski suggested that the position of attachment of the spermatozoa to the nurse cell, whether perpendicular or parallel to the rod, was characteristic of each species. Differences in the shape of nurse cells have been mentioned by Marcus & Marcus (1963) and Jordan & Ramorino (1975). In the present study the appearance of the nurse cells has been found to be a useful taxonomic character, often diagnostic of the species. In particular there is interspecific variation in the size and shape of the nurse cells, the size of the contained yolk granules, and the shape and number of the rods. There is also a rather wide range of variation within species and even within individuals. This does not greatly detract from the usefulness of the character, but means that spermatozoa of several individuals should be examined if the nurse cells are to be used to confirm identification. This variability is partly due to the not uncommon malformations of the nurse cells, which are hardly surprising in view of the abortive and variable processes described by Woodard (1942*a*) during the course of their development. Small and malformed nurse cells seem to become more common in the seminal vesicle outside the main breeding season. Most *Littoraria* species examined usually possess more or less distinct rods, although within certain species they are sometimes small or apparently absent. Although the position of attachment of the spermatozoa to the nurse cell was stressed by Borkowski (1971), in *Littoraria* this is variable, even within individuals, as also found by Marcus & Marcus (1963) in *L. flava*. Most commonly, attachment is perpendicular to the rod, but this orientation is seldom determined by any direct attachment of sperm to the rod itself (as observed by Reinke, 1912). Rather, the spermatozoa attach to the region of the cell occupied by the yolk granules, as evident from species in which granules do not entirely pack the cell (Fig. 5a,b).

Within the genus *Littoraria* the form of the nurse cells is not only a useful specific character, but is diagnostic of the new subgenus *Palustorina*. In the subgenera *Littoraria*, *Littorinopsis* and *Lamellilitorina* the nurse cells are round or oval, rarely more than 30 μm in length, and the rods, if present, usually extend the full length of the cells or even appear to project from them (Fig. 5a–d). In addition to the species described in detail in the present work, nurse cells of this form are known in *L. nebulosa* (Reinke, 1912; rods absent), *L. irrorata* (Woodard, 1942*a*; pers. obs.), *L. angulifera* (Lenderking, 1954), *L. flava* (Marcus & Marcus, 1963), *L. pintado*, *L. undulata*, *L. coccinea*, *L. zebra*, *L. fasciata* and *L. varia* (all pers. obs.). In contrast, in the species of the subgenus *Palustorina* the nurse cells are elongate or fusiform, 30 to 55 μm in length, and the rods and the yolk granules are concentrated at opposite ends of the cell. Most importantly, each nurse cell bears a single flagellum, 160 to 250 μm in length, arising from the end occupied by the rods (referred to as basal simply for convenience of description) (Fig. 5e,f). Flagellate nurse cells have not previously been reported in the family, and their occurrence supports the probable homology of nurse cells with the apyrene sperm of other prosobranchs (Fretter & Graham, 1962, p. 342; Buckland-Nicks & Chia, 1977). The eupyrene spermatozoa are attached at the opposite end of the nurse cell to the flagellum, their orientation either in the same or opposite direction to that of the flagellum. In either case the flagellum is unlikely to contribute significantly to locomotion of the nurse cell once eupyrene sperm are attached, since each of the many spermatozoa is comparable in length to the single flagellum and their combined beat must outweigh that of the latter. It is not known whether the flagellum is a compound structure, consisting of a number of fibres as in the atypical sperm of some other mesogastropod families such as Strombidae (Reinke, 1912) and Cerithiopsidae (Fretter & Graham, 1962, p. 340). This is likely, however, since the flagellum appears broad and ribbon-like near the base, and can occasionally be seen to fray into several strands. The mucronate tip conspicuous in the nurse cells of *L.* (*P.*) *conica* is reminiscent of the stalk by which nurse cells are attached to the germinal epithelium in the early stages of their formation (Reinke, 1912; Linke, 1933; Woodard, 1942*a*). The subgenera of *Littoraria* differ also in the lengths of the eupyrene sperm, 205 to 350 μm in *Palustorina* species, 100 to 187 μm in the subgenus *Littorinopsis*, while in species of the nominate subgenus lengths are more variable, from 60 to 303 μm.

Amongst other genera in the family Littorinidae, cytoplasmic rods occur in the nurse cells of most species of *Nodilittorina* which have been examined (Borkowski, 1971; Jordan

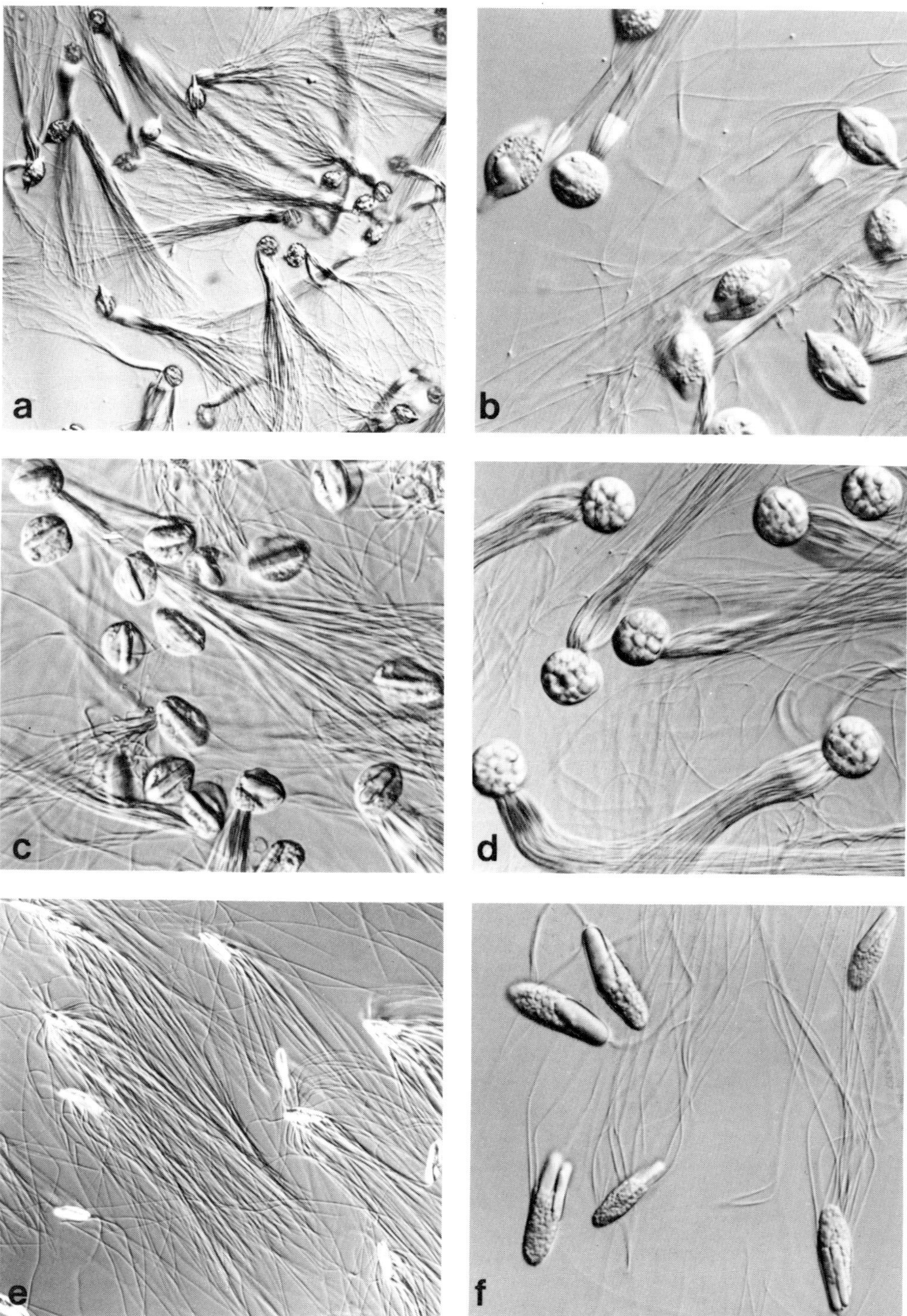


**Fig. 5** Spermatozeugmata of *Littoraria* species (differential interference contrast illumination): **(a,b)** *L.* (*Littorinopsis*) *scabra*, Magnetic I., Qld. ( × 190, × 490); **(c)** *L.* (*Littorinopsis*) *cingulata pristissini*, Denham, W. A. ( × 490); **(d)** *L.* (*Littorinopsis*) *philippiana*, Magnetic I., Qld. ( × 490); **(e,f)** *L.* (*Palustorina*) *articulata*, Magnetic I., Qld. ( × 190, × 490).

& Ramorino, 1975; pers. obs.) and in *Cenchritis muricatus* (Borkowski, 1971). Variation in the shape of the nurse cells in *Nodilittorina* is such that, as in *Littoraria*, the character may prove to be diagnostic of species. Accounts of the nurse cells of *Littorina* species (Reinke, 1912; Ankel, 1930; Linke, 1933; Buckland-Nicks & Chia, 1977) and of *Melarhaphe neritoides* (Battaglia, 1952) have failed to record the presence of rods in these genera. Only typical eupyrene sperm have been found in the seminal vesicle of *Bembicium nanum* and *B. auratum* (pers. obs.). This result requires confirmation, since Bedford (1965) reported nurse cells in *Bembicium nanum*, although without providing clear figures. It is tentatively suggested that the absence of nurse cells in *Bembicium* and of rods in *Littorina* can be regarded as primitive character states (Fig. 17). Confirmation must await description of sperm in other littorinacean groups. At least in a few *Littoraria* species and in *Melarhaphe neritoides*, rods appear to have been secondarily lost. The flagellate nurse cells of the subgenus *Palustorina* are apparently uniquely derived in the family.

### *Female reproductive tract*

In its basic plan the anatomy of the ovary and oviduct of *Littoraria* is closely similar to that of the European *Littorina* species described in detail by Linke (1933) and Fretter & Graham (1962) (Figs 7, 9). The branched lobules of the ovary ramify through the digestive gland, and from the gonad a thin, transparent oviduct runs just beneath the surface epithelium against the columellar pillar of the shell. This portion of the duct is of gonadial origin, and is followed by the short, convoluted and rather thick walled renal section. The organogenesis of the gonoduct of *Littorina saxatilis* has been described by Guyomarc'h-Cousin (1976). In several littorinids (*Littorina littorea*, Fretter & Graham, 1962, p. 45; *Littoraria angulifera*, *L. flava*, *Nodilittorina lineolata*, Marcus & Marcus, 1963; *L. melanostoma*, Berry & Chew, 1973) a small gonopericardial duct has been observed, joining the oviduct at the proximal end of the renal section. However, no pericardial connection was found during dissections of *Littoraria* species in the present study. Close to the seminal receptacle the renal section enters the pallial oviduct, which is derived from the mantle wall. This final section is concerned with transport and storage of sperm and with production of egg capsules, and opens to the mantle cavity at a small, distal aperture.

The structure of the pallial oviduct is complex and can best be understood by reference to the schematic drawing of Fig. 6a. The basic form is a laterally flattened tube, giving off a blind sac, the bursa copulatrix, near the anterior opening. Posteriorly, a narrow duct leads to the seminal receptacle and is joined by the renal oviduct. Internally, both grooves of the slit-like lumen are ciliated, and the dorsal groove is surrounded by a lobe of glandular tissue. From the functional account of the system in *Littorina littorea* given by Fretter & Graham (1962), it is evident that during copulation sperm are deposited in the bursa and subsequently pass along the ventral sperm groove to the receptacle for storage. Eggs enter the pallial oviduct close to the receptacle, and it is probably at this point that fertilization occurs. By a combination of ciliary and muscular action, eggs are passed along the dorsal egg groove and during their passage are successively coated by the products of glands differentiated along the length of the groove.

Before describing the structure of the pallial oviduct in detail, it is necessary to review the nomenclature of the glandular elements, which has been inconsistently applied by previous authors (Table 2). Existing accounts of the littorinid oviduct refer only to the

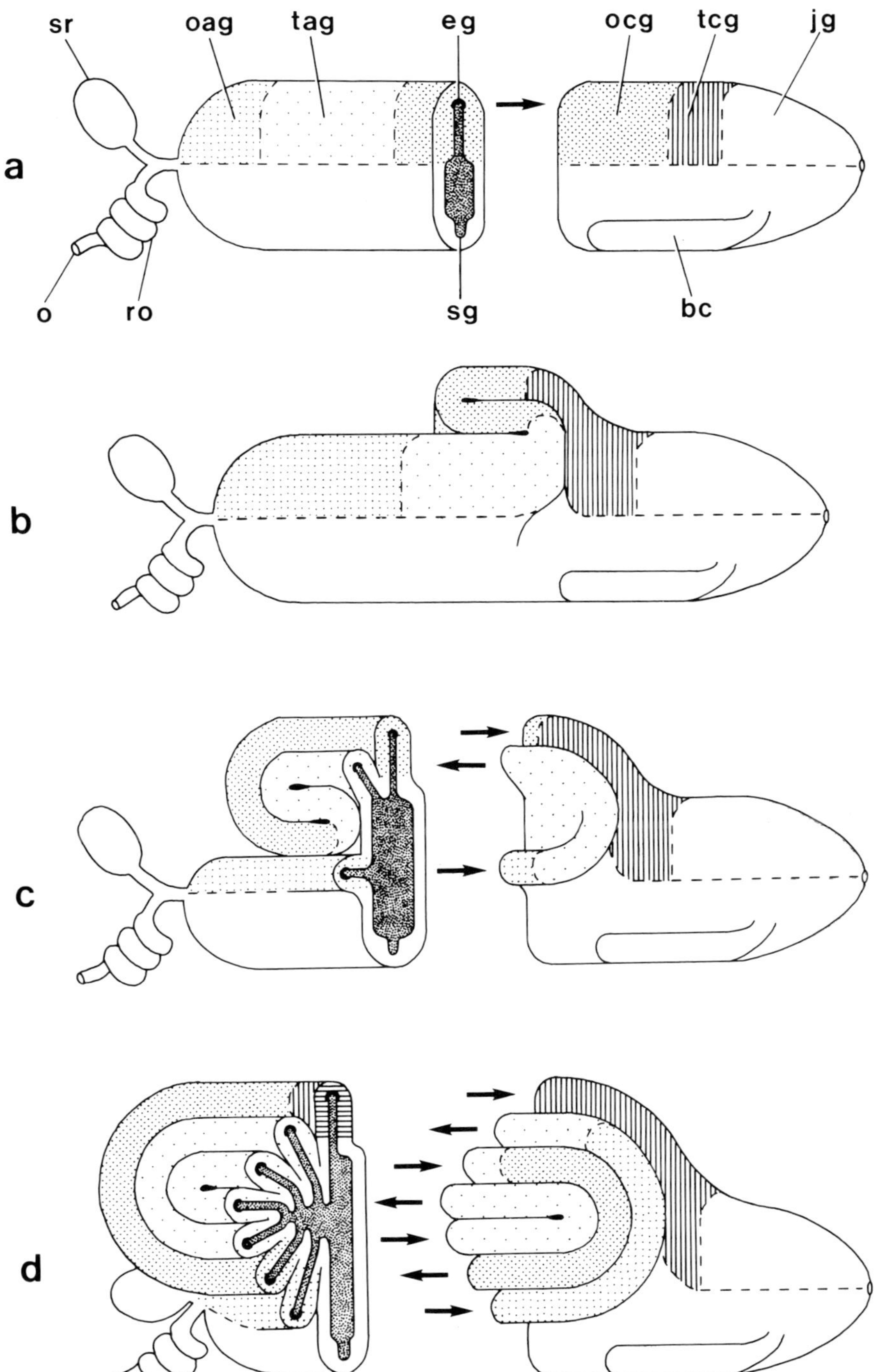


**Fig. 6** Schematic diagram explaining the form of the pallial oviduct in oviparous *Littoraria* species; **(a)** to **(d)** represent hypothetical stages showing progressive lengthening of egg groove by twisting into a double spiral (see text for details); **bc,** bursa copulatrix; **eg,** egg groove; **jg,** jelly gland; **o,** ovarian oviduct; **oag,** opaque albumen gland; **ocg,** opaque capsule gland; **ro,** renal oviduct; **sg,** sperm groove; **sr,** seminal receptacle; **tag,** translucent albumen gland; **tcg,** translucent capsule gland; arrows indicate pathway of eggs.

**Table 2** Nomenclature of the glandular components of the pallial oviduct in the family Littorinidae.

| Author | Species | Albumen gland opaque | Albumen gland translucent | Capsule gland opaque | Capsule gland translucent | Jelly gland |
|---|---|---|---|---|---|---|
| Linke (1933) | *Littorina obtusata* | Eiweissdrüse | | Kapseldrüse | | Gallertdrüse |
| | *Littorina littorea* | Eiweissdrüse | | Kapseldrüse | | Gallertdrüse |
| | *Littorina 'saxatilis'* | Eiweissdrüse | | Kapseldrüse | | Brutraum |
| Fretter & Graham (1962) | *Littorina littorea* | albumen gland | | capsule gland (part) | covering gland | capsule gland (part) |
| Marcus & Marcus (1963) | *Littoraria flava* | Kapseldrüse (part) | | Kapseldrüse (part) | | |
| | *Littoraria angulifera* | Kapseldrüse (part) | | Kapseldrüse (part) | | |
| Berry & Chew (1973) | *Littoraria melanostoma* | albumen gland | | covering gland | capsule gland (part) | capsule gland (part) |
| Sacchi (1975) | *Littorina nigrolineata* | albumen gland | | capsule gland | | jelly gland |
| | *Littorina saxatilis* | albumen gland | | capsule gland | | brood pouch |
| Hannaford Ellis (1979) | *Littorina arcana* | opaque albumen gland | translucent albumen gland | capsule gland | | jelly gland |
| | *Littorina rudis* | opaque albumen gland | translucent albumen gland | capsule gland | | brood pouch |
| Fretter (1980) | *Littorina* spp. with benthic egg mass | opaque albumen gland | translucent albumen gland | membrane gland | | jelly gland |
| | Ovoviviparous *Littorina* spp. | opaque albumen gland | translucent albumen gland | membrane gland | | brood pouch |
| | *Littorina littorea* | opaque albumen gland | translucent albumen gland | jelly gland | shell gland | capsule gland |

European *Littorina* species and to several species of *Littoraria* and *Bembicium*. During the present study, oviducts of species from ten genera (p. 5) have been dissected and it has been found that in all but *Bembicium* the same three glandular types can be distinguished. Proximally, the albumen gland appears white, pale grey or fawn in fresh or preserved material, and as first noted by Hannaford Ellis (1979) it is differentiated into an initial opaque and a subsequent translucent part. This is followed by the gland variously named capsule, covering or membrane gland. In most species the greater part of this gland is opaque pink, or sometimes cream, while the segment closest to the egg groove, and often the entire distal part of the gland, is a translucent dark red or orange brown. These two components have been distinguished by different names by several authors (Table 2), and are absent in *Bembicium*. However, since the term capsule gland is in general use in other prosobranch groups, has historically been applied to both these glandular components in littorinids, and since the two components are intimately associated, the terms opaque and translucent capsule glands will be adopted here. The third and final section of the pallial oviduct has been variously modified in littorinids according to the method of development and although the possible homology of structures in this position in the various littorinid genera is uncertain, they will be referred to as the jelly gland.

The names given to the various glands of the pallial oviduct reflect their supposed contributions to the coverings of the eggs, although there has as yet been no detailed histological study relating structures of the egg capsule to the glands producing them. Berry & Chew (1973) reported that the staining reactions of the egg albumen, covering and capsule in *L. melanostoma* corresponded to those of the glands of the same names (Table 2), although neither the composition of the egg capsule, nor the structure of the final parts of the oviduct, were described in detail. The probable functions of the glands can, perhaps, be deduced from the correlation of oviducal structure with the type of spawn produced. In *Littorina littorea* and *Melarhaphe neritoides*, both producing pelagic capsules (Linke, 1933; Lebour, 1935; Fretter & Graham, 1962), the egg is coated with albumen, surrounded by a covering membrane, and embedded in a viscous fluid within the firm outer capsule. In these two species all glandular types are present; the opaque capsule gland occupies a relatively large volume of the oviduct (pers. obs.) and, as suggested by Linke (1933), probably produces the most voluminous component of the spawn, the viscous fluid. In contrast, in *Littorina arcana* and *Littorina obtusata*, which produce a benthic gelatinous spawn (Hannaford Ellis, 1979; Fretter, 1980), the opaque capsule gland is considerably smaller in relation to the albumen gland. The reduction of the opaque capsule gland has proceeded even further in the ovoviviparous *Littorina saxatilis* (Fretter, 1980) and in none of these three species has a translucent portion of the capsule gland been described. Fretter & Graham (1962, pp. 389, 425) reported that embryos of *Littorina saxatilis* and *Littorina obtusata* are enclosed in thin egg coverings, but that no capsule is present. It seems possible, therefore, that the translucent capsule gland is concerned with production of the outer egg capsule, while the opaque capsule gland produces the material between the capsule and the albumen layer, including the so-called egg covering. Although, as discussed below, capsule glands are absent in the ovoviviparous species of *Littoraria*, the embryos are still enclosed within a covering, the origin of which is unknown. Linke (1933) suggested that the capsule may be produced by the ovipositor. However, this seems unlikely, since Fretter & Graham (1962, p. 387) found fully formed capsules within the oviduct of *Melarhaphe neritoides*, while in *Littoraria* species an ovipositor is apparently lacking. The secretion of the ovipositor may merely harden the capsule (Linke, 1933; Fretter & Graham, 1962, p. 47).

In species of *Littorina* (Linke, 1933; Hannaford Ellis, 1979; Fretter, 1980) and *Bembicium* (Anderson, 1958; Bedford, 1965) which produce benthic gelatinous egg masses, the final section of the pallial oviduct is greatly enlarged by glandular folds thrown into the lumen and is named the jelly gland (Table 2). In the ovoviviparous *Littorina saxatilis* this section is also folded internally and functions as the brood pouch. As suggested by Hannaford Ellis (1979), brood pouch and jelly gland are homologous in the genus *Littorina*. In *Littorina littorea*, *Melarhaphe neritoides* and the other littorinid genera examined during the present study which are known to produce pelagic capsules, this final glandular section is relatively poorly developed. It follows the translucent capsule gland and is visible only as a small whitish lobe around the egg groove, sometimes with a pale yellow or brown proximal region. From its position, it is to be expected that this final glandular element will produce material found externally to the egg capsule. Such a substance has indeed been described in *Littorina sitkana* Philippi (in which capsules are embedded in a benthic gelatinous egg mass; Buckland-Nicks *et al.*, 1973), in *Littorina keenae* Rosewater (=*planaxis* Philippi, in which capsules are initially embedded in a pelagic, gelatinous egg mass; Schmitt, 1979) and in *Littoraria nebulosa* (in which capsules are shed in mucous strings;

Bandel, 1974). In each case the gelatinous material soon disintegrates to release the capsules and perhaps for this reason has not been observed in other species. It is therefore tentatively suggested that the final glandular component in species with pelagic capsules is similar in function to the jelly gland of species with benthic spawn. The name jelly gland will be used here and the homology thereby implied is strengthened by the occurrence of all three conditions of the gland within the genus *Littorina*. However, confirmation must await further investigation.

Figure 6a illustrates in schematic form the arrangement of the components of the pallial oviduct in littorinids which produce pelagic egg capsules. In fact the egg groove and its surrounding glands do not follow a straight course as shown in this figure. In all littorinid genera so far examined the pathway of the eggs has been greatly lengthened and the volume of the glands increased by throwing the egg groove into a series of loops and spirals lying on the right side of the organ. The pattern of the convolutions appears to be characteristic of each of the genera examined (Fig. 11). In *Littoraria* the characteristic pattern is a doubly wound spiral, which can be visualized as an elongated loop wound back upon itself (Fig. 6b–d). The sperm groove, however, continues to follow a straight course to the seminal receptacle along the morphologically ventral side of the lumen. Throughout its spiral route the egg passage remains an open groove, its lumen continuous with that running the length of the oviduct, and in cross section the common lumen of the spiral whorls can be seen. Although in most previous accounts of the littorinid oviduct the external spiral form has been illustrated, the internal structure has been correctly interpreted only by Berry & Chew (1973) in *Littoraria melanostoma*. Since, however, their account was in some respects incomplete, the oviduct of *L. melanostoma* will be redescribed here, as a typical example of the genus. Amongst other species of *Littoraria*, oviducts have previously been figured only for *L. angulifera* and *L. flava* from the western Atlantic (Marcus & Marcus, 1963).

Species of *Littoraria* in which development has been observed either release pelagic egg capsules or are ovoviviparous, releasing early veliger larvae after brooding the eggs between the lamellae of the gills in the mantle cavity. As a result of the present study it has been found that the method of development is clearly indicated by the structure of the pallial oviduct. Only two species of the *scabra* group, *L. articulata* (pers. obs.) and *L. melanostoma* (Berry & Chew, 1973) are definitely known to spawn pelagic egg capsules. In both species the spiral portion of the oviduct is large, composed of 4½ to 6½ whorls, along which all the glandular types defined above can be distinguished. Six other species of the *scabra* group possess a pallial oviduct of this type (*L. vespacea*, *L. albicans*, *L. conica*, *L. carinifera*, *L. sulculosa*, *L. strigata*); veligers have never been found in their mantle cavities and they can safely be assumed to be oviparous also. This prediction is strengthened by the discovery of similar oviducts in all other *Littoraria* species of which egg capsules have been described in the literature (*L. undulata*, *L. coccinea*, *L. nebulosa*, *L. flava*, *L. irrorata*, *L. pintado*, p. 46). In the remaining ten species of the *scabra* group of which the anatomy is known, the spiral portion of the oviduct is reduced, consisting of 2½ or 3½ whorls, and both opaque and translucent capsule glands are absent. These species are ovoviviparous and retention of embryos in the mantle cavity has been observed in eight of them (*L. scabra*, *L. lutea*, *L. pallescens*, *L. philippiana*, *L. intermedia*, *L. subvittata*, *L. filosa*, *L. luteola*).

The pallial oviduct of *L.* (*Palustorina*) *melanostoma* is representative of the oviparous group, which comprises all members of the subgenera *Littoraria*, *Palustorina* and *Lamellil-*

*itorina*. When examined externally from the right side (Fig. 7a,c) the pallial oviduct is seen to consist of a proximal swelling marked by a conspicuous spiral line of black pigment. This spiral portion partly overlies the rectum, and is followed by a straight portion which terminates at a pore opening into the mantle cavity to the right of the anal papilla. The internal structure can be understood by cutting successive transverse sections and examining these with a dissecting microscope (Fig. 8). These may be compared with the diagrams of Fig. 6. Within the straight portion the lumen is a laterally flattened slit, with a ventral pigmented sperm groove and a dorsal pigmented groove for the passage of eggs. Distally, the bursa copulatrix is represented only by a ventral chamber communicating with the lumen of the oviduct. The bursa only separates as a distinct sac posteriorly, at the level of the most anterior part of the spiral oviduct, and continues posteriorly just to the right of the sperm groove. (The bursa opens in a more anterior position in the subgenus *Lamellilitorina* and in some species of the subgenus *Littoraria*, as discussed below.) Running beneath the spiral portion, the sperm groove continues posteriorly as a short duct leading to the seminal receptacle. The convoluted renal oviduct enters the pallial section just anterior to the receptacle, from which point the morphologically dorsal egg groove leads forward into the spiral portion. Viewed from the right side, the egg groove runs in an anticlockwise direction through three revolutions to the apex of the spiral, and continues through three descending clockwise whorls, which alternate with the ascending whorls of the groove. Leaving the spiral on the dorsal side, the descending groove runs forward as the dorsal egg groove of the straight portion of the duct.

Within the spiral section the lining of the egg groove is darkly pigmented and all whorls of the groove communicate by a common central lumen, although several serial sections must be examined to demonstrate all interconnections. Nevertheless, the opening of the groove is constricted by the glandular folds between which it runs, so that the egg passage may function effectively as a closed tube. The surrounding glandular tissues can be identified by their coloration even in unstained fresh or preserved material. For the first one third of a revolution the groove runs through opaque white tissue of the albumen gland. The subsequent portion of the albumen gland is translucent pale grey or fawn, and surrounds the $2\frac{1}{2}$ whorls ascending to the apex, as well as the first whorl of the descending spiral. The following descending whorls are surrounded by the capsule glands, pale pink opaque tissue predominating for the first $1\frac{1}{2}$ whorls and dark red translucent tissue for the last whorl of the spiral, extending around the lowermost chamber. A small amount of translucent whitish jelly gland surrounds the egg groove in the straight portion of the oviduct.

Identification of the glandular components was confirmed by histological techniques. The initial opaque white portion of the albumen gland stains dark pink in haematoxylin and eosin (HE), blue in Mallory-Heidenhain trichrome (MHT) and blue and purple (indicating acidic and mixed mucins) using the alcian blue and periodic acid-Schiff (ABPAS) technique. The following translucent section of the gland stains pale pink in HE, very pale blue or colourless in MHT, and largely magenta but with some blue areas (mainly neutral mucins) in ABPAS. Similar MHT staining reactions of the albumen gland were described by Hannaford Ellis (1979) in two *Littorina* species. The opaque capsule gland is packed with a granular secretion, staining dark pink in HE, red in MHT and magenta to purple in ABPAS (indicating neutral and mixed mucins). The translucent capsule gland stains pink in HE, blue in MHT and purple in ABPAS. Berry & Chew (1973) reported

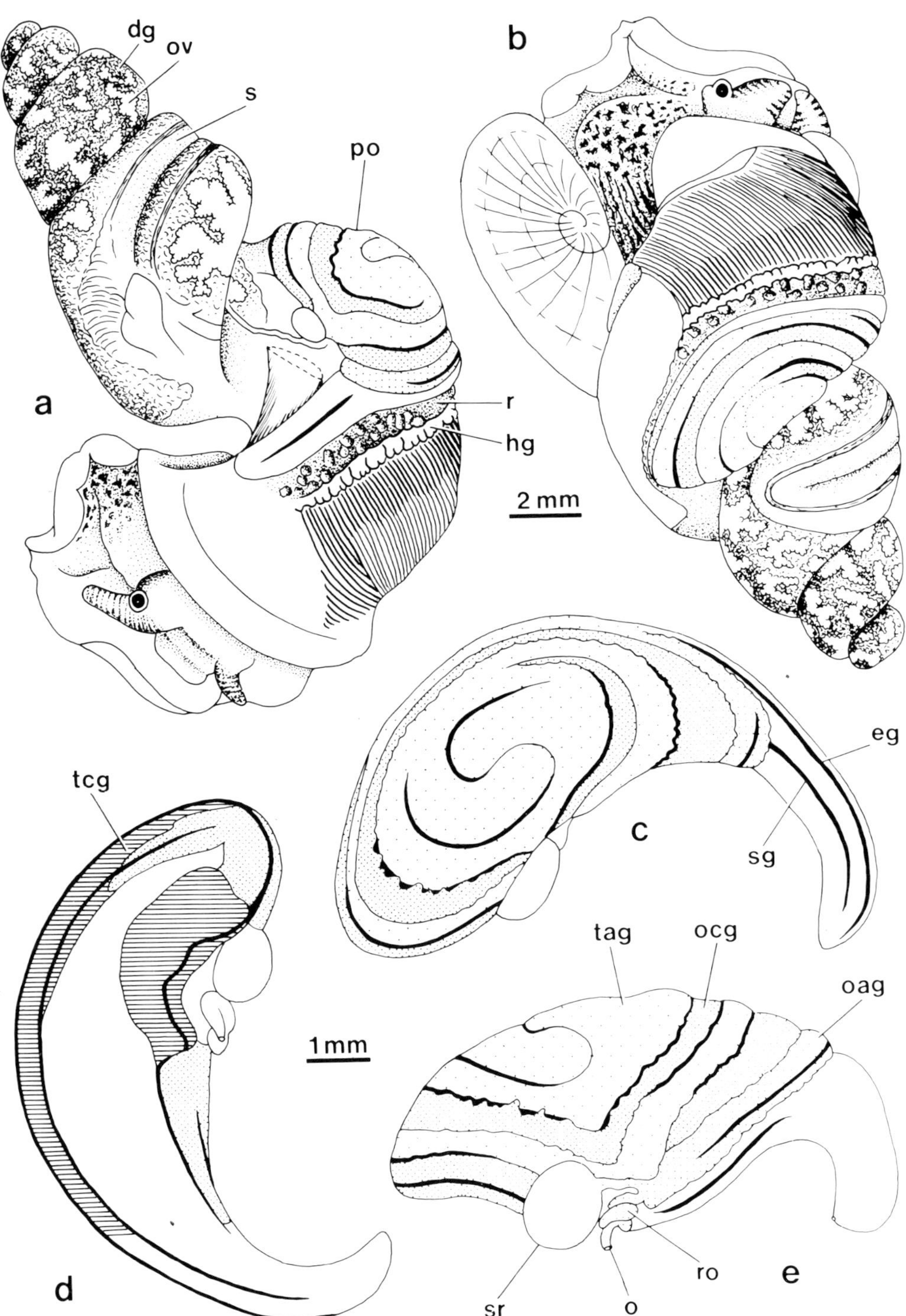


**Fig. 7** *Littoraria* (*Palustorina*) *melanostoma*: female reproductive tract; **(a,b)** animal removed from shell; **dg**, digestive gland; **hg**, hypobranchial gland; **ov**, ovary; **po**, pallial oviduct; **r**, rectum; **s**, stomach; **(c,d,e)** lateral, medial and ventral views of pallial oviduct; abbreviations and shading as in Fig. 6.

similar trichrome staining reactions for the glandular tissues in this species. Within the whorls of the opaque capsule gland a small patch of cells near the base of the egg groove shows the reactions of the translucent albumen gland, while an adjacent area is an extension of the translucent capsule gland. The close association of the two regions of the capsule gland is further demonstrated by the appearance of an area of the opaque tissue type within the translucent region of the gland posterior to the seminal receptacle. This complex pattern is found in the other oviparous species also. The staining reactions of the jelly gland are the same as those of the translucent albumen gland. In each type of gland the subepithelial cells are arranged in indistinct, closely packed lobules, each delimited by a very thin connective tissue envelope, with a single layer of secretory cells surrounding the narrow central lumen. This structure has been illustrated by Linke (1933, figs 57, 62) and Fretter & Graham (1962, p. 335). The arrangement applies also to the jelly gland, which is thus in contrast to the glandular epithelial folds of this region in *Littorina* (Linke, 1933) and *Bembicium* (pers. obs.)

The lumen of the pallial oviduct is lined throughout by an epithelium, supported by a thin basement membrane which clearly separates it from the glandular subepithelial cells. In the final and largest chamber of the spiral the epithelium is ciliated and columnar, with cells up to 70 μm tall, and is thrown into closely packed folds up to 150 μm in height. In the remaining spiral portions of the oviduct ciliated columnar epithelium is found only within the grooves, while the lining of the common central lumen is low and lacks cilia. Most of the ciliated cells within the grooves contain granules of black pigment, which is responsible for the black spiral clearly visible on the right side of the intact oviduct. The secretion of the subepithelial glandular cells reaches the spiral groove through fine cytoplasmic extensions terminating between the ciliated epithelial cells. Consequently the epithelial layer appears to show the same staining reactions as the glandular tissue beneath it, although close examination shows that the epithelial cells are not themselves glandular. The secretory pathway is shown most clearly by the granular secretion of the opaque capsule gland. In the final chamber of the spiral there is no glandular tissue beneath the epithelium. Here the surface area is much increased by folding of the epithelial layer and the cells contain clear vesicles. These cells stain strongly magenta in ABPAS and may well serve a secretory function. The structure and staining reaction of the epithelium of the bursa are similar. In contrast, in the sperm groove the epithelium contains numerous goblet-shaped cells, colourless in MHT, but bright blue (indicating acidic mucins) in ABPAS, and these are clearly secretory. A layer of circular muscle 20 to 80 μm in thickness underlies the epithelium in the final chamber, while a thinner layer surrounds bursa and oviduct in the straight section.

In the ovoviviparous species, comprising the subgenus *Littorinopsis*, the spiral form of the pallial oviduct is similar to that of the oviparous species, although the glandular region is relatively smaller in size and consists of fewer whorls. The spiral structure is usually less clear externally owing to the small amount of black pigment in the ciliated spiral groove. In comparison with the oviparous species of *Palustorina* the bursa copulatrix is situated further anteriorly, lying beside the straight section of the oviduct, which it joins near the opening to the mantle cavity. Throughout its length the pallial oviduct runs along the right side of the rectum, and the straight portion terminates in a raised papilla to the right of the anus. This papilla is presumably concerned with positioning the ova in the mantle cavity for brooding, and is absent in the oviparous species. *L.* (*Littorinopsis*) *scabra* is

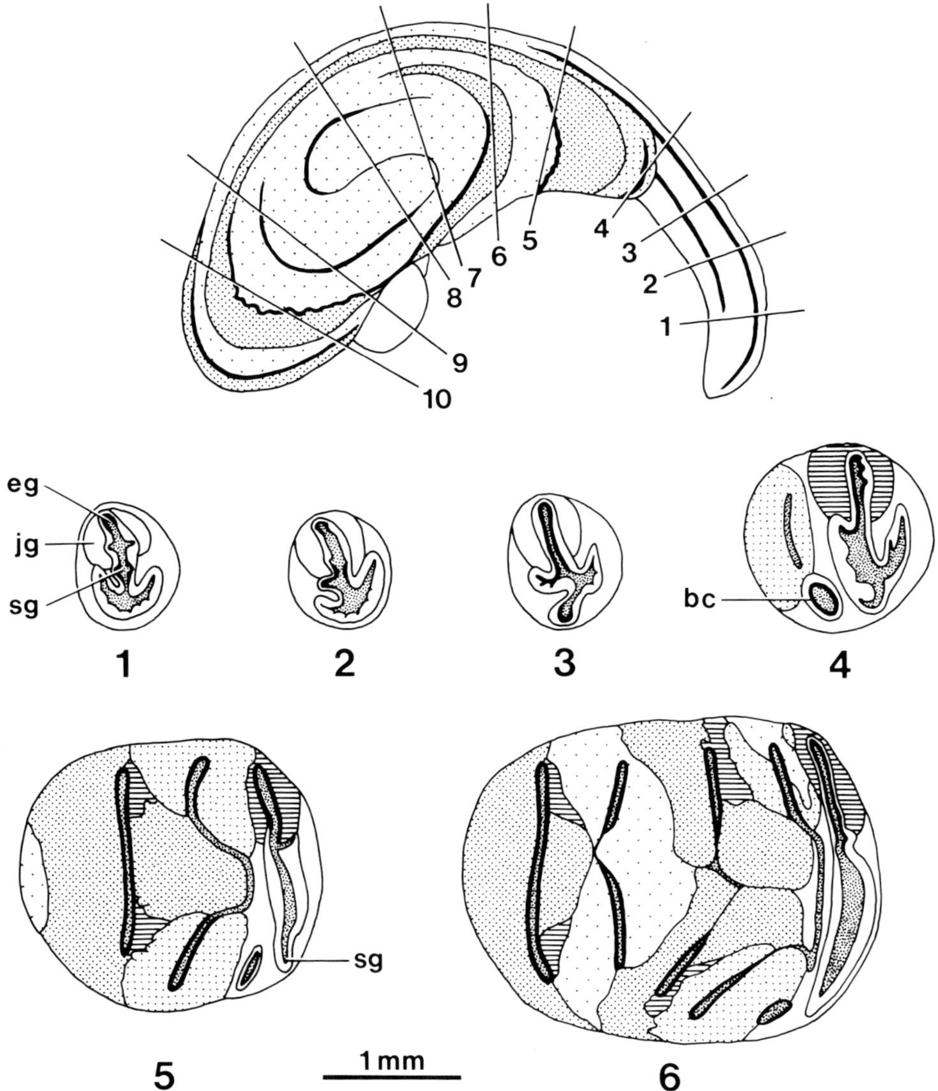


**Fig. 8** *Littoraria* (*Palustorina*) *melanostoma*: serial sections of pallial oviduct; abbreviations and shading as in Fig. 6.

representative of the ovoviviparous group (Fig. 9). In this species the opaque region of the albumen gland occupies the first anticlockwise ascending whorl of the spiral and is followed by the translucent albumen gland which continues for half a revolution to the apex, and through a further two descending clockwise whorls into the straight portion of the oviduct (Fig. 10). The two parts of the albumen gland are similar in appearance and staining reactions to the homologous glands in *L.* (*Palustorina*) *melanostoma*. The small amount of glandular tissue around the egg groove running through the straight section appears to be the reduced jelly gland; staining reactions are similar to those of the translucent albumen gland, although in ABPAS the tissue stains red rather than magenta.

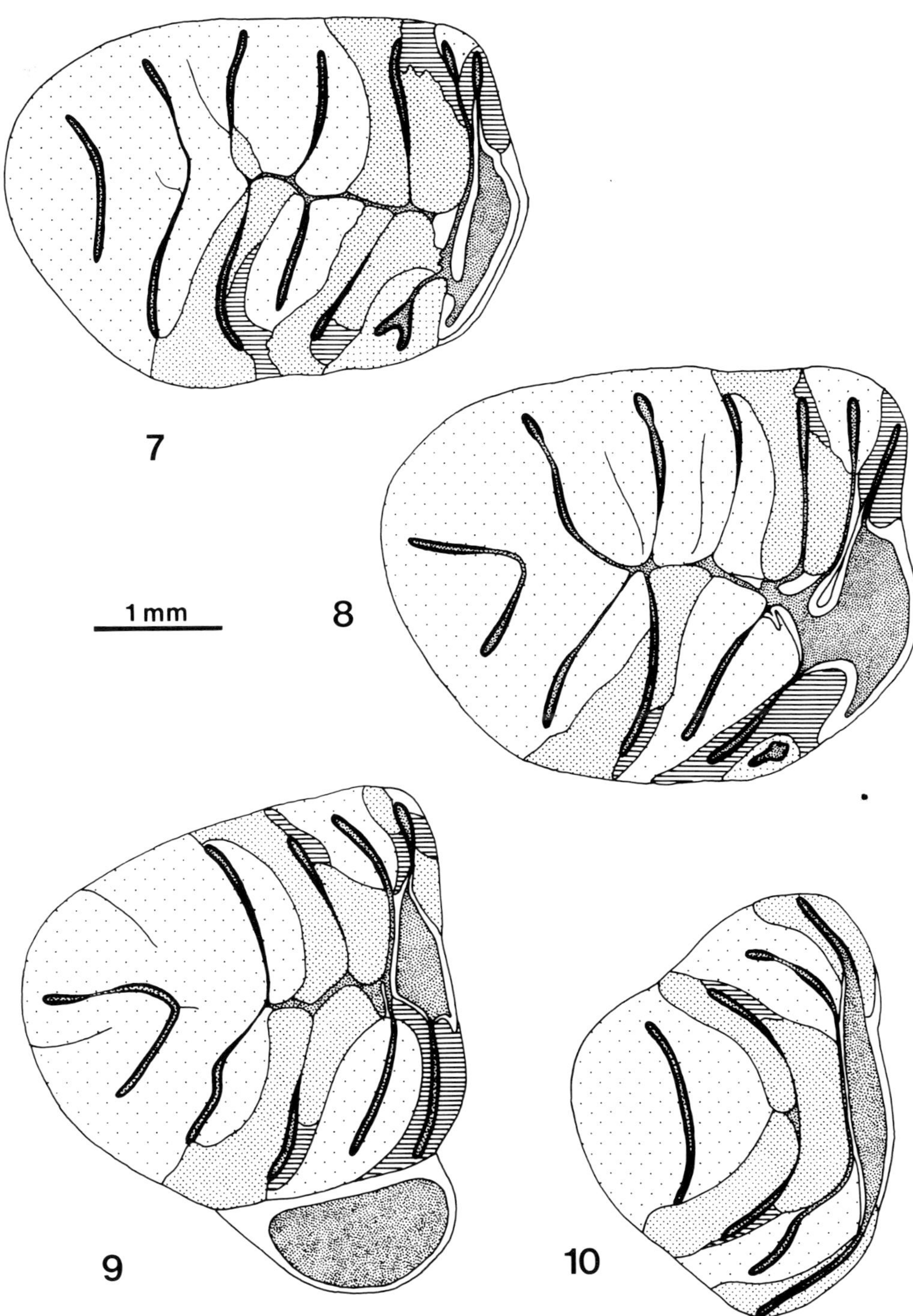


**Fig. 8** continued.

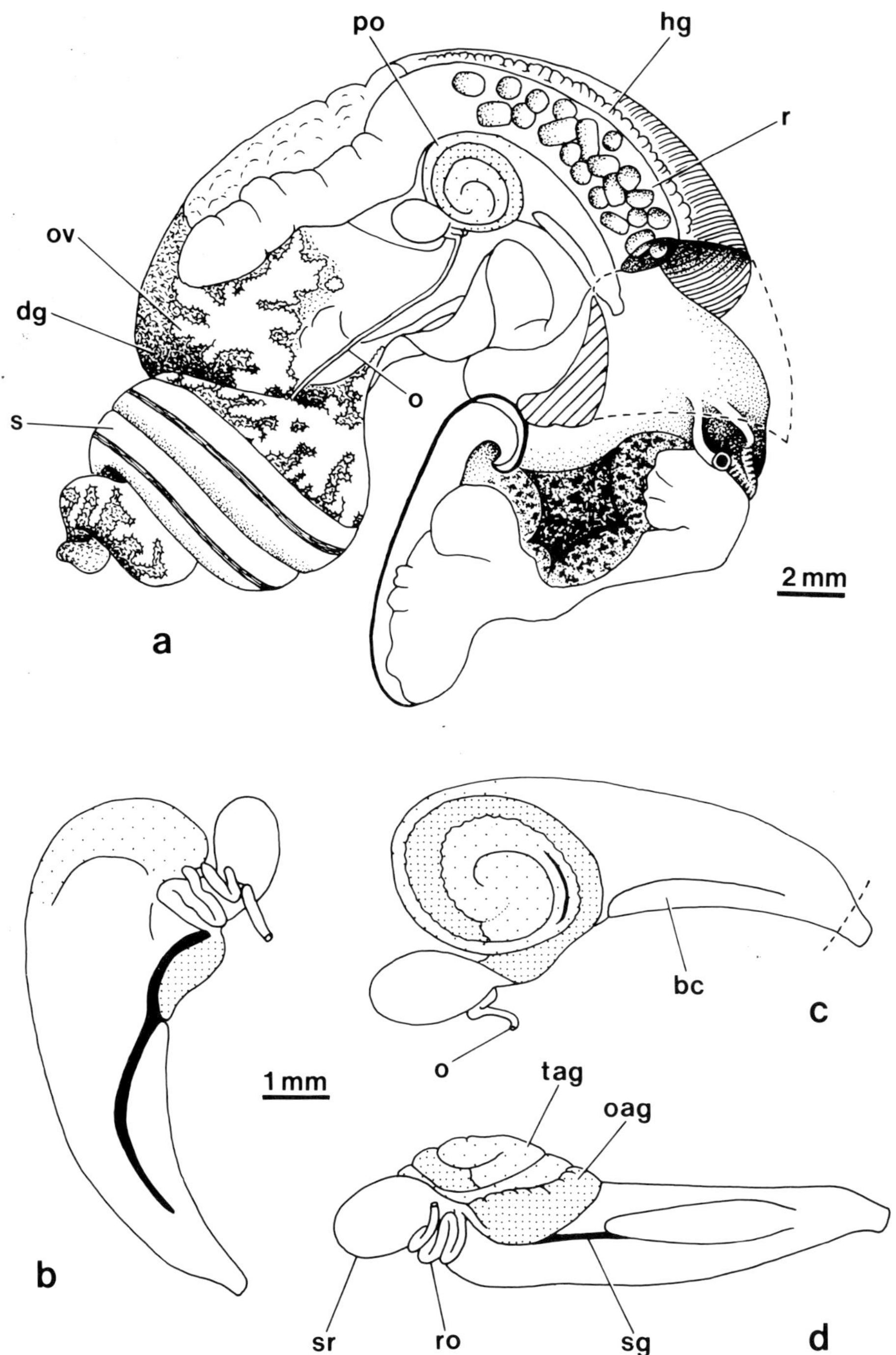


**Fig. 9** *Littoraria* (*Littorinopsis*) *scabra:* female reproductive tract; **(a)** animal removed from shell; **(b,c,d)** medial, lateral and ventral views of pallial oviduct; abbreviations and shading as in Figs. 6,7; dashed lines indicate cut mantle tissue.

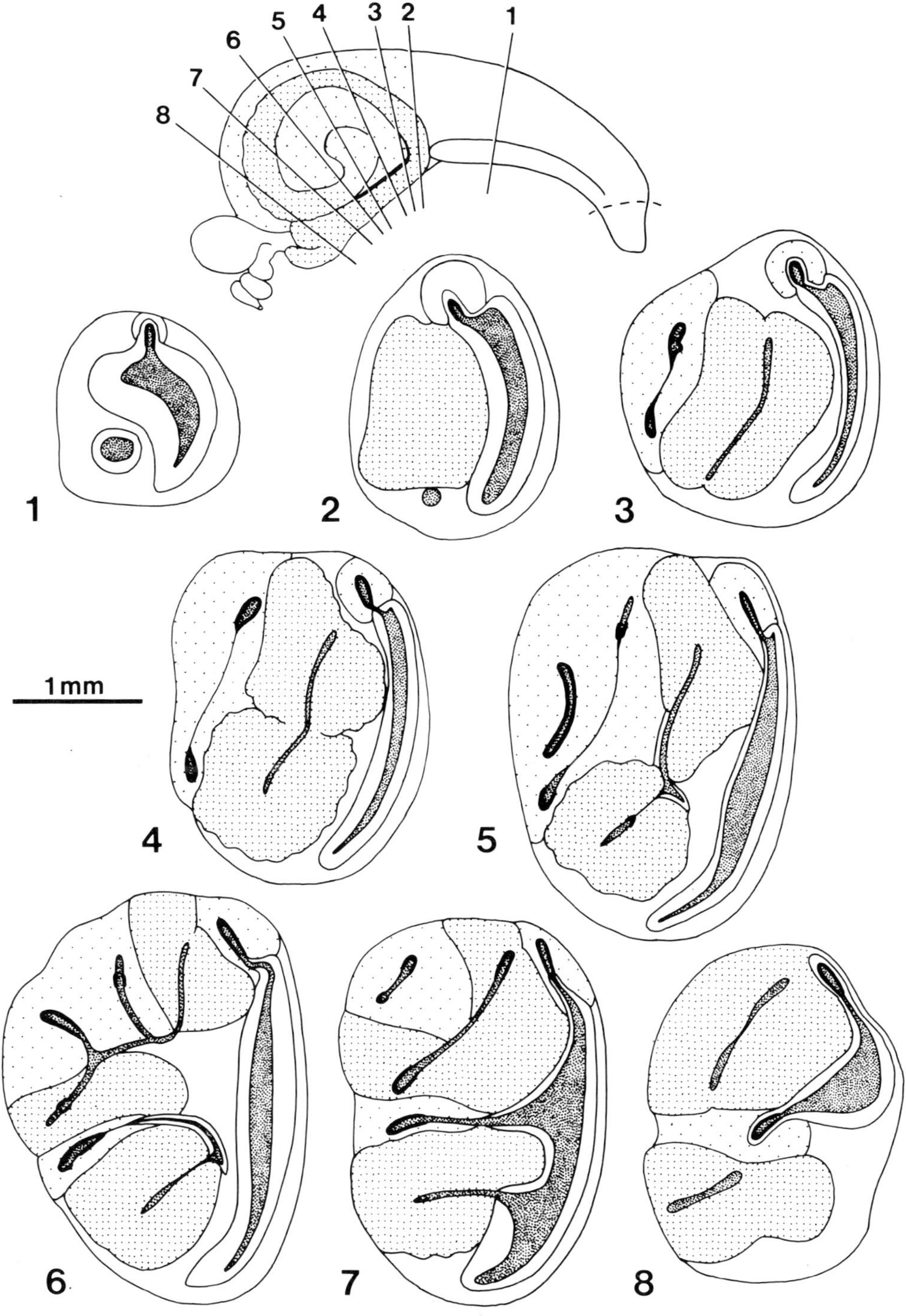


**Fig. 10** *Littoraria (Littorinopsis) scabra:* serial sections of pallial oviduct; shading as in Fig. 6.

Both types of capsule gland are absent. The epithelial lining of the pallial oviduct is closely similar to that described for *L. melanostoma*, except for the fact that the goblet cells staining blue in ABPAS (indicating acidic mucins), while still concentrated in the sperm groove, are also found in smaller numbers through the convoluted epithelium of the final chamber and straight section. Despite the absence of capsule glands, embryos and veligers are enclosed within a spherical covering while retained in the mantle cavity; the origin of this covering layer is unknown.

It is possible to speculate upon the evolution of the pallial oviduct in the family Littorinidae. It is likely that the pallial oviduct of prosobranchs arose as a straight, elongate gutter in the pallial wall (e.g. Fretter & Graham, 1962; Ponder, 1968) which in most groups has closed ventrally to form a tube, as represented diagrammatically in Fig. 6a. Amongst the various littorinid genera examined during the present study, certain species of *Nodilittorina* (*N. angustior*, *N. australis*, *N. dilatata*, *N. hawaiiensis*, *N. millegrana*, comprising 'group 1' in Fig. 11), of *Fossarilittorina*, and also *Echininus antoni*, approximate most closely to this hypothetical ancestral condition. In these forms the albumen gland occupies a proximal loop of the egg passage, while capsule glands and jelly gland follow an almost straight course to the genital opening. Although there is as yet no evidence to support the suggestion that this form of the oviduct is in fact primitive for the family, it is nevertheless possible to arrange the oviducts of other species in a morphological sequence suggesting how progressive elaboration of a simple oviduct might have occurred (Fig. 11). The single spiral form found in *Littoraria* is not unique to the genus for very similar oviducts are seen in *Cenchritis muricatus* and *Tectarius grandinatus*. The spiral oviduct of *Bembicium* is only superficially similar, and lacks capsule glands (pers. obs.; see also Anderson, 1958; Bedford, 1965). The condition in *Echininus cumingi* and *Nodilittorina ziczac* is intermediate between that of *Tectarius grandinatus* and the simplest form of the oviduct. In some species of *Nodilittorina* the capsule gland is thrown into a loop (*N. pyramidalis*, *N. unifasciata*, comprising 'group 2' in Fig. 11), while in others both capsule gland and jelly gland are each thrown into a loop (*N. knysnaensis*, *N. praetermissa*, comprising 'group 3' in Fig. 11) and in *Melarhaphe neritoides* the jelly gland is elaborated into a spiral. The condition in *Littorina* is distinctive, with albumen and capsule glands each forming a spiral, while the jelly gland follows a relatively straight course. The sequence of Fig. 11 should not be interpreted as indicative of evolutionary relationships between the species illustrated, but it does suggest that derivation of the oviduct of *Littoraria* can most reasonably be postulated from the condition found in the genera *Nodilittorina*, *Fossarilittorina*, *Echininus*, *Tectarius* and *Cenchritis*, rather than from the specialized forms found in modern species of *Littorina* and *Melarhaphe*.

Amongst the oviparous species of *Littoraria*, the simplest oviduct is that of *L. pintado*, with only $3\frac{1}{2}$ whorls in the spiral portion. In other respects also this species displays characters regarded as primitive in this genus (Fig. 18). The most elaborate oviducts are found in the Eastern Pacific species *L. zebra* and *L. fasciata*, the latter showing a spiral of $9\frac{1}{2}$ whorls. The degree of separation of the bursa from the straight section of the oviduct appears to be another character of phylogenetic significance in the genus. In the subgenus *Littorinopsis* all species of which the anatomy is known show a bursa of the type described above in *L. scabra*. This condition, in which the bursa joins the oviduct close to the genital aperture, is also found in five species of the subgenus *Littoraria* (Fig. 18). In contrast, in other members of this subgenus and in *Palustorina*, the bursa is not fully separated from

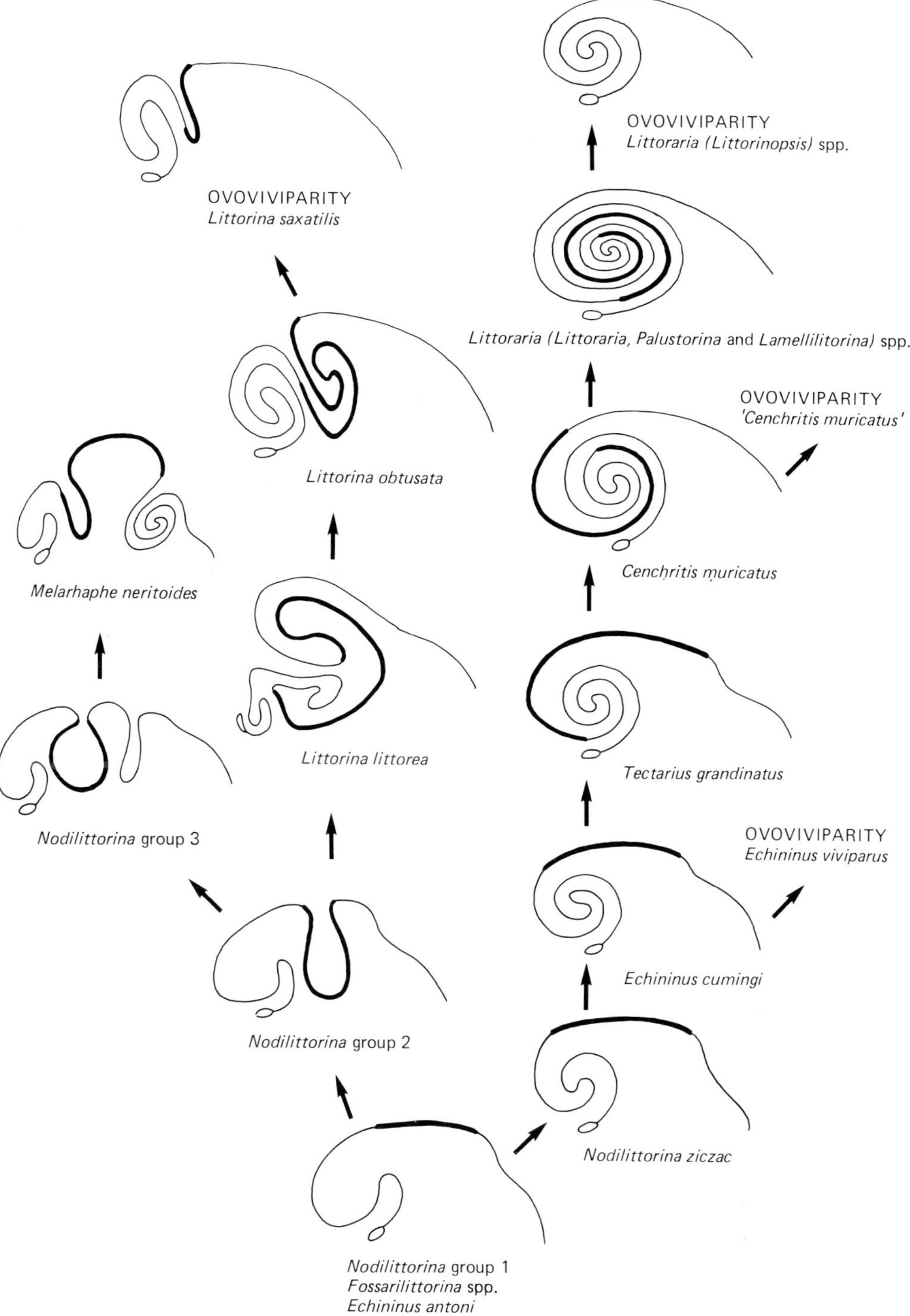


**Fig. 11** Diagrammatic representations of the pallial oviducts of some genera of Littorinidae, arranged in morphological sequence, showing how progressive elaboration of the path of the egg groove may have occurred; opaque capsule gland indicated by thick line; see text for details.

the oviduct until a point close to the anterior part of the spiral section, and this condition appears to be primitive in the genus *Littoraria*. Amongst the other littorinid groups examined, the bursa shows the more posterior opening in most but not all species of *Nodilittorina* and in *Littorina littorea*. Marcus & Marcus (1963) described two bursae in *Nodilittorina lineolata* (as *Littorina ziczac*, but see Bandel & Kadolsky, 1982), but this observation is probably erroneous.

The subgeneric division of *Littoraria* proposed here (p. 76) separates the ovoviviparous species as the subgenus *Littorinopsis*. Within the genus, as in the family as a whole, oviparity is certainly the ancestral condition. Little structural modification of the pallial oviduct has occurred with the development of ovoviviparity; the oviduct has merely been simplified by the loss of capsule glands and reduction in number of whorls of the spiral groove. Since the form of the oviduct, and also the method of brooding in the mantle cavity to the early veliger stage, are so similar in the species of *Littorinopsis*, they may be a monophyletic group (Fig. 18). The anatomy of *L. aberrans* from the eastern Pacific is as yet unknown, and so, consequently, is its subgeneric assignment, but in view of the character of its protoconch (p. 14) it would appear to undergo nonplanktotrophic development. The thin and colour polymorphic shell of this species suggests a high supratidal zonation, so that ovoviviparity (rather than a benthic egg mass) might be predicted, but whether embryos are brooded in the oviduct or in the mantle cavity remains to be discovered.

It may be noted that the ovoviviparous habit has evolved independently in several other littorinid groups. In *Littorina* it has been described in the species *saxatilis* and *neglecta* (e.g. Fretter, 1980), in which embryos are retained within the brood pouch, formed from the distal section of the pallial oviduct, until after metamorphosis (Linke, 1933; Fretter & Graham, 1962, p. 343). Rosewater (1982) described *Echininus viviparus*, which broods its young in a similar fashion, although not closely related to *Littorina* (Fig. 17). The brooding of embryos to the early veliger stage between the lamellae of the gills has only been described in the subgenus *Littorinopsis*. However, Bandel (1974) noted the release of veligers by *Cenchritis muricatus*, although other authors have found this species to release egg capsules (Lebour, 1945; Borkowski, 1971).

Variation of the type of larval development within a single species (as reviewed by Robertson, 1974, and Jablonski & Lutz, 1983) is extremely rare in gastropod molluscs. Many reported cases have subsequently been explained by the existence of unrecognized sibling species (Jablonski & Lutz, 1983). Several littorinids have been cited as examples of species with variable development. In *Littorina 'saxatilis'* (e.g. Fretter & Graham, 1962; Robertson, 1974) the ovoviviparous and oviparous forms are now recognized as distinct species, partly on the basis of oviducal anatomy (Heller, 1975*a*; Sacchi, 1975; Hannaford Ellis, 1979). Although the taxonomic significance of developmental type and the associated oviducal anatomy is still disputed in the case of the pair of species *Littorina saxatilis* and *Littorina arcana* (Caugant & Bergerard, 1980; Smith, 1982), differences in allele frequency have been demonstrated in sympatric populations (Ward & Warwick, 1980), supporting the case for a specific difference. A report of a short pelagic stage in *Littorina 'obtusata'* by Tattersall (1920) has never been confirmed, for although two closely related species are now recognized (Sacchi & Rastelli, 1967), subsequent observations have been of nonplanktotrophic development in benthic egg masses. The variation in development of *Cenchritis muricatus* noted by Bandel (1974) suggests that the form requires taxonomic study. In the genus *Littoraria* the following species have been reported to release both veligers and earlier developmental stages: *L. pallescens* (Abe, 1942, as *Melaraphe scabra*; c.f. pers. obs. of veligers in mantle cavity); *L. angulifera* (Lenderking, 1954; c.f. Gallagher & Reid, 1974) and *L. intermedia* (Struhsaker, 1966, as *Littorina scabra*). As suggested by Struhsaker (1966), release of embryos rather than veligers may be an abnormal response to laboratory conditions. In the present study females of ovoviviparous species have occasionally been found to release embryos prematurely upon disturbance. In each of these three species the embryonic stages shed were devoid of egg capsules, so that this habit represents only a very minor modification of the normal ovoviviparous habit. Anatomical examination has provided no suggestion that any species of *Littoraria* is able to show both oviparous and ovoviviparous modes of development. In conclusion, the release of veligers as opposed to egg capsules, and the associated differences in anatomy of the oviducts described above, must be accepted as evidence of at least specific difference in *Littoraria*, as in other littorinids.

In the present study the form of the pallial oviduct has been used principally to define generic and subgeneric groups and to suggest the method of development in species for which it is unknown. Although not one of the most useful characters at the specific level, variation between species of the same subgenus can be observed in such details as the number of whorls of the spiral section, extent of the glandular components, and relative lengths of the bursa, spiral and straight portions. Usually the details of the arrangement of the spiral egg groove and relative sizes of glandular lobes, as seen in transverse section, are not of great importance, for these vary somewhat between individuals and are dependent upon the exact plane of the section. Use of the anatomy of the pallial oviduct as a taxonomic character is complicated by the reduction in size and darkening in colour of the organ, to grey or brown, which takes place outside the breeding season. However, these changes are largely a consequence of a decrease in thickness of the glandular walls of the duct; the form and structure do not change. Likewise in females infected with parasitic trematodes the gonad and oviduct may be reduced in size, but trematode infections are rare in *Littoraria* species.

### *Egg capsules*

As discussed in the preceding section, species of *Littoraria* in which spawning has been described produce either pelagic egg capsules or release veligers after a period of development in the mantle cavity. *L. aberrans* is an apparent exception, with some form of nonplanktotrophic development. Spawning and development of the western Atlantic *L. angulifera* have been investigated by Lebour (1945), Lenderking (1954) and Gallagher & Reid (1974), and of *L. irrorata* by Bingham (1972*b*) and Gallagher & Reid (1974), while the egg capsules alone have been described in *L. flava* (Marcus & Marcus, 1963) and *L. nebulosa* (Bandel, 1974). Of the Indo-Pacific species only the following have been examined in any detail: *L. pallescens* (Abe, 1942, as *Melaraphe scabra*); *L. pintado* and *L. intermedia* (Struhsaker, 1966, the latter as *Littorina scabra*); *L. melanostoma* (Berry & Chew, 1973); and *L. luteola* (Muggeridge, 1979, as *Littorina scabra*). The egg capsules of *L. coccinea* and *L. undulata* have been illustrated by Rosewater (1970) and of *L. articulata* by Kojima (1958*c*, as *L. strigata*). From these accounts the duration of brooding in ovoviviparous species varies from four days in the tropical *L. angulifera* to seventeen days in the temperate

*L. luteola*, while capsules of *L. irrorata* hatch in one to two days, *L. pintado* in three, and *L. melanostoma* in seven days. Gallagher & Reid (1974) suggested a planktotrophic life of eight to ten weeks before settlement in *L. angulifera*.

In the ovoviviparous species of *Littoraria* the embryo is contained within an egg covering alone and no capsule is produced. Amongst the oviparous members of the genus, the egg capsules have been described in eight species, as illustrated in Fig. 12. In each case the capsule is a more or less symmetrical biconvex disc, ranging from 150 to 400 μm in diameter, and typically containing a single ovum. The capsule is surrounded by a circumferential flange or lamella, which in *L. coccinea* and *L. glabrata* (=*L. kraussi*) is turned down to form a flotation skirt (Rosewater, 1970). The capsule of *L. pintado* (Struhsaker, 1966) differs somewhat from the others in the genus and is considered below.

The utility of the form of the egg capsule as a taxonomic character has been disputed in the past (Borkowski, 1975). However, revised species concepts based upon detailed anatomical studies have shown that in several groups of littorinids the form of the egg capsule is a useful diagnostic character (Murray, 1979; Bandel & Kadolsky, 1982). In the genus *Littoraria* the published drawings suggest that egg capsule form might be species specific in many cases, although at least in *L. articulata* there appears to be some variation within a species (p. 206).

The range of capsule types produced by members of the family Littorinidae has been reviewed most recently by Bandel (1974) and by Jordan & Ramorino (1975). Systematic trends in capsule form above the species level have been denied (Rosewater, 1970), but in the light of new generic concepts (Bandel & Kadolsky, 1982) some generalizations can be made. The capsules of oviparous species of the genus *Littoraria* are rather distinctive (Fig. 12), although similar to those described in *Cenchritis muricatus* (Lebour, 1945; Borkowski, 1975; figures given by these authors differ, as illustrated in Fig. 12). Species of *Littorina* with planktotrophic development show capsules of similar form again (e.g. Buckland-Nicks *et al.*, 1973; Kojima, 1957, 1958*b*; Linke, 1933; Murray, 1979; Schmitt, 1979; illustrated in Fig. 12), although usually of larger size and frequently enclosing several eggs. The capsules of *Nodilittorina* contain single eggs and are more elaborate, with a cupola sculptured by concentric rings, the edge of the cupola overhanging the lower half of the capsule as a peripheral band or skirt (e.g. Struhsaker, 1966; Borkowski, 1971; Pilkington, 1971; Bandel, 1974; Jordan & Ramorino, 1975; Bandel & Kadolsky, 1982). A similar shape has been reported in *Echininus antoni* (=*E. 'nodulosus'*, Bandel & Kadolsky, 1982) and in *Fossarilittorina meleagris* (Borkowski, 1971; illustrated in Fig. 12).

These trends in capsule form have to a certain extent been overlooked, because in several littorinid groups closely related species may show a range of developmental types, as exemplified by the genus *Littorina*, with species producing pelagic capsules, benthic egg masses or retaining embryos in the oviduct. Littorinids known to produce a benthic gelatinous spawn without individual egg capsules include six species of the genus *Littorina* (Kojima, 1958*a*; Buckland-Nicks *et al*, 1973; Fretter, 1980), three of *Bembicium* (H. Anderson, 1958; D.T. Anderson, 1961, 1962) and one of *Risellopsis* (Pilkington, 1974). This type of spawn has been described also in several Antarctic littorinaceans (Picken, 1979) and in members of the family Lacunidae from Europe (Fretter & Graham, 1962, p. 388). On the basis of the spawn in Lacunidae, 'the most primitive family of the Littorinacea', Fretter (1980) suggested that production of this type of benthic gelatinous

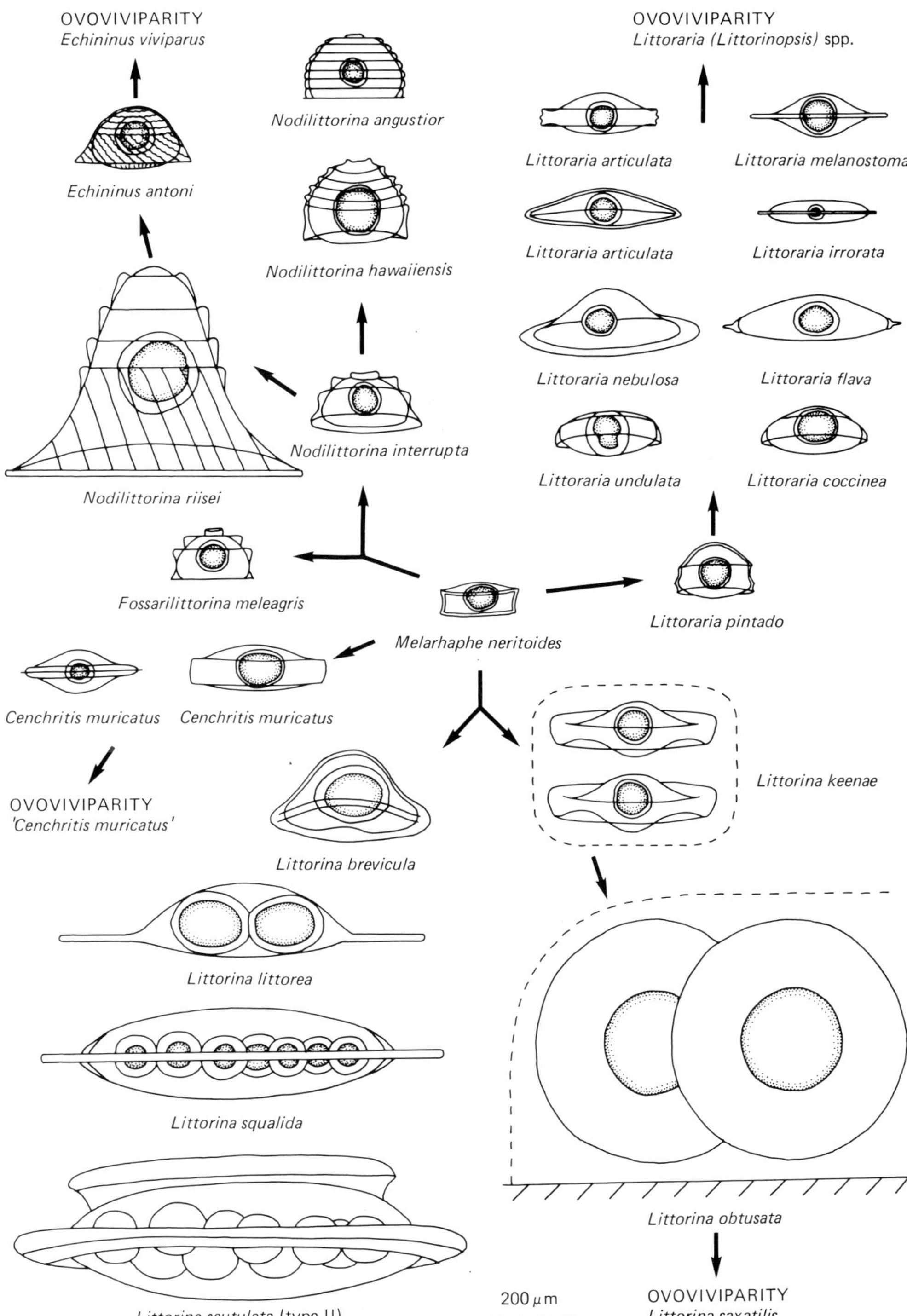


**Fig. 12** Egg capsules of Littorinidae, arranged to show possible derivation of the forms characteristic of several genera from a simple pelagic capsule (represented by that of *Melarhaphe neritoides*). (After Bandel, 1974; Bandel & Kadolsky, 1982; Berry & Chew, 1973; Bingham, 1972*b*; Borkowski, 1971; Kojima, 1957, 1958*b,c*,; Lebour, 1935, 1945; Linke, 1933; Marcus & Marcus, 1963; Murray, 1979; Rosewater, 1970; Schmitt, 1979; Struhsaker, 1966).

egg mass was the primitive condition in the Littorinidae. If these families are sister groups in the sense of Hennig (1966), this is a reasonable assumption. However, their relationship is uncertain and Ponder (1976) has questioned whether the two groups should be maintained as distinct families. It is also noteworthy that benthic egg masses are only known to occur in littorinaceans from cold or temperate latitudes. Various prosobranch and other invertebrate groups show a tendency to eliminate planktonic developmental stages at high latitudes (review in Picken, 1979), and the distribution of benthic spawn in littorinaceans may be the result of adaptation to similar ecological conditions rather than retention of an ancestral character.

If a benthic spawn is indeed primitive in the Littorinidae, then as suggested by Fretter (1980) the pelagic capsule of, for example, *Littorina littorea* could be derived from the ancestral state, as represented by *Littorina obtusata*, via the intermediate condition shown by *Littorina mandshurica* Schrenk (Kojima, 1958*c*) and *Littorina squalida* Broderip & Sowerby (Kojima, 1958*b*), in which a large capsule contains up to fourteen eggs. Reduction in the number of eggs to one, and modification of the capsule shape, could then have given rise to the condition of *Melarhaphe neritoides*, from which other forms could be derived. One difficulty in this scheme is that the similarity between a benthic gelatinous egg mass lacking capsules and the pelagic capsules of *Littorina* species containing many eggs is deceptive. If the interpretation of the function of the glands of the pallial oviduct given in the preceding section is correct, then the jelly of the benthic egg mas is produced by the jelly gland, while that in the capsule is a product of the opaque capsule gland. Accordingly, derivation of the pelagic capsule from the gelatinous spawn would require the following modifications of the oviduct: increase in size of the opaque capsule gland, development of a new gland (the translucent capsule gland, to secrete the capsule) and reduction in size of the jelly gland.

In view of the characteristics of the spawn of *Littorina sitkana* (Buckland-Nicks *et al.*, 1973) and of *Littorina keenae* (Murray, 1979, as *Littorina planaxis*), a new interpretation can be offered. If release of pelagic capsules were the ancestral state, elaboration of the jelly gland would have allowed production of a spawn mass with embedded capsules. In *Littorina keenae* the spawn mass is pelagic and soon disintegrates to release the individual capsules, while in *Littorina sitkana* it is attached to the substrate and benthic juveniles emerge. When hatching takes place within the spawn mass, the need for a protective egg capsule must be reduced and loss of the translucent capsule gland may follow, thereby deriving a spawn like that of *Littorina obtusata*, without individual egg capsules (Goodwin, 1979). In support of this suggestion it can be argued that, as discussed above, benthic development may be a response to environmental conditions at high latitudes. Further, in most genera a few species show capsule forms which could have been derived from a generalized ancestral type. This type may be represented by *Melarhaphe neritoides*, with a simple 'pill box'-shaped capsule, closely similar to that of *Cenchritis muricatus* (Lebour, 1945), and requiring only slight modification by development of the lower rim as a narrow skirt, and duplication of the upper rim as rings of sculpture on a cupola, to derive capsules like those of *Littoraria pintado* (Struhsaker, 1966), *Fossarilittorina meleagris* (Borkowski, 1971) and *Nodilittorina interrupta* (C.B. Adams in Philippi) (Bandel, 1974, as '*Littorina* sp.', but see Bandel & Kadolsky, 1982). This scheme is illustrated in Fig. 12.

Although this second scheme is the more attractive on the basis of available evidence, it is highly speculative. Much more information is required, especially concerning the anatomy and relationships of the lacunids and of the small littorinacean forms from the Antarctic. The two schemes may not even be mutually exclusive, for the pelagic capsule might be ancestral with respect to the genera discussed, and yet derived in the superfamily as a whole.

Whichever scheme is considered more likely, it appears that within the genus *Littoraria* the rather *Nodilittorina*-like capsule of *L. pintado* may be ancestral and the symmetrically biconvex capsule forms derived. This suggestion is in agreement with the cladogram of Fig. 18. The scheme of Bandel (1974) shows similarities to Fig. 12, but places the ovoviviparous species without capsules in a central position. This cannot be accepted, since these species are undoubtedly derived from closely related oviparous species. Ovoviviparous species of *Littorina* were probably derived from those with large jelly glands producing benthic spawn (Fretter, 1980), while in the genus *Littoraria* the cladogram of Fig. 18 suggests derivation of ovoviviparous species from those producing symmetrically biconvex capsules containing single eggs. Ovoviviparous species of *Echininus* and *Cenchritis* were presumably likewise derived from ancestors with pelagic capsules.

## *Radula*

The littorinid radula is of the generalized taenioglossate type. Each transverse tooth row consists of a central rachidian, flanked on each side by a lateral and an inner and outer marginal tooth, all bearing posteriorly directed cusps. The range of variation within the family in the form of the radular teeth can be appreciated from the figures of Rosewater (1970, 1972) and Bandel (1974). The function of the littorinid radula has been described by Ankel (1936, 1938) and the mechanical relationships of the teeth by Bandel & Kadolsky (1982). Rosewater (1980*a*) has classified littorinid radulae into five major types, on the basis of the form of the rachidian tooth, and has correlated each type with a particular habitat and food source. The importance of radular characters in taxonomic studies has been emphasized by Rosewater (1970, 1972, 1981), but he has made use of them only at subgeneric and higher levels of classification. Recently, Bandel & Kadolsky (1982) have discriminated between species of *Nodilittorina* solely on the basis of radular characters.

In all twenty species of the *scabra* group the radulae were found to be of the type defined by Rosewater (1980*a*) as 'hooded', that is with a frontal plate anterior to the cusps of the rachidian tooth. Rosewater has suggested that this radular type may be adapted for feeding upon the algal flora of mangroves, driftwood and marsh grass. It is indeed striking that the majority of species known to possess radulae of the hooded type are found upon these substrates, and in addition to the species of the *scabra* group there can be listed the following: *L. angulifera* (Marcus & Marcus, 1963; Bandel, 1974; Rosewater, 1980*b*), *L. nebulosa* (Troschel, 1858; Bandel, 1974), *L. irrorata* (Troschel, 1858; Allen, 1953), *L. flava* (Marcus & Marcus, 1963), *L. aberrans* (Rosewater, 1980*b*), *L. varia*, *L. zebra* and *L. fasciata* (all Rosewater, 1980*a*). Nevertheless, species with hooded radulae occur also on rocky shores, for example *L. undulata* (Adam & Leloup, 1938; Rosewater, 1970) and *L. cingulifera* (Rosewater, 1981). Amongst the species living upon mangrove trees several can be found also upon sheltered rocks (e.g. *L. articulata*, *L. strigata*). Since all the species known to possess a hooded radula are members of the genus *Littoraria*, and because they are not entirely restricted to mangrove and similar habitats, it seems likely that this specialized radular type is not only of ecological, but also of phylogenetic significance, supporting classification of these species in a single generic group.

The radulae of *L. pintado*, *L. coccinea*, *L. glabrata* and *L. mauritiana*, all illustrated or described by Rosewater (1970), are apparently not hooded. Since a hooded rachidian is unknown in other littorinid genera, the species of *Littoraria* lacking this feature appear to retain the ancestral character state (Fig. 18). Within the family Littorinidae there has been a tendency towards reduction of the width of the rachidian tooth in several lines (Bandel, 1974; Bandel & Kadolsky, 1982). A wide rachidian, referred to as 'rhomboidal' by Rosewater (1980*a*), appears to be the ancestral type and is found in the genera *Littorina*, *Laevilitorina*, *Pellilitorina* (as reviewed by Rosewater, 1980*a*) and in some species of the Lacunidae also (Troschel, 1858).

Amongst the species of *Littoraria* described here, two different forms of the hooded radula can be recognized, which for convenience will be referred to as 'saw-toothed' and 'chisel-toothed'. The radulae of *L. scabra* (Fig. 28e,f) and *L. cingulata cingulata* (Fig. 57b) may be used as examples of the saw-toothed form; the rachidian tooth has three well developed cusps, the larger central cusp being pointed or spade shaped; the paired teeth bear cusps of almost equilaterally triangular shape and all of roughly similar size; cusps on the lateral teeth are five in number, of which the central one is the largest, in front of which is a more or less prominent gap, and the other cusps on the laterals are also clearly visible; the inner marginal has four subequal cusps; the outer marginal bears up to six small cusps, of which the outermost is the largest. In the chisel-toothed form, typified by *L. melanostoma* (Fig. 74b), the number of cutting edges is much reduced; the rachidian bears a single, broad, straight, cutting edge formed from the widened central and two small lateral cusps; the cusps of the paired teeth are obliquely triangular, with their points directed towards the midline, so that the cutting edges are effectively long blades rather than pointed teeth; the lateral bears five cusps, but only two of these are large, with a wide gap between, the inner one and outer two cusps being reduced to denticles; the inner marginal has four cusps, of which again only the central two are large; the outer marginal bears only two large cusps. There are, however, intermediates between the extreme saw- and chisel-toothed types. These two main radular forms lend at least partial support to the subgeneric classification adopted here. The chisel-toothed type as described is found only in *L.* (*Palustorina*) *melanostoma* and *L.* (*P.*) *conica*, while other species in the subgenus show radulae of intermediate type. In the subgenus *Littorinopsis* the majority show typical saw-toothed radulae, with the exceptions of *L. pallescens* and *L. lutea* which are intermediate. *L.* (*Lamellilitorina*) *albicans* is of the chisel-toothed type, but for its unique rachidian and somewhat toothed outer marginal. *L.* (*Littoraria*) *vespacea* is of the saw-toothed type. The two main morphological types do not correspond with habitat differences, since both include leaf and trunk dwelling species.

From the present study it appears that radular characters are of use mainly in separating groups of species, which from other evidence seem to be closely related. Taxonomic studies of the genus *Littorina* (e.g. Heller, 1975*a*; Raffaelli, 1979) have likewise not revealed useful differences between closely related species. Bandel (1974) was able to distinguish between the radulae of eighteen Atlantic species of the family, although these included members of seven genera, while Bandel & Kadolsky (1982) separated the radulae of twelve species of the genus *Nodilittorina* from the Atlantic and defined some species solely on this basis. Small differences can sometimes be found between species in each of the categories of the hooded radula described above. Characteristics of the rachidian tooth are especially obvious, for example that of *L. albicans* is unique in bearing three equal cusps, while in three species of the subgenus *Littorinopsis* (*L. filosa, L. cingulata, L. subvittata*) the central cusp is rather elongated. Other subtle differences seem to exist in the details of cusp sizes and shapes, but these are difficult to measure or to describe. It is possible that accurate measurement of cusps with the light microscope (as made by Borkowski, 1975) might permit finer discrimination, but are beyond the scope of this work.

There are several difficulties to be considered when comparing the details of the radular cusps. Although all radulae were mounted flat and viewed from above it was found that even slight differences in the orientation of individual teeth altered their appearance considerably. Furthermore, the natural orientation of the teeth obscured certain details of the cusps. These difficulties are demonstrated in Fig. 28e,f, showing standard and side views of the same radula. Light microscopy is more suitable for detailed comparisons of cusp shapes (Hickman, 1977). Because of these difficulties it is hard to judge the extent of variation between the four specimens examined for each species, but in general the range is apparently small. Of previous studies of littorinid radulae only those of Borkowski (1975) and Goodwin & Fish (1977) have made attempts to describe intraspecific variation and both showed a considerable range. Borkowski also pointed out possible regional variations from comparisons with published figures. However, so much depends on tooth orientation that it is very difficult to make comparisons between the figures given by different authors, and especially to compare camera lucida drawings with scanning electron micrographs. All specimens examined here were taken from large adults, since changes in tooth form with size have been recorded by Borkowski (1975) and Raffaelli (1979).

Radulae of *L. 'scabra'* have been figured by several authors, but because of the similarities within the group it is not usually possible to assign the figures to the correct species in the absence of other information. In two cases radulae have been used in discussions of the classification of the *scabra* complex. Adam & Leloup (1938) figured the tooth row of *L. scabra s. s.* (as 'forme typique') and of *L. pallescens* (as *Littorina scabra* var. *filosa*). Despite the obvious differences between the figures, these authors commented that the radulae were 'almost completely identical'. Rosewater (1980*b*) gave scanning electron micrographs of *L. scabra s. s., L. angulifera* and *L. aberrans*; of which the first two were closely similar, while the last showed certain differences. Nevertheless, in considering the three species as subspecies of *L. scabra*, Rosewater stressed their similarities in his descriptions. In conclusion, if radular characters are to be used for taxonomy at the specific level in this family, then differences finer than those admitted by most previous authors must be sought, and validated by adequate studies of intraspecific variation.

Littorinids from mangroves are known to have relatively shorter radulae than those of rocky shore species (e.g. Peile, 1937; Marcus & Marcus, 1963), presumably because of the need for more rapid tooth replacement when the radula abrades a harder surface. The present study supports this observation, radular lengths varying from 0.58 to 1.35 times the height of the shell, which may be compared with a mean value of 2.2 in the rock dwelling *Littorina saxatilis* complex (James, 1968) and an extreme value of 8.2 in '*Littorina ziczac*' (Peile, 1937, i.e. *Nodilittorina* sp.). Furthermore, the three species found mainly upon the leaves of *Avicennia* mangrove trees showed amongst the shortest radulae (mean relative lengths: *L. filosa* 0.77; *L. albicans* 0.67; *L. luteola* 0.58), while five out of six species occurring largely on *Rhizophora* trunks had radulae approximately equal to or exceeding the length of the shell. Relative radular length may have some value as a taxonomic character, but although quoted in the descriptions has not been used as such in the present monograph. James (1968) found a change in relative radular length with age, although Allen (1953) found no correlation between lengths of the shell and radula.

### *Alimentary system*

The alimentary system of *Littoraria* species (as represented by *L. scabra*, Fig. 13a) is in most respects similar to that of *Littorina littorea* as described by Johansson (1939) and Fretter & Graham (1962, pp. 28–32). Werner (1950, 1951) described the morphology and histology of the tract of *L. irrorata*, but his morphological account was inaccurate in several respects. In the foregut (Fig. 13b,c) a pair of pouches arises from the ventral side of the anterior oesophagus just behind the red buccal mass. These oesophageal pouches are usually bilobed, the anterior lobe being the larger, so that they may appear four in number. They are hollow, separated from the food channel by darkly pigmented longitudinal folds, and are themselves pale, or sometimes black in darkly pigmented animals. In most species the pouches are thin walled, but in *L. carinifera* and *L. conica* they are thick walled, red, enlarged and apparently glandular. The paired salivary glands are pale pink in colour and lie in a dorsal position between the buccal mass and the glandular mid-oesophagus; they are constricted into two parts by the nerve ring around the oesophagus. The details of the oesophageal pouches and salivary glands thus differ slightly from the condition in *Littorina litttorea*, as described by Fretter & Graham (1962), in which the openings of the pouches are constricted and the salivary glands lie entirely posterior to the nerve ring.

Johansson (1939) and Fretter & Graham (1962, p. 30) have described the littorinid stomach in detail, and that of *L. flava* was illustrated by Marcus & Marcus (1963). Stomachs of all the *Littoraria* species examined agree with these accounts. By careful dissection, details of ciliary tracts are visible from the outside. There are three ducts from the digestive gland and the cuticle forming the gastric shield is thickened in two places (Fig. 14). Shapes of the ciliary sorting areas and of the shield show individual variation and no consistent differences between species were detected. In a few species (e.g. *L. melanostoma*, *L. articulata*, *L. strigata*) the blind upper end of the stomach is relatively rather shorter, but this character is not correlated with habitat.

### *Pallial complex*

The ctenidium of *Littoraria* species consists of triangular leaflets attached to the left side of the roof of the mantle cavity (Fig. 3), numbering 60 to 80 in the smaller species (such as *L. strigata*, *L. articulata*, *L. vespacea*) and 100 to 120 in the larger species. Anatomical details have been given by Marcus & Marcus (1963) for *L. angulifera*. The leaflets extend only about halfway across the roof of the mantle cavity and all but the most anterior leaflets are continued as folds as far as the hypobranchial gland adjacent to the rectum. The left edge of each leaflet is white and a little thickened, while the remaining area of the gill, including the folds, is usually black or grey, or white in entirely unpigmented animals. The proportions of the individual leaflets show some small variations between

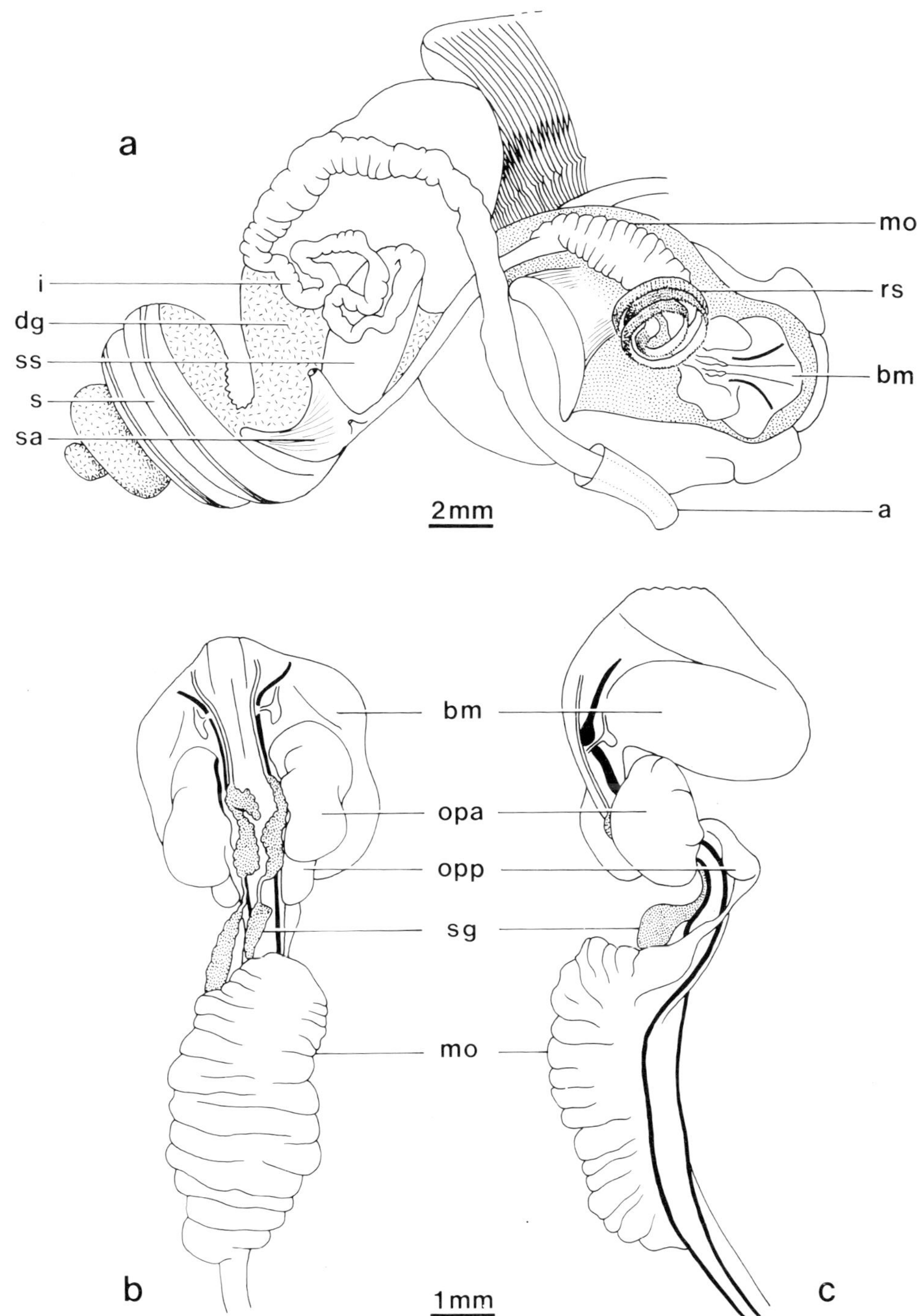


**Fig. 13** *Littoraria* (*Littorinopsis*) *scabra:* **(a)** dissection of alimentary system; **(b,c)** dorsal and lateral views of foregut; **a**, anus; **bm**, buccal mass; **dg**, digestive gland; **i**, intestine; **mo**, glandular mid-oesophagus; **opa**, anterior lobe of oesophageal pouch; **opp**, posterior lobe of oesophageal pouch; **rs**, radular sac; **s**, proximal region of stomach; **sa**, ciliary sorting area; **sg**, salivary gland; **ss**, distal style sac of stomach.

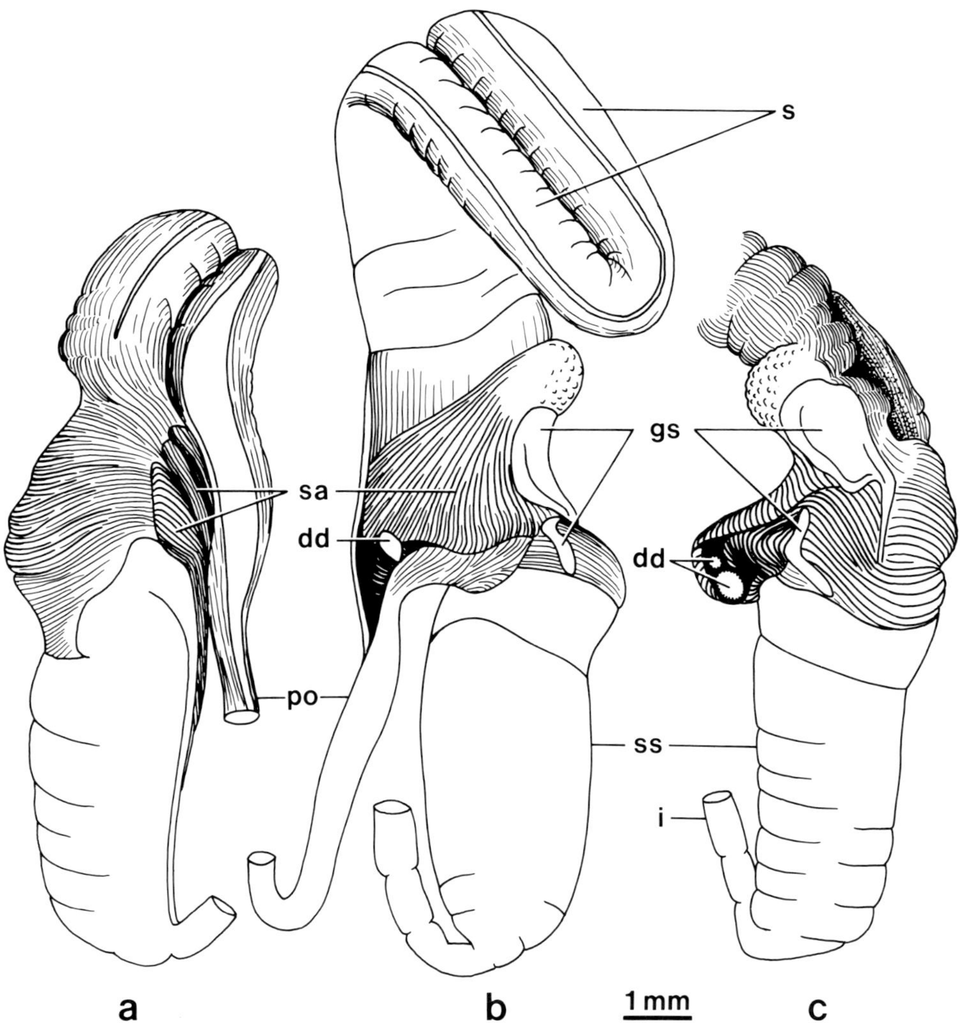


**Fig. 14** *Littoraria (Littorinopsis) scabra:* stomach dissected away from digestive gland; **(a,b,c)** outer, inner and dorsal views; **dd**, ducts of digestive gland; **i**, intestine; **gs**, gastric shield; **po**, posterior oesophagus; **s**, proximal region of stomach; **sa**, ciliary sorting area; **ss**, distal style sac of stomach.

species, but in the subgenus *Littorinopsis* most species show proportions similar to the following for *L. cingulata:* left edge 0.5 mm, right edge 1.2 mm, height of leaflet 0.4 mm, length of fold 2.0 mm. In the subgenus *Palustorina* the leaflets occupy a somewhat greater width of the mantle cavity, as in *L. melanostoma:* left edge 1.3 mm, right edge 1.8 mm, height of leaflet 0.7 mm, length of fold 1.8 mm. In the ovoviviparous species the developing eggs are retained in the mantle cavity, embedded in a thin sheet of mucus over and between the leaflets and folds. At least in some species (e.g. *L. pallescens*, *L. lutea*) the leaflets of females brooding larvae are about twice the usual height.

Several authors have suggested that the littorinids living at supratidal levels on mangrove trees and rocky shores represent an intermediate stage in the evolution of terrestrial from marine gastropods, being adapted for aerial respiration by reduction in the size of the gill leaflets (Quoy & Gaimard, 1833, p. 476; Prashad, 1925; Risbec, 1942; Berry, 1963). The gill leaflets of *Littoraria* do indeed seem smaller than those of the intertidal *Littorina littorea* (illustrated by Johansson, 1939), but differences from other species are less marked. Remmert (1969) examined the gills of European *Littorina* species and found that the apparent decrease in gill size upshore was explained simply by the positive allometric growth of the gills in combination with the smaller size of the high level species. This cannot account for the relatively smaller gills of *Littoraria* species, which cover a wide size range and show no obvious correlation between gill proportions and adult shell size, at least when comparisons are made between species. No correlation was found between an index of relative gill area and the habitat of the species, so that gill area is not reduced further in the *Littoraria* species living at the higher tidal levels. Since the members of the *scabra* group climb upwards in order to remain above the water level throughout the tidal cycle (p. 59), respiration must take place in air in all species, explaining the similarity in gill structure.

On the roof of the mantle cavity, between the gill folds and the rectum, lies the hypobranchial gland. In comparison with the intertidal *Littorina littorea* (Johansson, 1939; Fretter & Graham, 1962, p. 22) this is much reduced in *Littoraria* species. The gland is largest in *L. carinifera* (to 2.2 mm wide) and is also relatively large in *L. conica*, *L. articulata*, *L. strigata* and *L. vespacea*. In most species the gland is from 0.5 to 0.8 mm in width, folded, and white, grey or black in colour. In *L. vespacea* it is yellow and conspicuous.

The mantle itself is usually grey to whitish in colour, but in *L. albicans* the entire outer surface is bright yellow.

# HABITAT AND BEHAVIOUR

## Habitat and Zonation

The majority of the members of the *L. scabra* group, and some other *Littoraria* species (p. 72) show an obligate association with mangrove trees or salt marsh vegetation. Amongst other littorinids, few species occur in these habitats. These include an undescribed species of *Nodilittorina* (in South East Asia), *Bembicium auratum* (in eastern Australia), a species of *Peasiella* (in northern Australia) and several members of the genus *Littorina* (in the northern Atlantic).

In South East Asia as many as ten *Littoraria* species can be found at a single locality, as in the mangrove forests of Singapore. Each species occupies a characteristic zone or habitat, although overlap between species is considerable. The zonation pattern is at least partially correlated with that of the mangrove tree species, which is itself usually well developed. Macnae (1968) described mangrove zonation in terms of the dominant tree in each association. From the seaward edge of the forest he recognized the following zones:

1. *Sonneratia* zone
2. Seaward *Avicennia* zone
3. Zone of *Rhizophora* forests
4. Zone of *Bruguiera* forests
5. Zone of *Ceriops* thickets
6. Landward fringe

The zonation of *Littoraria* species will be described by reference to this system. The following brief account is based on personal observation (Reid, in prep.) and, for tidal levels and distributional data on mangroves, on the review by Macnae (1968).

Various environmental factors are known to influence mangrove zonation, but tidal inundation and rainfall are amongst the most significant from the present point of view. An outer *Sonneratia* zone is present only on tropical coasts and trees may grow down to the level of the lowest high water mark. Neither the flaky bark nor smooth leaves seem to offer very suitable habitats for *Littoraria* and numbers of individuals are few. Trees of the genus *Avicennia* occur sporadically throughout the forest, but are most abundant in a zone at the seaward edge, extending down to mean sea level. *Littoraria* species are always most common near the seaward edges of swamps, where *Avicennia* and *Rhizophora* trees provide the two most important habitats. Certain *Littoraria* species are more abundant on one or other so that the *Avicennia* and *Rhizophora* zones usually support distinctly different faunas. Leaf dwelling species are especially characteristic of *Avicennia* trees, possibly because the undersides of the leaves bear hairs which are grazed by the snails. Of all the mangrove genera, *Avicennia* extends furthest into the temperate zones and therefore provides the main habitat for such subtropical species as *L. luteola* and *L. cingulata pristissini*. The greatest area of the mangrove forest is often dominated by dense stands of *Rhizophora* trees which occupy levels inundated by all medium high tides. Several species of *Littoraria* occur on the trunks and stilt roots of *Rhizophora*, but are not common on the leaves, which are smooth and often carried on branches above the levels occupied by the snails. *Littoraria* species are only common in the outer parts of *Rhizophora* forests.

Both *Avicennia* and *Rhizophora* zones are present on most mangrove shores, but development of the zones behind these is variable, depending largely upon climate. In the wet tropics the full zonation is usually present. The *Bruguiera* forests are well developed, although the *Ceriops* zone is sometimes represented only be scattered trees within the *Bruguiera* zone. At the back of the swamp the landward fringe supports the greatest diversity of mangrove trees, including species of *Bruguiera*, *Avicennia*, *Xylocarpus*, *Excoecaria*, *Lumnitzera* and, in the Indo-Malayan region, the mangrove palm *Nypa*. This zone merges into rain forest behind the swamp and is flooded only by extreme high tides. Few *Littoraria* species occur in the *Bruguiera* and landward zones, but *L. carinifera* and *L. conica* are typically found there. In regions of seasonal or low rainfall the *Bruguiera* and landward zones are reduced in extent, while the *Ceriops* thickets occupy a wide area at a level inundated by all spring tides. In arid areas the *Bruguiera* zone is absent, bare salt flats appear in the *Ceriops* zone and the landward fringe is replaced by a belt of samphires (e.g. *Arthrocnemum* species). Under these climatic conditions *Littoraria* are rare or absent in and behind the *Ceriops* thickets.

Within the mangrove forest, habitats can be classified according to tree species, substrate and tidal level. At least three species, *L. filosa*, *L. luteola* and *L. albicans*, are know to occur mainly upon the leaves of *Avicennia*, while *L. pallescens* is found on leaves of *Avicennia*, *Rhizophora* and other trees. Species living on trunks and stilt roots may be more common on *Rhizophora* than on *Avicennia* (e.g. *L. scabra*, *L. intermedia*), but are generally less specific in their habitat than the leaf dwelling species. Since the snails are dispersed by a pelagic larval stage, even isolated mangrove trees on rocky and coral shores usually support populations.

Although over most of the geographical range occupied by *Littoraria* the dominant type of intertidal halophytic association is a mangrove forest, in some areas samphires and other saltmarsh plants occur at suitable levels and may be colonized by snails. For example, in the arid region of Shark Bay, Western Australia, *L. cingulata pristissini* occurs amongst samphires and on the temperate south coast of New South Wales *L. luteola* may be found in salt marshes as well as mangroves. Driftwood and wharf pilings may also be suitable substrates on sheltered coasts. Several species, while typically found upon mangrove trunks, also occur on sheltered rocks, for example *L. articulata*, *L. strigata*, *L. intermedia* and *L. sulculosa*. However, in all cases *Littoraria* species require a hard substrate and are never found on the mud between mangrove trees.

On a larger scale, an interesting distinction can be made between continental and oceanic species. The former are common in the wide mangrove fringes of large land masses, in deltaic mangroves, in areas with turbid water and muddy soil, and occur not only at the seaward edge but also within the forest. In contrast, oceanic species are found in clear water situations, on the trees of narrow mangrove fringes on rocky or coral shores and on oceanic islands. On continental margins oceanic species are restricted to promontories and offshore islands, and in wide mangrove forests are found only on the outermost trees at the seaward edge. Similar distinctions have been made in other molluscan groups by Abbott (1960) and Taylor (1971*a*). The explanation of these distribution patterns is at present obscure. Possible considerations include the higher nutrient levels near continental margins, or the effects of freshwater runoff, although neither would appear to explain the local distribution of oceanic species on mainland coasts. The most obvious features of the environment which can be correlated with the absence of oceanic species are water turbidity and a muddy substrate. It is difficult to envisage a direct effect of sediment upon adult snails living above the water level, but an influence upon their algal food or upon larval settlement is not unlikely. Taylor (1971*a*) related the distinction between oceanic and continental distributions to suppression of pelagic larval stages in continental species, but Abbott (1960) found that, as in *Littoraria* species, there was no correlation of distribution with method of development, size, abundance or microhabitat. At least three species belong to the oceanic group, *L. scabra*, *L. pallescens* and *L. intermedia*. *L. subvittata* is probably also an oceanic species. The distribution of *L. lutea* is poorly known, but appears intermediate between the two types. The contrast between an oceanic and a continental distribution is very clearly shown by a comparison of the records of *L. philippiana* and *L. pallescens* from eastern Australia; the former is common in the mangroves of the mainland, the latter restricted to promontories and offshore islands.

The arboreal mangrove fauna shows both a vertical zonation up the trees and a horizontal zonation through the forest from seaward to landward (Berry, 1963; Macnae, 1968). The vertical zonation of *Littoraria* is most apparent near the seaward edge of the forest, where species and individuals are most numerous. In the systematic descriptions, horizontal zonation is indicated by reference to the zonation of the mangrove trees, while vertical zonation is roughly indicated by specifying heights above ground level at which snails are found during low tide periods.

That mangrove littorinid species are zoned to different heights on the trees is well known, and is a frequently quoted example of a vertical zonation pattern in the mangrove community (Berry, 1963, 1972; Macnae, 1968, p. 165; Vermeij, 1973*a*; Sasekumar, 1974). These previous authors, working in Malaysia, have given upper limits of zonation in the following order, from lowest to highest: *L. undulata*, *L. carinifera*, *L. scabra*, *L. melanostoma*. These observations are, however, inaccurate both in regard to nomenclature and to the zonation implied. *L. 'undulata'* refers to *L. articulata* (and probably also to *L. strigata* and *L. vespacea*); *L. 'carinifera'* may include *L. conica* as well as *L. carinifera s. s.*; while *L. 'scabra'* refers to *L. intermedia* and perhaps also to *L. pallescens*. Furthermore, in stable or accreting mangroves these species show pronounced horizontal zonation so that their vertical distribution cannot easily be compared. Only on eroding shores, as examined by Berry (1963, 1972) and perhaps by others, will all occur on the outermost *Rhizophora* trees behind the seaward mud cliff. Regarding horizontal zonation, previous authors have mostly noted simply that littorinids penetrate the mangrove forest, but become scarce away from the seaward edge. Only Nielsen (1976) and Frith *et al.* (1976) have observed that *L. carinifera* is in fact typically found in the landward zone of the swamp.

A mangrove at the mouth of the Sarawak River in Borneo may be used as an example of *Littoraria* zonation. Here six species occurred, zoned from the seaward edge in the order: *L. articulata* and *L. vespacea*, *L. albicans*, *L. melanostoma*, *L. carinifera*, *L. conica*; while the maximum tidal levels attained increased in the order: *L. articulata*, *L. vespacea*, *L. melanostoma*, *L. carinifera*, *L. conica*, *L. albicans*. At this and other localities it is evident that *L. carinifera* and *L. conica* are species of the landward zone, usually occurring at higher levels than *L. melanostoma*, in contrast with most previous reports.

## Behaviour

Vertical zonation patterns are complicated by the mobility of the snails, for all *Littoraria* species will climb to avoid submersion by a rising tide (Abe, 1942; Lenderking, 1954; Nielsen, 1976; Reid, in prep.). In addition the snails occupy higher levels on the trees, even during low tide, at spring tide periods (Berry & Chew, 1973; Reid, in prep.). Species of lower tidal levels, such as *L. intermedia* and *L. articulata*, will often descend following the water level on a falling tide. Of the high level species, *L. filosa* can be found up to 3.0 m above the ground (*Avicennia* zone, Magnetic Island, Queensland), *L. albicans* to 3.5 m (*Avicennia* zone, Santubong, Sarawak), *L. pallescens* to 4.5 m (*Rhizophora* forest, Phuket Island, Thailand) and *L. philippiana* to 5.2 m (*Rhizophora* forest, Magnetic Island, Queensland). These snails of the highest vertical levels are often beyond the reach of the tide, although individuals will occasionally descend to the water level to release larvae and perhaps to obtain moisture. Such species seem to be influenced more by rainfall, dew and humidity than by the tide. They occupy lower levels at the edge than in the humid interior of the forest, and move to higher levels during wet weather. Horizontal zonation is unaffected by tidal cycles or by rainfall, since all *Littoraria* species seem unable to crawl upon soft mud and therefore can only move horizontally where stilt roots or branches of adjacent trees are in contact.

Under dry conditions most littorinids withdraw into the shell, remaining attached to the substrate by mucus. The formation of the mucous holdfast of *L. irrorata* has been described by Bingham (1972*c*) and is similar in the *scabra* group. Unless the humidity is low, the mucous holdfast remains fluid, and in some species, notably *L. scabra*, the secretion is so copious that if dislodged a snail may remain suspended by a thick string of mucus, which it subsequently ascends to regain its position (Boschma, 1948; pers. obs.). This observation has given rise to the myths that thread spinning is defensive (Macnae, 1968) or even assists in climbing from branch to branch (Allan, 1959). In fact the function of the holdfast is to conserve moisture, which would be lost by evaporation if attachment depended upon extension of the foot (Vermeij, 1973*a*).

The diets of *Littoraria* species have not been investigated in any detail. As mentioned above, leaf dwelling species graze leaf hairs from the undersides of *Avicennia* leaves, but will only rarely rasp holes through the leaf lamina. Leaf hairs do not seem to be an essential component of the diet, since no species is entirely restricted to *Avicennia*. The thick leaves of *Rhizophora* and other mangroves are not eaten. On trunks and stilt roots the snails graze the surface layer, presumably digesting diatoms and other algae.

## Copulation

The behaviour of *Littoraria* species during copulation is the same as that recorded in other littorinids (Linke, 1933; Abe, 1942; Gibson, 1964; Bingham, 1972*b*; Gallagher & Reid, 1974). Males search for females, mounting the shell of any other individual encountered, of which they are only able to determine the sex by attempting to insert the penis into the bursa copulatrix (Gibson, 1964; Muggeridge, 1979). If the shell beneath is a male, the pair soon separates. During copulation the male attaches to the right side of the anterior end of the female shell and inserts the penis under the outer lip of the aperture. The act of copulation is of short duration, but in *Littoraria* species the male may remain attached

**Table 3** Pairs in copulation position, recorded at Magnetic Island, Queensland, Australia (September, 1980 to September, 1981).

| Female | Male | | | | |
|---|---|---|---|---|---|
| | *L. filosa* | *L. philippiana* | *L. scabra* | *L. intermedia* | *L. articulata* |
| *L. filosa* | 174 | 1 | 10 | 1 | |
| *L. philippiana* | 8 | 27 | 3 | | |
| *L. scabra* | 2 | | 37 | | |
| *L. intermedia* | | | | 93 | 8 |
| *L. articulata* | | | | 10 | 824 |

to the female shell in the copulation position for several hours. Under dry conditions both animals withdraw into their shells and attachment is by mucus alone. Copulation only occurs under moist conditions, during or after high tide, after rain, or in the early morning. Bingham (1972*b*) found that high temperatures stimulated copulation in *L. irrorata*.

At Magnetic Island, Queensland, copulation was recorded over a whole year during a study of reproductive patterns in *Littoraria* species (Reid, in prep.). The results are presented in Table 3. All pairs seen in the copulation position were recorded; in the species *L. filosa*, *L. philippiana*, *L. scabra* and *L. intermedia* the penis was inserted under the female shell in 37% of the cases observed, and in only 2.9% of these copulating pairs were both partners males. Out of a total of 1198 pairs, in only 43 (3.6%) did the individuals belong to different species. This provides evidence that the species as defined by morphological criteria are indeed true biological species.

It could be argued that the zonation of the species or separate breeding seasons might be responsible for the apparent assortative mating. In fact, the zones of *L. intermedia* and *L. articulata* show very broad overlap, as do those of the three remaining species, and even within each of these groups the deviation from random mating is clear. Although breeding seasons do differ in duration, all the species show maximum reproductive activity during the wet summer months. In view of the indiscriminate behaviour of males in searching for mates, it is not surprising that some attempts at interspecific copulation were found. One male *L. scabra* was even observed attempting to copulate with a *Nerita articulata*.

It is not known whether transfer of sperm occurs during interspecific mating, but no possible hybrids with intermediate shell or anatomical characters have been discovered. Rosewater (1970, 1981) has suggested that closely related littorinids may hybridize, but this seems extremely unlikely. If hybrids were frequently produced and were fertile, then a complete range of intermediates would be expected. If infertile, a hybrid of intermediate phenotype might appear to be a distinct species. However, all the five species listed in Table 3 (besides most of the ovoviviparous species described in the present study) have

**Table 4** Pairs in copulation position, recorded at Broome, Western Australia (3 November, 1981).

| Female | Male | | | |
|---|---|---|---|---|
| | *L. filosa* | *L. cingulata cingulata* | *L. sulculosa* | *L. articulata* |
| *L. filosa* | 36 | 4 | | |
| *L. cingulata cingulata* | 2 | 55 | | |
| *L. sulculosa* | | 1 | 2 | |
| *L. articulata* | | | | 1 |

been seen to produce progeny which develop at least to the veliger stage. Interspecific mating has been recorded in several studies of the reproduction of co-occurring littorinids (Struhsaker, 1966; Gallagher & Reid, 1974). Raffaelli (1977) found that amongst copulating pairs of *Littorina saxatilis* and *Littorina nigrolineata* almost half were combinations other than male–female intraspecific pairs.

Copulating pairs were also recorded at Broome, Western Australia, where four species occurred together in broadly overlapping zones (Table 4). Here the frequency of interspecific pairing was 6.9%.

# BIOGEOGRAPHY

## Patterns of Distribution

The genus *Littoraria* is largely confined to the tropics. Although several members extend into subtropical regions, the only species with predominantly subtropical or temperate distributions are *L. irrorata* in the western Atlantic (to 39°N) and *L. luteola* in Australia (to 37°S). This restriction to the tropics is not a consequence of dependence upon the mangrove habitat, which is itself largely tropical and subtropical in distribution. In subtropical regions *Littoraria* species regularly occur in salt marshes, while in Australasia the distribution of mangroves of the genus *Avicennia* extends to higher latitudes (Chapman, 1976) than that of *Littoraria* species. Furthermore, species of *Littoraria* from rocky shores are also absent from temperate zones.

The subgenus *Littoraria* is of worldwide distribution. Species of *Littorinopsis* are largely confined to the Indo-Pacific, with only a single representative in the tropical Atlantic. The subgenera *Palustorina* and *Lamellilitorina* are restricted to Indo-Malaya and Australia (list p. 72).

To a large extent the distributions reported here support the subdivisions of the Indo-Pacific biogeographic region proposed by Macnae (1968) on the basis of the endemic fauna of mangroves (Table 5). Two centres of endemicity are especially obvious. Firstly the Malayan Peninsula, eastern Sumatra, western Borneo and southern Vietnam, which is also the focus of highest species richness. Secondly, the north-western coast of Australia,

**Table 5** Comparison of distributions of mangrove-associated *Littoraria* species with the subdivisions of the Indo-Pacific province proposed by Macnae (1968) on the basis of mangrove faunas.

| Division | Endemic species | Species shared with one other division |
|---|---|---|
| 1. West Indian Ocean | | *L. subvittata* |
| 2. West central | *L. delicatula* | *L. subvittata*<br>*L. conica* |
| 3. East central | *L. vespacea*<br>*L. albicans* | *L. ardouiniana*<br>*L. conica* |
| 4. North-east | *L. flammea* | *L. ardouiniana* |
| 5. East Borneo, Celebes | | |
| 6. Qld., New Guinea | *L. philippiana*<br>*L. luteola* | *L. filosa* |
| 7. N.W. Australia | *L. cingulata cingulata*<br>*L. cingulata pristissini*<br>*L. sulculosa* | *L. filosa* |
| 8. West Pacific islands | | |

The eight remaining species are distributed over three or more divisions.

where two species and one subspecies are endemic, while two of the four species shared with other divisions (*L. articulata* and *L. filosa*) show rather distinctive shell types not found elsewhere. The isolation of this division is further emphasized by the absence of the widely distributed *L. intermedia*, and by the presence of two further endemic littorinids, *Nodilittorina australis* (Gray) (probably not specifically distinct from *N. nodosa* (Gray)) and *Tectarius rusticus* (Philippi) (Rosewater, 1970, 1972). The eastern limit of the north-western Australian endemic littorinids falls between Cape Londonderry and Darwin, while the western limit for the eastern Australian endemic *L. luteola* is the Torres Strait and for *L. intermedia* is close to Darwin. These limits correspond with the fact that the Arafura Sea was land during low sea level periods in the Pleistocene, and before that was dry back until the late Tertiary, while the Torres Strait became a seaway only in middle to late Pleistocene times (Doutch, 1972).

Largely on the basis of the distribution of shorefishes, Springer (1982) has proposed that the Pacific Plate should be recognized as a major biogeographic subunit of the Indo-Pacific province. Springer has demonstrated that many fish and other organisms show distributions restricted to either the oceanic or continental lithospheric plates of the Indo-Pacific, and that 20% of the shorefish species occurring on the Pacific Plate are endemic. Springer explained such patterns by proposing a present or historical barrier to dispersal at the western margin of the Pacific Plate, but pointed out that to confirm his vicariance hypothesis it would be necessary to demonstrate sister group relationships between widespread Pacific Plate endemics and species restricted to adjacent continental plates. Such cladistic data were not available to Springer, but are presented here for the genus *Littoraria*. Of the twenty-five Indo-Pacific species of the genus, only *L. coccinea* could be regarded as a possible Pacific Plate endemic, although its distribution extends across the western margin of the plate as far as Fiji, New Caledonia and Queensland (but not into the Indian Ocean, *contra* Rosewater, 1970). On the basis of anatomical and conchological similarities, the sister species of *L. coccinea* is believed to be *L. glabrata* (=*kraussi*), endemic to the Indian Ocean. The distribution of *L. pintado* and the variation of *L. carinifera*, as discussed below, also suggest a barrier to dispersal between the Indian and Pacific Oceans. Of the remaining widespread, pan-Indo-Pacific species of the genus (*L. undulata*, *L. scabra*, *L. pallescens*, *L. intermedia*), all extend far into the Pacific across the western boundary of the Pacific Plate. Sixteen of the twenty species of the *scabra* group are restricted to continental plate areas in the central Indo-Pacific, but none shows an eastern limit of distribution corresponding to the western margin of the Pacific Plate. Distributions of *L. carinifera* (Fig. 85) and *L. lutea* (Fig. 33) do, however, correspond rather precisely with the margins of the southern part of the Eurasian Plate. The distributions of *Littoraria* species do not, therefore, lend support to Springer's vicariance hypothesis.

An ecological explanation of the real distinction between the faunas of the Pacific Plate and of its adjacent continental margins, in terms of distribution of high islands and estuarine habitats, was only briefly considered, and set aside, by Springer (1982, p. 123). Nevertheless, in the case of *Littoraria* species, local distribution patterns strongly suggest such an ecological interpretation (p. 57). On a larger scale, distributions of the pan-Indo-Pacific *Littoraria* species can also be adequately explained in terms of habitat availability. Of these widespread species, only *L. pallescens* is entirely restricted to mangrove trees, and its distribution in the western Pacific (Fig. 38) corresponds very precisely with that of mangroves (Macnae, 1968; McCoy & Heck, 1976; Chapman, 1976). *L. scabra* shows a similar distribution, but can occasionally be found on other maritime trees and on driftwood, and probably in consequence there are a few records from the Line and Tuamotu Islands, and also from the Hawaiian Islands, where mangroves have only been introduced this century (Wester, 1981). *L. intermedia* is common on both mangroves and sheltered rocky shores, and *L. undulata* on rocky shores and driftwood, and both species are widely distributed across the Pacific (Rosewater, 1970; Fig. 47 herein).

In very few cases do the distribution patterns show sudden geographical replacement which could suggest competitive interactions between species. One possible case is the absence of the otherwise widely distributed *L. intermedia* from north-western Australia, where two endemic species occur, but as mentioned above this is more likely to be a result of geological history. Another possible example concerns the two species found on *Avicennia* foliage in eastern Australia; *L. luteola* is abundant in New South Wales and rare in Queensland, while the reverse is true for *L. filosa*, but this is a case of gradual, rather than sudden, replacement. It is interesting that at most localities in the Indo-Pacific region only one leaf dwelling species occurs, or only one occurs commonly. Thus the distributions of *L. filosa*, *L. luteola*, *L. ardouiniana*, *L. delicatula* and *L. albicans* are almost mutually exclusive, and all occur in continental situations. The remaining species known to inhabit foliage is *L. pallescens*, with a wide distribution across the Indo-Pacific, but since it is restricted to oceanic situations it seldom occurs together with the others.

A curious disjunct distribution is shown by the rock dwelling species *L. pintado*. The typical form of the species occurs in the tropical north-western Pacific (including the Hawaiian, Ryukyu and Caroline Islands) and in the western Indian Ocean (Mascarene Islands, Madagascar, South Africa and also Somalia, Mienis, 1973). Despite much collecting in the central Indo-Pacific, the only records from intervening localities are single shells from Bombay and Bali (both MCZ) which are of doubtful provenance. Shells are identical from the two main areas of distribution, although preserved animals have not been seen from the Indian Ocean. Rosewater (1970) recognized a subspecies *L. pintado schmitti* (Bartsch & Rehder) from Clipperton Island in the Eastern Pacific and mentioned that *L. pullata* (Carpenter) from Baja California was an analogue of *L. pintado*. *L. schmitti* and *L. pullata* differ from *L. pintado* only in their more heavily pigmented shells and specimens of *L. pullata* are anatomically identical to *L. pintado* from the western Pacific. Consequently both are here regarded as conspecific with *L. pintado*, and the species becomes the only member of the family known to occur on both sides of the Pacific Ocean. It is of interest that *L. pintado* retains many characters regarded as ancestral within the genus (p. 79), and conceivably may be of more ancient origin than other members of the group. Amongst other species of *Littoraria* from the Indo-Pacific only *L. intermedia* might possibly extend into the Eastern Pacific province, a single shell being recorded from the Galapagos Islands. Few other molluscs are known to cross the Eastern Pacific barrier (Emerson, 1967).

Vermeij (1972*a*, 1973*b*) has suggested that amongst molluscs of mangroves and other intertidal habitats, those of the supralittoral zone are more restricted in distribution as a result of their adaptation to local atmospheric conditions, which are more geographically variable than hydrological conditions. The high level, mostly leaf dwelling, species of *Littoraria* are indeed sometimes narrowly distributed (e.g. *L. albicans*, *L. luteola*, *L. delicatula*), although *L. pallescens* is widespread, and there is no clear correlation between zonation level and distribution in the genus as a whole. The two species typical of the landward fringes of swamps, *L. carinifera* and *L. conica*, may be more dependent upon atmospheric moisture than other species, since by reason of their position they cannot descend the trees to reach the water level for most of the lunar cycle. These species are restricted to the high rainfall areas of Indo-Malaya, and species are absent from this zone elsewhere. While this might be a case of climatic limitation, it would seem that Vermeij's suggestion might not be generally applicable to these highly mobile snails which show vertical migration with the tidal cycle.

The most striking correlate of distribution is the occurrence of species in oceanic or continental situations. Of the *Littoraria* species associated with mangroves, only three (*L. scabra, L. intermedia, L. pallescens*) occur across most of the Indo-Pacific province, and all are typical of oceanic habitats. Species of continental shores appear to be more restricted. A similar correlation has been noted by Abbott (1960) in the genus *Strombus*. This effect might be simply the result of species on promontories and islands having access to stronger currents for dispersal. Alternatively, the explanation of Vermeij (1972*a*, 1973*b*) might apply in modified form, since the oceanic habitat might be more geographically uniform than the continental margins. Continental species could then be restricted by their adaptation to local environmental conditions, which may be variable on a large scale.

## Dispersal

There is no direct evidence of the means and distance of dispersal of *Littoraria* species, but some deductions can be made. In the genus *Littorina* the degree of intraspecific variability in shell characters (James, 1968; Heller, 1975*a*) and enzyme patterns (Berger, 1973; Wilkins & O'Regan, 1980) has been correlated with the potential for larval dispersal suggested by the method of development. *Littorina littorea* releases pelagic egg capsules which hatch into veliger larvae and this presumably widely dispersed species shows rather constant characters. In contrast, the ovoviviparous species of the *Littorina saxatilis* complex lack a planktonic phase and show considerable intra- and inter-population variability. Populations of poorly dispersed species are thought to become precisely adapted to local conditions, explaining variability between populations. In similar fashion, Rosewater (1970) explained the interpopulation variability of *L. 'scabra'* in terms of restriction of gene flow caused by ovoviviparous reproduction. It is now clear that most of the supposed variation between populations of *L. 'scabra'* is in fact variation between species. Within species, shell form is rather constant and although several show striking colour polymorphism, this is either an adaptive genetic trait constant over wide areas, or is in other cases influenced by environmental effects (or possibly by selection) from one tree to the next (p. 17). Where intraspecific variation in shell form or colour is significant, it is over a considerable geographical range, as discussed below. Therefore shell characters provide no evidence for small scale geographical isolation between populations.

In the two oviparous *Littoraria* species from mangroves of which the egg capsule has been described (p. 46), the diameter of the ovum is 76 to 140 $\mu$m, while in the ovoviviparous species the maximum dimension of the larval shell when spawned from the mantle cavity of the parent is 100 to 140 $\mu$m. In all the *Littoraria* species from mangroves, with the exception of *L. albicans*, the height of the shell at settlement and metamorphosis is 320 to 415 $\mu$m (from measurements of protoconchs of adults and of newly settled snails collected in the field). These measurements suggest a considerable period of planktotrophic growth between spawning and settlement in both oviparous and ovoviviparous species.

The only tropical littorinid for which data on length of larval life are available is *Nodilittorina hawaiiensis* (Struhsaker & Costlow, 1968, as *Littorina picta*), which grows from an 80 μm ovum at spawning to a 250 μm larva at metamorphosis in 24 days, under laboratory conditions. Gallagher & Reid (1979) have estimated the pelagic life of *L. angulifera* as eight to ten weeks, based on the delay between spawning and peak recruitment. The protoconch of *L. angulifera* is closely similar to those of other members of the subgenus *Littorinopsis*, and a similar length of larval life might be predicted for these.

Scheltema (1971) has estimated the maximum time required for transport of larvae across the Atlantic in ocean currents as 60 days from east to west and 96 days in the reverse direction, so that the estimated pelagic life of *L. angulifera* may be just sufficient to permit transatlantic drift. *L. angulifera* is found on both sides of the tropical Atlantic (Rosewater & Vermeij, 1972) and the eastern and western populations are not morphologically distinct, so that larval dispersal might occur between them. Despite the probability of long distance dispersal, Gaines *et al.* (1974) have reported significant genetic differences between populations of *L. angulifera* on mangrove islands less than 300 m apart. As observed by these authors this effect was probably due to selection rather than limited dispersal since there was no association between inter-island distance and heterogeneity. Rosewater (1963, 1970, 1981) pointed out that the association of the *scabra* group with mangroves provides the opportunity for dispersal of adults by rafting on floating vegetation, as has been observed by Marcus & Marcus (1963) in *L. angulifera*. All the above evidence suggests that *Littoraria* species have the potential for wide dispersal.

Within the genus *Littoraria* there is no correlation between the range of the distribution and the method of development, whether oviparous or ovoviviparous. This is not unexpected, since both types apparently involve a lengthy planktotrophic phase. In fact, contrary to the usual association of viviparity with restricted distribution, it is three ovoviviparous members of the subgenus *Littorinopsis* (*L. scabra*, *L. intermedia*, *L. pallescens*) which are the most widely distributed of the species associated with mangroves, while the oviparous subgenus *Palustorina* shows the more restricted distribution. However, within each subgenus there are examples of wide and narrow geographical ranges. The apparently oviparous *L. albicans* is unique in the genus by virtue of its large protoconch (610–660 μm), which could suggest a long planktotrophic life. Nevertheless, this species is restricted to a small area of Indonesia.

## Variation and Speciation

Morphological characters of *Littoraria* species are relatively constant over wide geographical areas, as expected in view of their presumably wide dispersal. For example, *L. scabra* shows one of the most extensive ranges, and specimens from Hawaii are indistinguishable from those from South Africa. In such uniform and widespread species it is difficult to envisage how sufficient isolation might be achieved to permit speciation. In fact some species of the genus show geographical variation on a large scale, which suggests how peripheral differentiation and vicariance may together lead to speciation in widely dispersing forms (Abbott, 1960; Schuto, 1974).

Clinal variation is shown by several species, for example *L. filosa* and *L. articulata*, which show different shell forms at the extremes of their 6500 km range around the tropical Australian coast. A cline has been described in *Nodilittorina africana* (Philippi) around the coast of South Africa by Hughes (1979*a*). *L. pallescens* shows distinctive shell types in the Arafura Sea and in Malaysia, while across the great range of *L. intermedia* shells from Hawaii and Polynesia can be distinguished from those from the rest of the Indo-Pacific. Thus even where ranges are continuous, regional differentiation can occur over sufficient distance. In such cases, differentiation might proceed to speciation if local extinction or geological events caused vicariance. In the case of *L. carinifera* the distinctive Pacific and Indian Ocean forms are in contact only through the Straits of Malacca. In north-western Australia there is some evidence that an originally continuous population of *L. cingulata* has become divided into geographically isolated northern and southern populations, which are sufficiently distinct to be recognized as subspecies. Amongst the rock dwelling species, *L. coccinea* is restricted to the Pacific Ocean (as far west as the Philippines) and *L. glabrata* to the Indian Ocean (as far east as the Cocos-Keeling Islands); although recognized as distinct species, shell and anatomical characters suggest a very close relationship. It may be noted that vicariance does not always lead to differentiation; for instance the Indian and Pacific Ocean forms of *L. pintado* show identical shell characters. Whether differentiation has proceeded to the point of speciation can only be demonstrated if the forms are sympatric over part of their range and remain distinct in the region of overlap. Such is the case for the closely related pair of species *L. articulata* and *L. strigata*. The three species *L. intermedia*, *L. philippiana* and *L. subvittata* may also have originated from a common ancestral species and now show partial overlap of their ranges.

## Regional Diversity

The present taxonomic treatment increases the number of recognized species of littorinids in mangrove forests of the Indo-Pacific province from three (Rosewater, 1970) to twenty. Of this number, up to ten species may be found together in a single mangrove swamp, as at Singapore. This change in classification does not significantly alter the pattern of regional diversity of the mangrove fauna as a whole (Vermeij, 1973*b*), but does considerably change the pattern of distribution of species amongst the genera of mangrove molluscs. Such diversity within a single genus is by no means unusual in the tropics, but stands out as unique amongst the larger mangrove molluscs. Within the family Littorinidae the sympatric occurrence of several congeneric species is usual, for example seven species of *Littorina* (and one of *Melarhaphe*) are found on rocky shores in Britain (Raffaelli, 1982).

Within the genus *Littoraria* as a whole, the worldwide pattern of species richness (Fig. 15) shows a maximum diversity in the central Indo-Pacific, with progressively fewer species in the Eastern Pacific, the western Atlantic and the eastern Atlantic. A similar pattern, with maximum diversity in the central Indo-Pacific, is well known in many marine organisms, such as hermatypic corals (Stehli & Wells, 1971; Rosen, 1981), bivalves (Stehli *et al.*, 1967) and in mangroves and sea grasses (McCoy & Heck, 1976). The various explanations that have been advanced to account for this pattern have been reviewed by Rosen (1981).

The diversity of the mangrove associated species of *Littoraria* (comprising the twenty members of the *scabra* group) is illustrated in Fig. 16. Of the other marine groups mentioned above, the contours of species richness show the closest correlation with those of mangrove tree genera (McCoy & Heck, 1976). Diversity of *Littoraria* species is therefore

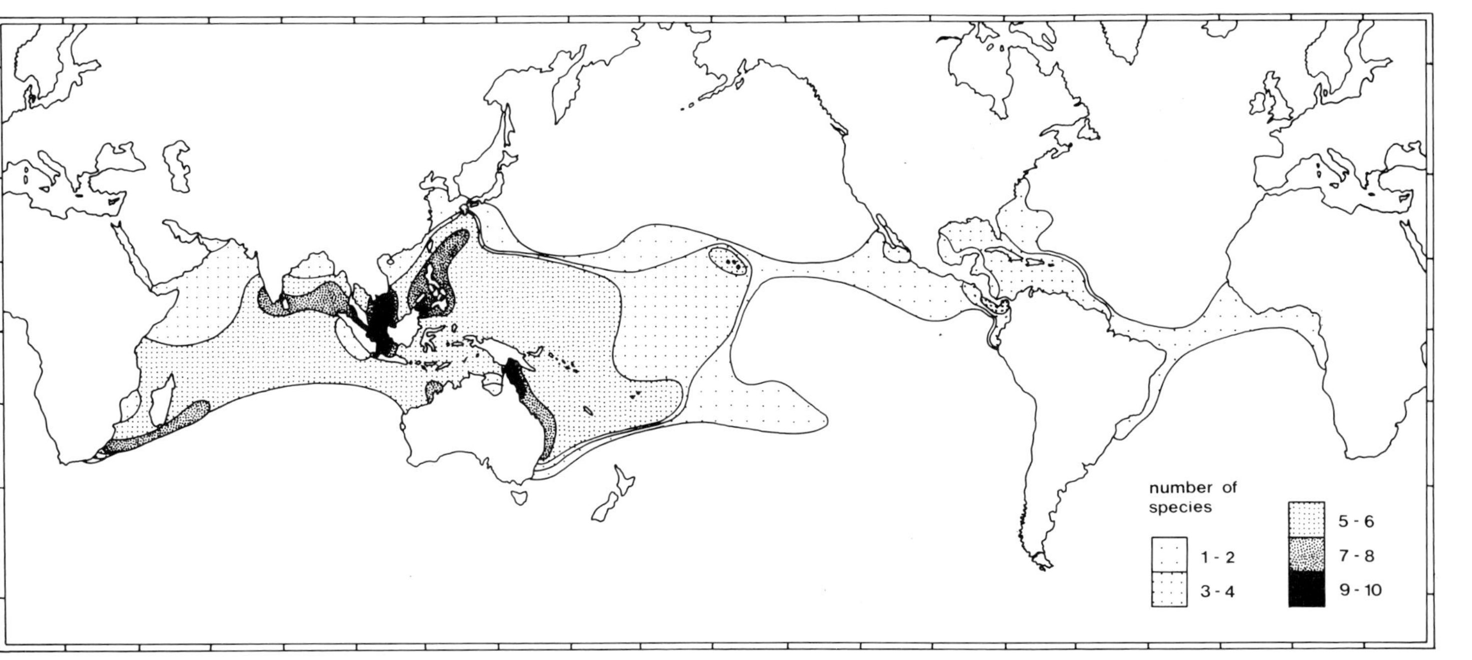


**Fig. 15** Worldwide contour map of species richness in the genus *Littoraria* (compiled from 2700 distribution records, representing all 36 known species).

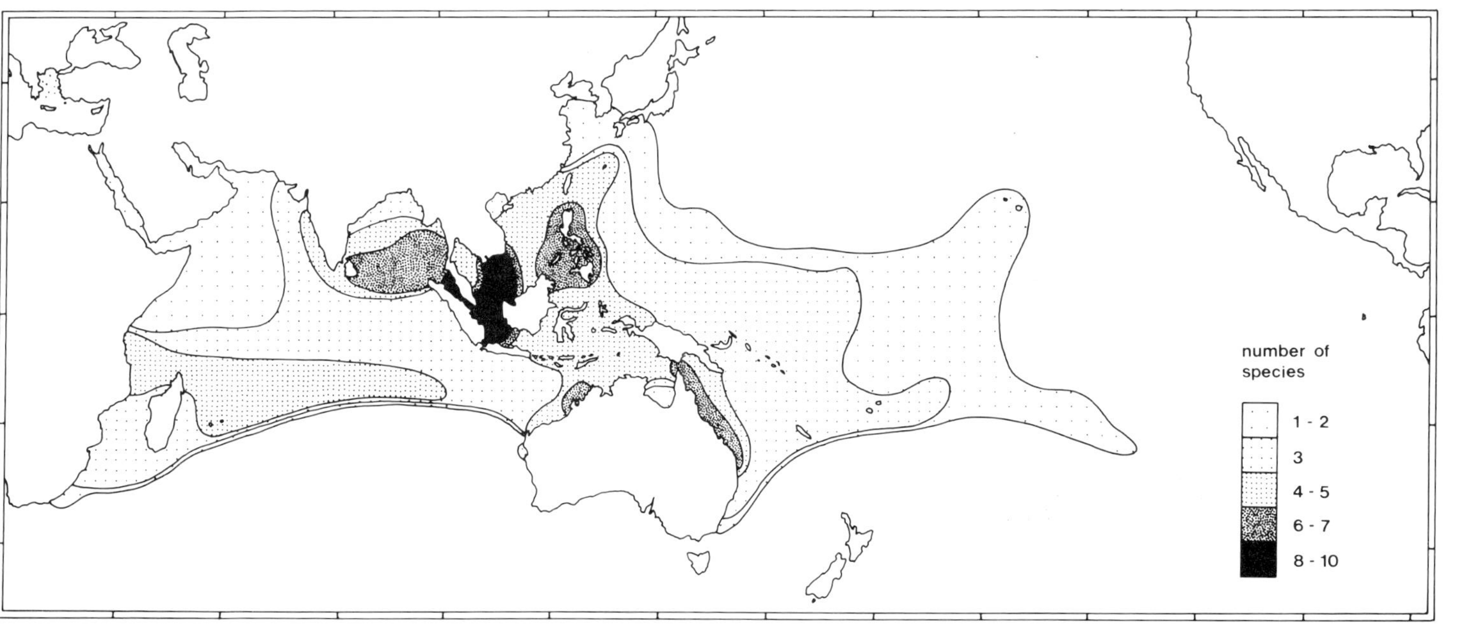


**Fig. 16** Contour map of species richness of the 20 members of the genus *Littoraria* associated with mangroves in the Indo-Pacific province (compiled from 1900 distribution records).

correlated with a measure of habitat diversity. However, the relationship is probably not a causal one, but rather may be the result of similar factors acting to produce and to maintain high species numbers in the two groups in the same environment. This seems likely, since species of *Littoraria* are seldom restricted to a single tree species, but occupy habitats characterized by substrate and tidal level. The effect of tree species is evident only in the rather distinct faunas of *Avicennia* and *Rhizophora* zones, but both these genera are very widely distributed throughout the Indo-Pacific (Chapman, 1976).

In comparison with the diversity patterns of corals and seagrasses, the centre of *Littoraria* species richness is located further westwards and does not encompass the archipelagic region of the western Pacific. In part this reflects the continental distribution of mangroves and the oceanic distribution of corals. In the latter group the importance of a large area of suitable habitat and possibilities of isolation, both provided by archipelagos, have been emphasized as explanations for the observed pattern of diversity (Rosen, 1981). The diversity pattern of the gastropod genus *Strombus* (Abbott, 1960) is similar to that shown by corals. In the genus *Littoraria* the very highest numbers of co-occurring species are found in the area of Malaysia, eastern Sumatra, northern Java, western Borneo and southern Vietnam, on the shores of continental land masses, rather than in the central region of Indonesia where continental islands are small and numerous. Since the larvae of *Littoraria* species are probably widely dispersed, archipelagic areas do not provide possibilities for isolation and speciation. In the past the Indo-Malayan centre of diversity has been regarded as a centre from which species have radiated (Ekman, 1953; Stehli & Wells, 1971). However, at least in certain groups, this is a region of accumulation, rather than (or as well as) of generation of species (Taylor, 1971; Kay, 1984). This appears to be the case in the genus *Littoraria*, in which differentiation and speciation appear to be taking place in the more peripheral areas of the Indo-Pacific province.

Accumulation of species in the Indo-Malayan region may have been favoured by the large area of the mangrove habitat which is available on the shores of the large continental land masses, which permits co-occurrence of many species. The effect of predation may also be significant. Vermeij (1978) has argued that predation in marine environments is more intense at lower latitudes. The major predators of *Littoraria* species are crabs, and at several localities it has been observed that densities of these snails are greatest in small patches of mangroves or on isolated trees, where tree climbing crabs appear to be scarce. Although no data are available on the local or geographical distribution of crabs, the possibility exists that intense predation in large mangrove forests in equatorial areas may contribute to the maintenance of high species richness in the genus by reducing population size and competitive effects, or by influencing zonation of the snails and overlap between their habitats.

# PHYLOGENY AND GENERIC CLASSIFICATION

## Status of the genus *Littoraria*

Until recently, an emphasis on shell characters in the classification of the family Littorinidae led to the inclusion of all smooth and spirally sculptured forms in the single genus *Littorina*, while species with nodulose sculpture were placed in *Nodilittorina*, *Tectarius* or *Echininus*. Even on the basis of shell characters, the group of large, often thin and colourful shells from mangrove habitats was generally recognized as distinct, and following Mörch (1876) was classified as the subgenus *Littorinopsis* (e.g. von Martens, 1897; Thiele, 1929). *Littorinopsis* was raised to generic rank by Cossman (1916), who was followed by Wenz (1938) and by authors in Japan (e.g. Kuroda & Habe, 1952).

The association of the generic name *Melarhaphe* Menke with the species here placed in *Littoraria* can be traced to the Adams brothers (1858). Tryon (1887) gave *Littorina scabra* as the typical example of his section *Melaraphe*, although the type species is in fact *Turbo neritoides* L. (see Rosewater, 1966, for a discussion of *Melarhaphe*). This error was followed by a number of authors.

The generic name *Littoraria* was first published in an index (Griffith & Pidgeon, 1834) and was neglected until used as a subgenus of *Littorina* (and a senior synonym of *Littorinopsis*) by Bequaert (1943). *Littoraria* has subsequently been accepted as a full genus by Japanese taxonomists (e.g. Azuma, 1960; Habe, 1964; Higo, 1973). The type of the genus is *Turbo zebra* Donovan, from the tropical Eastern Pacific, of which the anatomy has been hitherto unknown, which may explain the reluctance of some authors to use the generic name. Rosewater (1970, 1981) defined subgenera of *Littorina* partly on the basis of characters of the penis and radula, but still stressed features of the shell. Consequently, he recognized the subgenus *Littoraria* for robust, rather smooth, unicoloured or axially striped shells, which are oviparous and usually found on rocks. The subgenus *Littorinopsis* was retained for species with rather thin, spirally ribbed or carinate, spirally spotted shells, often associated with shore vegetation and mostly thought to be ovoviviparous. Nevertheless, penial and radular characters do not distinguish between these two groups. Placing primary importance on these two anatomical features, Bandel & Kadolsky (1982) united the two as the genus *Littoraria*. From a consideration of the morphological evidence discussed in detail in previous sections, it is clear that the species of *Littoraria* form a natural group, equivalent in rank to the more familiar genera *Littorina* and *Nodilittorina*. Dissection of *L. zebra*, the type species (Fig. 4l), has confirmed that the generic name *Littoraria* is applicable to the group, for previously only shell (Fig. 99g) and radula (Rosewater, 1980a) of this species were known. The genus is formally defined in the taxonomic section (p. 82).

Bandel & Kadolsky (1982) have proposed a reclassification of the family, using characters of the penis, radula, spawn and shell. The present study, which includes examination of new taxonomic characters of the pallial oviduct and sperm nurse cells, and a reappraisal of information on spawn and penial shape, almost entirely supports the scheme of these authors. The characters of phylogenetic significance in *Littoraria* and other genera are summarized as a cladogram (Fig. 17, Table 6), which complements the table of characters given by Bandel & Kadolsky. Only one modification of their scheme is made, that *Nodilittorina antoni* is considered a member of the genus *Echininus*. The familiar genus *Littorina* should be restricted to those northern temperate and arctic species which show a pallial oviduct with two consecutive spiral loops (of albumen followed by capsule gland), an open prostate, and a penis usually with several penial glands and lacking a glandular disc.

The following list includes all the genera and the subgeneric combinations which have been used by previous authors for species of *Littoraria*.

### *Generic combinations use for* Littoraria *species*

*Helix*—Linnaeus, 1758 [in part; not *Helix* Linnaeus, 1758]
*Buccinum*—Gmelin, 1791 [in part; not *Buccinum* Linnaeus, 1758]
*Phasianella*—Lamarck, 1822; Menke, 1830 [both in part; not *Phasianella* Lamarck, 1804]
*Turbo*—Donovan, 1825; Schumacher, 1838 [in part; not *Turbo* Linnaeus, 1758]
*Littorina*—Lesson, 1831; Quoy & Gaimard, 1833; Philippi, 1846; Reeve, 1857; Nevill, 1885; von Martens, 1887; Annandale & Prashad, 1919; Whipple, 1965; Kay, 1979 [all in part; not *Littorina* Férussac, 1822, type species *Turbo littoreus* Linnaeus, see Melville, 1980]
*Littorina* (*Littorina*)—Bequaert, 1943 [in part; not *Littorina* Férussac]
*Littoraria* Griffith & Pidgeon, 1834; Azuma, 1960; Habe, 1964; Habe & Kosuge, 1966; Higo, 1973; Yoo, 1976; Bandel & Kadolsky, 1982
*Littorina* (*Littoraria*)—Bequaert, 1943; Shikama & Horikoshi, 1963; Rosewater, 1970; Rosewater, 1981
*Litorina*—Philippi, 1847–48; von Martens, 1871; Lischke, 1871; Weinkauff, 1878, 1882 [all in part; unjustified emendation of *Littorina* Férussac, attributed to Menke, 1828, by Bequaert, 1943]
*Littorina* (*Melaraphe*)—Adams & Adams, 1858 [in part]; Tryon, 1887; Melvill & Standen, 1901; Dautzenberg & Fischer, 1905; Dautzenberg, 1929; Bequaert, 1943 [in part]; Biggs, 1958 [not *Melarhaphe* Menke, 1828, type species *Turbo neritoides* Linnaeus, see Rosewater, 1966]
*Melaraphe* (*Litt.*)—Dunker, 1871 [not *Melarhaphe* Menke]
*Littorina* (*Melaraphis*)—Tapparone-Canefri, 1874; Tryon, 1883 [not *Melaraphis* Philippi, 1836 = *Melarhaphe* Menke]
*Littorina* (*Malaraphe*)—Casto de Elera, 1896 [error for *Melarhaphe* Menke]
*Melarhaphe*—Hedley, 1918*a*; Yen, 1942 [in part; not *Melarhaphe* Menke]
*Melarapha*—Iredale & McMichael, 1962 [in part; see Rosewater, 1966]
*Littorina* (*Littorinopsis*) Mörch, 1876; Fischer, 1887; von Martens, 1897; Schepman, 1909; Prashad, 1921; Oostingh, 1927; Thiele, 1929; Adam & Leloup, 1938; Rosewater, 1970; Rosewater, 1981
*Littorinopsis*—Cossmann, 1916; Wenz, 1938; Kuroda & Habe, 1952; Oyama & Takemura, 1961; Brandt, 1974
*Melaraphe* (*Littorinopsis*)—Hirase, 1934; Abe, 1942
*Leptopoma?*—Heude, 1885 [not *Leptopoma* Pfeiffer, 1847]
*Littorina* (*Lamellilitorina*) Tryon, 1887
*Littorinopsis* (*Lamellilitorina*)—Wenz, 1938

### *List of recognized Recent taxa of* Littoraria

In the following list common synonyms and key taxonomic references are provided for those species not described in detail in the present monograph. Habitat and geographical distribution are indicated.

Subgenus *Littoraria* Griffith & Pidgeon, 1834
*zebra* (Donovan, 1825) [=*pulchra* (Sowerby, 1832)]; Keen (1971); mangroves; tropical E. Pacific
*fasciata* (Gray, 1839); Keen (1971); mangroves; tropical E. Pacific

*varia* (Sowerby, 1832); Keen (1971); mangroves; tropical E. Pacific
*irrorata* (Say, 1822); Bequaert (1943); salt marsh; NW. Atlantic
*vespacea* n. sp.; mangroves; Malaysia, Indo-China
*flava* (King & Broderip, 1832); Bequaert (1943; as *nebulosa* subsp.); mangroves and rocks; Brazil, Antilles
*tessellata* (Philippi, 1847); Bandel & Kadolsky (1982); mangroves and rocks; Caribbean
[n. sp. Reid (in prep.); salt marsh, mangroves; tropical E. Pacific]
*undulata* (Gray, 1839); Rosewater (1970); rocks, driftwood; Indo-Pacific
*nebulosa* (Lamarck, 1822); Bequaert (1943; in part); driftwood, mangroves; Caribbean
*cingulifera* (Dunker, 1845); Rosewater (1981); mangroves, rocks; W. Africa
*coccinea* (Gmelin, 1791) [=*obesa* (Sowerby, 1832)]; Rosewater (1970); rocks, driftwood; W. Pacific, Polynesia
*glabrata* (Philippi, 1846) [= *kraussi* (Rosewater, 1970)]; Rosewater (1970); rocks; Indian Ocean
*mauritiana* (Lamarck, 1822); Rosewater (1970); rocks; SW Indian Ocean
*pintado* (Wood, 1828) [=*schmitti* (Bartsch & Rehder, 1939); *pullata* (Carpenter, 1864)]; Rosewater (1970); rocks; W. Indian Ocean, W. and E. Pacific

Subgenus *Lamellilitorina* Tryon, 1887
*albicans* (Metcalfe, 1852); mangroves; Borneo

Subgenus unknown
*aberrans* (Philippi, 1846); Rosewater (1980*b*); mangroves; tropical E. Pacific

Subgenus *Littorinopsis* Mörch, 1876
*angulifera* (Lamarck, 1822) [=*ahenea* (Reeve, 1857)]; Bequaert (1943); Rosewater (1981); mangroves; tropical E. and W. Atlantic
*scabra* (Linnaeus, 1758); mangroves; Indo-Pacific
*lutea* (Philippi, 1847); mangroves; Indonesia, Philippines
*pallescens* (Philippi, 1846); mangroves; Indo-Pacific
*philippiana* (Reeve, 1857); mangroves; E. Australia
*intermedia* (Philippi, 1846); mangroves, rocks; Indo-Pacific
*subvittata* n. sp.; mangroves, rocks; W. Indian Ocean
*filosa* (Sowerby, 1832); mangroves; Australia, Sunda Is
*cingulata cingulata* (Philippi, 1846); mangroves; NW. Australia
*cingulata pristissini* n. subsp.; mangroves, salt marsh; Shark Bay, W. Australia
*luteola* (Quoy & Gaimard, 1833); mangroves, salt marsh; E. Australia
*ardouiniana* (Heude, 1885); mangroves; S. China Sea
*delicatula* (Nevill, 1885); mangroves; Bay of Bengal

Subgenus *Palustorina* n. subgen.
*melanostoma* (Gray, 1839); mangroves, salt marsh; Indo-Malaya
*flammea* (Philippi, 1847); China
*conica* (Philippi, 1846); mangroves; Malaysia, Indonesia
*carinifera* (Menke, 1830); mangroves; Indo-Malaya
*sulculosa* (Philippi, 1846); mangroves, rocks; NW. Australia
*articulata* (Philippi, 1846); mangroves, rocks; Indo-Malaya, Australia
*strigata* (Philippi, 1846); mangroves, rocks; Indo-Malaya

## Relationships of the Genus

Throughout the discussions of morphological characters the attempt has been made to assess which pairs of character states are likely to be ancestral and which derived, or in the terminology of cladistic analysis, which are plesiomorphic and which apomorphic (Hennig, 1966). These character states are summarized in Tables 6 and 7. In the absence of information from ontogeny and from a fossil record, out-group comparison has been

**Table 6** Character states in the family Littorinidae.

| Character | Plesiomorphic | Apomorphic | Notes |
|---|---|---|---|
| 1. penial glands | absent | present | Secondary reversal of character presumed in *Littoraria, Melarhaphe, Fossarilittorina, Cenchritis.* |
| 2. sperm nurse cells | absent | present | Polarity uncertain, no evidence from out-groups. |
| 3. sperm nurse cell rods | absent | present | Polarity uncertain. Secondary loss presumed in *Melarhaphe* and a few *Littoraria* and *Nodilittorina* spp. Unknown in *Tectarius* and *Fossarilittorina.* |
| 4. two consecutive spiral loops in pallial oviduct, of albumen followed by capsule gland | no | yes | |
| 5. penial glandular disc | absent | present | |
| 6. operculum | paucispiral | meso- to multispiral | *Tectininus* considered a subgenus of *Echininus*, not of *Nodilittorina.* |
| 7. penial papillae | absent | present | |
| 8. penial sperm groove | open | closed | |
| 9. prostate gland | open | closed | |
| 10. jelly gland a spiral loop | no | yes | |
| 11. single spiral ($\geq 3\frac{1}{2}$ whorls) of pallial oviduct, incorporating capsule gland | no | yes | Secondary loss of capsule gland in ovoviviparous species. *Bembicium* lacks capsule glands. |

used to assess plesiomorphies and apomorphies (Wiley, 1981). Insufficient anatomical information is available for the small littorinid forms from southern oceans including *Laevilitorina, Rissolittorina* and *Rufolacuna*, and for members of the family Lacunidae (considered by Ponder, 1976, to be a subfamily of the Littorinidae). Therefore, in considering the relationships of the genus *Littoraria*, only the genera *Nodilittorina, Echininus, Tectarius, Melarhaphe, Fossarilittorina, Cenchritis, Peasiella, Littorina* and *Bembicium* have been compared. From the small amount of information available, however, it seems possible that these genera (with the exception of *Bembicium*) may bear closer phylogenetic relationship to each other than to the excluded genera.

In making decisions as to the plesiomorphic and apomorphic states of a character, the conclusion will usually be dependent upon the choice of the out-group with which to draw comparison. That is, a phylogeny must be assumed before it can be tested, which appears to be a circular argument. As discussed by Wiley (1981) this logical difficulty is overcome if sufficient characters are available so that phylogenetic hypotheses can be retested by others in a process of reciprocal illumination.

The character states of the littorinid genera are summarized as a cladogram in Fig. 17. The distribution of apomorphies shows that a great deal of parallel evolution has occurred amongst the ten genera, and that most of the branching points are defined by only one or two characters. For these reasons the construction of a dendrogram to indicate recency of common ancestry is highly speculative, and owes more to subjective weighting of characters and estimations of 'overall similarity' than to rigorous application of the criterion of parsimony. The resulting cladogram must be regarded only as an hypothesis, to be tested as further information becomes available. Nevertheless, the cladogram is a useful means of summarizing information visually. Bandel & Kadolsky (1982) have noted a high incidence of parallel evolution in the genus *Nodilittorina.*

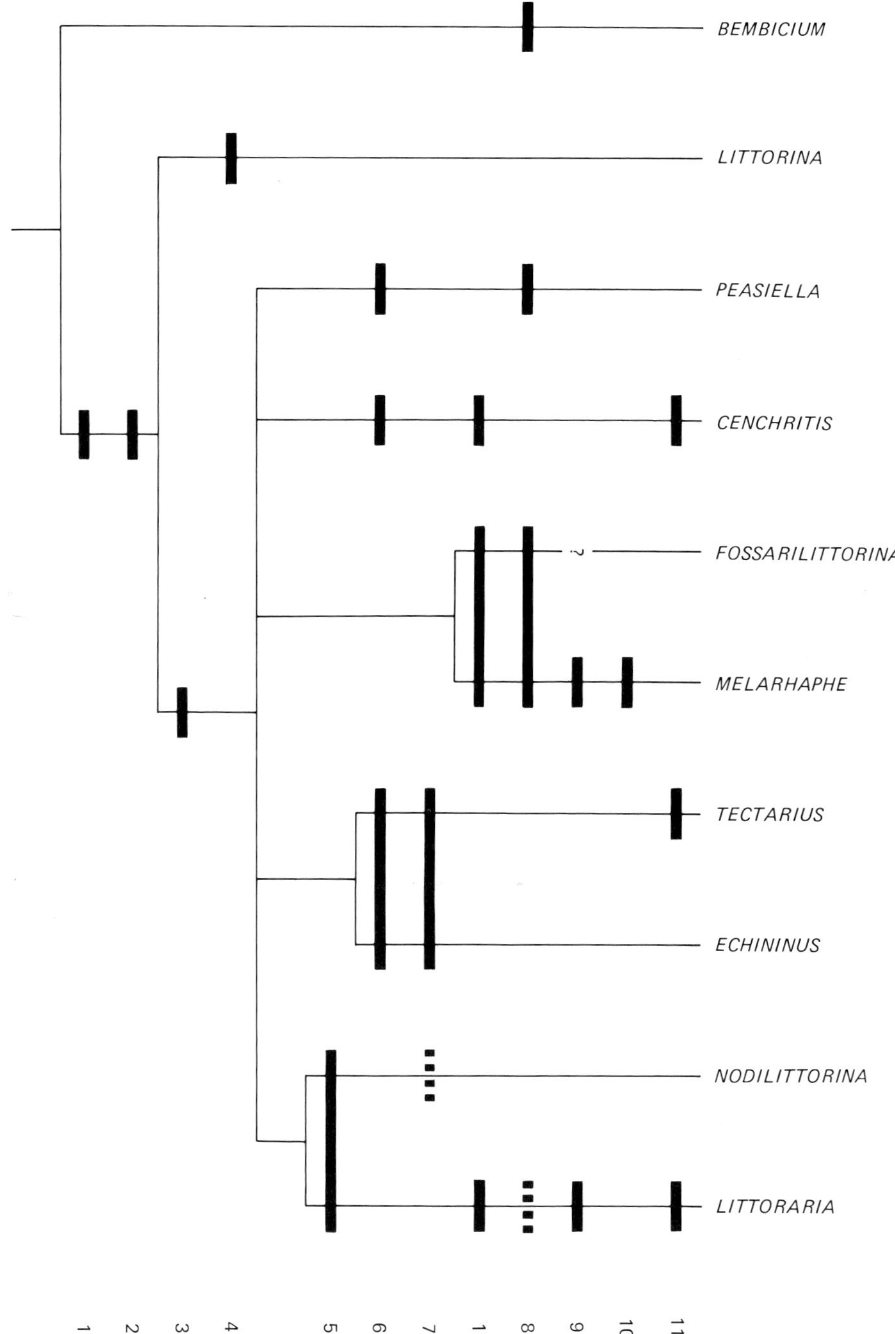


**Fig. 17** Cladogram representing an hypothesis of phylogenetic relationships amongst 10 genera of Littorinidae; numbers refer to the character states listed in Table 6; asterisk indicates a character reversal; solid bars indicate apomorphic states; broken bars indicate occurrence of both character states; query indicates character state unknown.

The genus *Bembicium* is rather poorly known, but peculiarities of penial form, lack of capsule glands and possible absence of sperm nurse cells, suggest that it is distantly related to the others. Of the eight remaining genera, *Littorina* is the most distantly related to *Littoraria*. The relationships of the other genera are less clear, and claims could be made for *Melarhaphe* and *Fossarilittorina*, *Cenchritis* or *Nodilittorina* as the sister group of *Littoraria*. For *Melarhaphe*, possible synapomorphies include the closed prostate and absence of penial glands, while the specialized pallial oviduct, lack of penial glandular disc, specialized radula and smooth shell militate against close relationship. *Fossarilittorina* is probably closely related to *Melarhaphe*, and the oviduct is less specialized. Evidence for *Cenchritis* includes the spiral oviduct, egg capsules, sperm nurse cells with rods, and lack of penial glands, while the shell, operculum, open prostate and absence of penial glandular disc are conspicuous differences from *Littoraria*. The evidence appears to favour *Nodilittorina* as the genus closest to *Littoraria*. An apparently unique synapomorphy is the penial glandular disc, while similarities of less certain value include the shape, sculpture and colour pattern of the shell of species of *Nodilittorina* which lack nodulose sculpture. Variability in numbers of penial glands in *Nodilittorina* suggests that these might readily be lost to derive the condition in *Littoraria*. Sperm nurse cells are similar in the two genera, as also in *Cenchritis*. The pallial oviduct is of sufficiently generalized structure in some species of *Nodilittorina* not to preclude the derivation of the spiral form shown by *Littoraria*. Egg capsules are divergently specialized in the two genera, although both could be derived from the generalized capsule form of *Melarhaphe neritoides*.

The only study to have applied techniques of biochemical taxonomy to a range of littorinid species from several genera is that of Jones (1972). This study compared the electrophoretic banding patterns shown by three proteins in twelve littorinids from Panama. However, the results were inconsistent and failed to demonstrate the groupings that might have been expected on the basis of the classification adopted in the present work.

## Subgeneric Classification

There is no doubt that the genus *Littoraria* is a monophyletic group, for all its members are characterized by the unique combination of four synapomorphies: the presence of the penial glandular disc, the single spiral loop of the pallial oviduct, the closed prostate and absence of penial glands. Within the genus, derived characters of the oviduct, developmental type, radula, penis and sperm nurse cells can be used in the reconstruction of a hypothetical phylogeny (Table 7, Fig. 18). Within the groups thus defined, shell characteristics are often similar, although impossible to describe in terms of simple character states. Geographical distribution and habitat also show some correlations with the species groups (see list p. 72).

The most distinctive group of species within the genus is here recognized as the new subgenus *Palustorina*. This group is defined by the unique synapomorphy of its flagellate sperm nurse cells. Unfortunately sperm nurse cells have not been described in all species of *Littoraria*, but the subgenus is also distinguished by a combination of penial form, position of the bursa copulatrix, and shell microsculpture, which are diagnostic when taken together. This subgenus is restricted to the Indo-Pacific province, and most of its seven members are known to occur exclusively in mangrove habitats, only two species being found also on sheltered rocky shores. Cossmann (1916) proposed the section *Touz-*

**Table 7** Character states in the genus *Littoraria*.

| Character | Plesiomorphic | Apomorphic | Notes |
|---|---|---|---|
| 1. penial glands | | | Absent in all *Littoraria* spp. |
| 5. penial glandular disc | | | Present in all *Littoraria* spp. |
| 8. penial sperm groove | open | closed | |
| 9. prostate gland | | | Closed in all *Littoraria* spp. |
| 11. single spiral of pallial oviduct, incorporating capsule glands | | | Present in all oviparous *Littoraria* spp.; secondary loss of capsule glands in ovoviviparous *Littoraria* spp. |
| 12. egg capsule with spiral rim above flotation skirt or lamella | yes | no | |
| 13. spiral whorls of pallial oviduct $\geqslant 3\frac{1}{2}$ | no | yes | Secondary reduction of number of whorls to $2\frac{1}{2}$ in a few ovoviviparous *Littoraria* spp. |
| 14. rachidian tooth of radula of 'hooded' type | no | yes | |
| 15. sperm nurse cells flagellate | no | yes | |
| 16. bursa opens in anterior position | no | yes | Condition variable within other littorinid genera. |
| 17. brooding of embryos in mantle cavity, loss of capsule glands | no | yes | |
| 18. planktonic protoconch $>610\,\mu m$ | no | yes | |
| 19. nonplanktotrophic development | no | yes | Indicated by protoconch of *L. aberrans*. |

*inia* of *Littorinopsis*, with *Phasianella prevostina* Basterot, from the Miocene of France, as the type species. Although Cossmann's figures of the species bear a superficial resemblance to *L. melanostoma*, examination of specimens in the BMNH has shown that '*Littorinopsis*' *prevostina* is not a member of the same subgenus, and probably not even of the genus *Littoraria*.

The twelve ovoviviparous species, known or suspected to brood embryos until the early veliger stage in the mantle cavity, are here recognized as the subgenus *Littorinopsis*. The anatomical difference between ovoviviparous and oviparous species is only a minor modification of the oviduct by the loss of capsule glands, which might, conceivably, have occurred several times during the evolution of the genus. Nevertheless, it seems useful to recognize the group as distinct. The penes are quite uniform, being bifurcate, with a round glandular disc. The shells of the group are also rather distinctive, showing spiral microsculpture in the grooves and being relatively thin in texture, while ten of the members are colour polymorphic. Probably associated with the thin and colourful shells is the habit of many of the species of living at high levels on the foliage of mangrove trees. Only two species occur occasionally on rock substrates. Eleven species occur in the Indo-Pacific and one in the Atlantic.

The subgeneric classification of *L. albicans* and *L. aberrans* is uncertain. *L. albicans* is anatomically close to members of the subgenus *Littoraria*, especially to the *L. zebra* group. However, the species is unique in its large protoconch of the planktotrophic type, equal cusps of the rachidian tooth, large number of primary grooves on the shell, and peculiar coloration of the head-foot. The shell microsculpture, varices and colour polymorphism are also unlike members of the subgenus *Littoraria*. For these reasons, the subgenus *Lamellilitorina* of Tryon (1887) is retained for this single species. In the absence of anatomical information, *L. aberrans* cannot yet be assigned to a subgenus; shell characters are similar to those of the thin shelled and colour polymorphic species of *Littorinopsis*, but

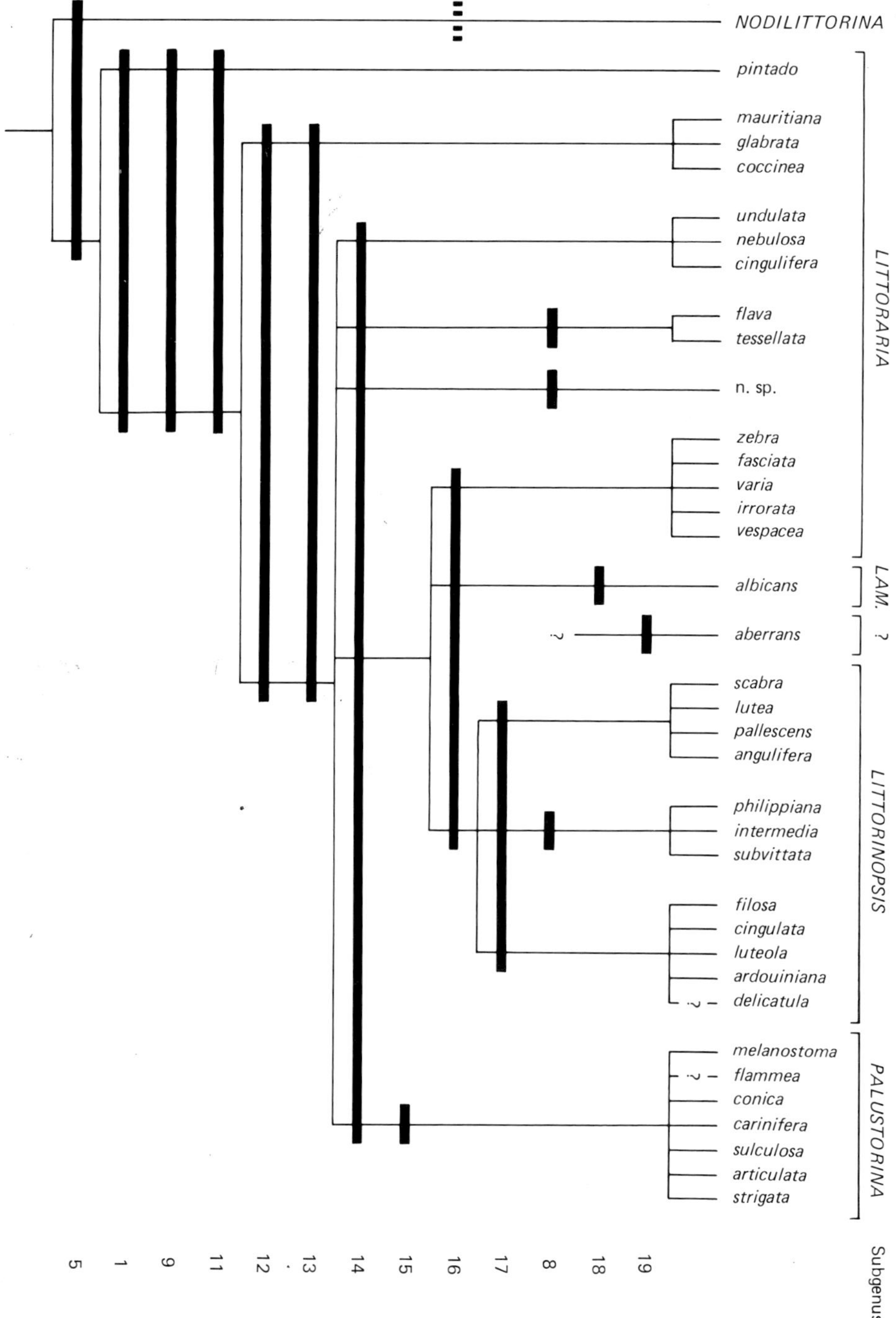


**Fig. 18** Cladogram of species of *Littoraria*, derived from the character states listed in Table 7. *Nodilittorina* is the inferred sister group of *Littoraria*. Terminal groupings are based upon close similarity in the form of the shell and or penis; solid bars indicate apomorphic character states; broken bars indicate the occurrence of both character states; queries indicate uncertainty of position of species of which anatomical characters are unknown; *Lam.*, subgenus *Lamellilitorina*. Note egg capsules and sperm are undescribed in some species (see text).

the protoconch shows that the species has nonplanktotrophic development, which is unique in the genus.

The remaining fifteen species are retained in the nominate subgenus. Although this is a paraphyletic and somewhat heterogeneous group, little would be gained by further subdivision. Shells in this group are neither thin nor colour polymorphic. Members occur in all the tropical provinces. *L. pintado* is of interest as the least specialized species in the genus, showing the greatest number of ancestral character states. The penial glandular disc is poorly differentiated, the pallial oviduct shows the least number of whorls amongst the oviparous species of *Littoraria*, the rachidian tooth of the radula is not hooded and the cupola of the egg capsule is sculptured by a single ring. In all these plesiomorphic states, *L. pintado* bears greater resemblance to species of *Nodilittorina* and other genera than is shown by any other *Littoraria* species. In addition, the shell is pale and the aperture dark brown with a somewhat paler basal stripe, which is also reminiscent of certain smooth shelled *Nodilittorina* species. In common with *L. coccinea*, *L. glabrata*, *L. mauritiana* and *L. undulata*, which also show few derived characters, *L. pintado* typically occurs on exposed rocky shores at high intertidal levels. This is certainly the ancestral habitat of the genus, and that in which most species of *Nodilittorina*, *Melarhaphe*, *Echininus*, *Tectarius*, *Cenchritis* and *Peasiella* also occur. Within the genus *Littoraria* there seems to have been an increasing specialization to the mangrove habitat in species showing more apomorphic characters (compare list of species p. 72 with Fig. 18). *L. pintado* is also remarkable for its wide and perhaps relict distribution (p. 64), which suggests that the species may be of greater age than its congeners.

# SYSTEMATIC DESCRIPTIONS

## Key to Shells

Shell variation in these species is such that a few rare and atypical shell forms may not conform to the diagnoses given, or may key out incorrectly. Reference should be made to the figures of shells, distribution maps, and, where possible, to anatomical characters. It should be noted that the primary grooves on the spire whorls should be counted on whorl four of the teleoconch or earlier; colour polymorphic shells may be yellow, pink, brown or patterned; columellar colour refers to the excavated area, rather than the pillar which is often white.

1 Columella narrow, rounded, not excavated . . . . . . . . . . 2
– Columella excavated or flattened, usually wide . . . . . . . . . . 8

2 Sculpture of 9–11 narrow carinae on last whorl; colour polymorphic . *L. filosa* (p. 140)
– Sculpture of low or rounded ribs, or numerous fine riblets . . . . . . 3

3 Ribs on last whorl numbering 40–78 . . . . . . . . . . 4
– Ribs on last whorl numbering 11–28 . . . . . . . . . . 5

4 Primary grooves on spire whorls numbering 10–13; secondary sculpture appears on whorl 7; microsculpture indistinct; colour polymorphic *L. ardouiniana* (p. 165)
– Primary grooves on spire whorls numbering 7–9; secondary sculpture appears on whorls 5–6; microsculpture usually of spiral striae in grooves or over whole surface; colour polymorphic . . . . . . . . *L. cingulata pristissini* (p. 152)

5 On last whorl grooves $\frac{1}{2}$–1 times rib width, containing strong spiral microsculpture; 11–13 prominent, rounded ribs on last whorl; primary grooves on spire whorls numbering 5–6; colour cream marbled with brown . . *L. cingulata cingulata* (p. 147)
– On last whorl grooves less than $\frac{1}{3}$ rib width; spiral microsculpture in grooves weak or absent; 15–28 ribs on last whorl; primary grooves on spire whorls numbering 6–10 . . . . . . . . . . . . . . . . . . 6

6 Spire outline straight sided, sutures not impressed; colour pale yellow with pattern of brown dots; parietal callus dark purple brown . . . . *L. melanostoma* (p. 174)
– Sutures impressed, spire whorls rounded . . . . . . . . . . 7

7 Sculpture on last whorl of rounded ribs, of which 2 at periphery are most prominent; microsculpture indistinct; colour polymorphic . . . . . *L. luteola* (p. 159)
– Sculpture on last whorl of low ribs of equal width; microsculpture of spiral striae on ribs and pits in grooves; colour pale with oblique, brown, axial stripes *L. flammea* (p. 180)

8 Primary grooves on spire whorls numbering 17–26; length of protoconch 0.6 mm; up to 20 varices on last whorl; colour polymorphic, fading to white *L. albicans* (p. 88)
– Primary grooves on spire whorls numbering 4–14; length of protoconch 0.4 mm or less . . . . . . . . . . . . . . . . 9

9 Spire outline almost straight sided, sutures not impressed; strong peripheral keel on last whorl; columella wide . . . . . . . . . . . . . . 10
– Sutures usually impressed and spire whorls rounded; if not, then columella narrow; last whorl not usually strongly keeled . . . . . . . . . . . 11

10 Sculpture of 50–70 fine ribs on last whorl; protoconch is a papilla on blunt apex of teleoconch; colour cream with irregular brown pattern . . . . ***L. conica*** (p. 182)
– Sculpture of 1–9 narrow carinae on last whorl, largest at periphery; colour grey with axial red brown lines . . . . . . . . . . . ***L. carinifera*** (p. 186)

11 Sculpture on last whorl of 9–11 prominent rounded ribs, separated by grooves 1–3 times rib width . . . . . . . . . . . . . . . 12
– Sculpture on last whorl of more numerous, small or low ribs, separated by grooves less than width of ribs . . . . . . . . . . . . . 13

12 Microsculpture of fine, axial lines in grooves; colour cream, with pale orange brown dashes or bands on ribs . . . . . . . . . . ***L. sulculosa*** (p. 194)
– Microsculpture of faint spiral lines in grooves, or absent; colour polymorphic ***L. pallescens*** (in part; p. 108)

13 Columella white and wide . . . . . . . . . . . . . 14
– Columella purple, brown or narrow . . . . . . . . . . . 16

14 Shell colour yellow or orange pink . . . . . . ***L. pallescens*** (in part; p. 108)
– Shell colour pale with more or less dense black or brown pattern . . . . . 15

15 Microsculpture of spiral striae in grooves; adult size 20–44 mm . . . ***L. scabra*** (p. 94)
– Microsculpture of axial striae in the wider grooves, or indistinct; adult size less than 20 mm . . . . . . . . . . . . . . ***L. articulata*** (in part; p. 200)

16 Sculpture of 13–18 prominent, narrow cords on last whorl . . . . . . 17
– Sculpture of more numerous, or low and rounded ribs . . . . . . . 18

17 Secondary sculpture conspicuous between primary cords on last whorl; colour usually brown, sometimes polymorphic . . . . . . . . ***L. philippiana*** (p. 118)
– Secondary sculpture faint or absent between primary cords on last whorl; colour polymorphic . . . . . . . . . . ***L. pallescens*** (in part; p. 108)

18 Ribs on last whorl numbering 35–60, of equal width . . . . . . . 19
– Ribs on last whorl usually numbering less than 35, or if not then secondary ribs only half width of primary ribs . . . . . . . . . . . . . 20

19 Columella very narrow, excavated; primary grooves on spire whorls numbering 11–14; sutures only slightly impressed; colour polymorphic . . . ***L. delicatula*** (p. 171)
– Columella of moderate width; primary grooves on spire whorls numbering 10–12; sutures impressed; colour pale orange brown with dark brown pattern more or less aligned into 8–12 oblique axial stripes . . . . . . ***L. subvittata*** (p. 135)

20 Rib at periphery of last whorl more prominent than others, marking a distinct keel, which is often emphasized by colour pattern; colour polymorphic . . . . . 21
– Last whorl rounded and hardly angled at periphery; peripheral rib not more prominent than the rest; colour cream with dark brown or black pattern . . 22

21 Ribs on last whorl numbering 33–50, comprising primary ribs separated by single secondary ribs of half their width; spiral bands of colour never present ***L. lutea*** (p. 103)
– Ribs on last whorl numbering 21–26; secondary sculpture usually absent or limited to a few inconspicuous riblets; spiral bands of colour may be present ***L. pallescens*** (in part; p. 108)

22 Colour pattern on last whorl of 13–20 axially aligned series of dashes, or axial stripes; posterior rib usually the most prominent, slightly pushed up towards suture; columella purple; spiral microsculpture seldom present on last whorl ***L. intermedia*** (p. 124)
– Colour pattern on last whorl of 6–15 more or less axially aligned series of dashes, or axial stripes (alignment may only be evident at sutures and periphery); if axial series number 12–15 then columella usually brown, not purple; ribs of approximately equal width; faint spiral striae usually visible on ribs . . . . . 23

23 Colour pattern on last whorl of 12–15 well aligned axial series of dashes; faint pale band on middle of base; columella very deeply excavated, brown or dull purple **L. vespacea** (p. 82)

– Colour pattern on last whorl of 6–11 axially aligned series of dashes, alignment often interrupted between suture and periphery; columella excavated, purple **L. articulata** or **L. strigata** (pp. 200, 209)

# Genus *LITTORARIA*

*Littoraria* Griffith & Pidgeon, 1834: 598 [type species by monotypy *Littorina pulchra* 'Gray' Sowerby, 1832 = *Turbo zebra* Donovan, 1825]

Diagnosis. Shell sculpture of spiral ribs and grooves, nodulose sculpture lacking; shell pattern usually of short spiral dashes aligned to form axial stripes. Operculum thin, paucispiral. Penial glands absent; penial glandular disc present; penial sperm groove open or closed; prostate a closed tube. Sperm nurse cells present, usually with rod-shaped inclusions, sometimes flagellate. Egg groove of pallial oviduct runs through a single spiral loop of 2½ to 9½ whorls, incorporating both albumen and capsule glands, or albumen glands alone in ovoviviparous species. Egg capsules of oviparous species planktonic, biconvex, with a circumferential lamella, containing a single ovum. In ovoviviparous species capsule glands are absent and embryos are brooded in mantle cavity and spawned at veliger stage; one species is suspected to undergo direct development. Radula often 'hooded', with frontal plate anterior to cusps of rachidian tooth. Distribution largely tropical and subtropical; habitat usually mangroves, driftwood and saltmarshes, sometimes rocky shores; zonation supralittoral.

## Subgenus *LITTORARIA*

Diagnosis. Shell solid; varices absent; microsculpture, if present, is of spiral striae, often over whole surface; shell coloration may be somewhat variable but rarely polymorphic. Penial base bifurcate or simple; penial sperm groove open or closed. Sperm nurse cells not flagellate. Bursa copulatrix may open at anterior or posterior end of straight section of pallial oviduct; capsule glands present. Development oviparous. Rachidian tooth of radula may or may not be hooded.

### *Littoraria (Littoraria) vespacea* n. sp.

Types. *Holotype:* BMNH 198345, Santubong, Sarawak, Borneo. *Paratypes:* BMNH, USNM, AMS.

Etymology. Latin: like a wasp, in reference to colour pattern of shell.

Diagnosis. Shell: small; solid; spire low; whorls rounded; columella wide, very deeply excavated, excavated area brown; primary grooves 8; 23–35 flattened ribs on last whorl, with narrow grooves between; microsculpture of faint spiral striae on ribs; colour white to yellow with pattern of black dashes entirely aligned to form axial stripes, numbering 12–15 on last whorl. Animal: penis not bifurcate, filament large, constricted at base; oviparous.

Shell. (Fig. 19). *Shape.* Height 10–25 mm. Teleoconch 5–6 whorls. Shell of moderate thickness, solid. Spire relatively low, outline slightly convex; whorls well rounded; sutures impressed. Peripheral keel absent. Adult lip slightly thickened, not flared. Varices absent. Columella wide, so deeply

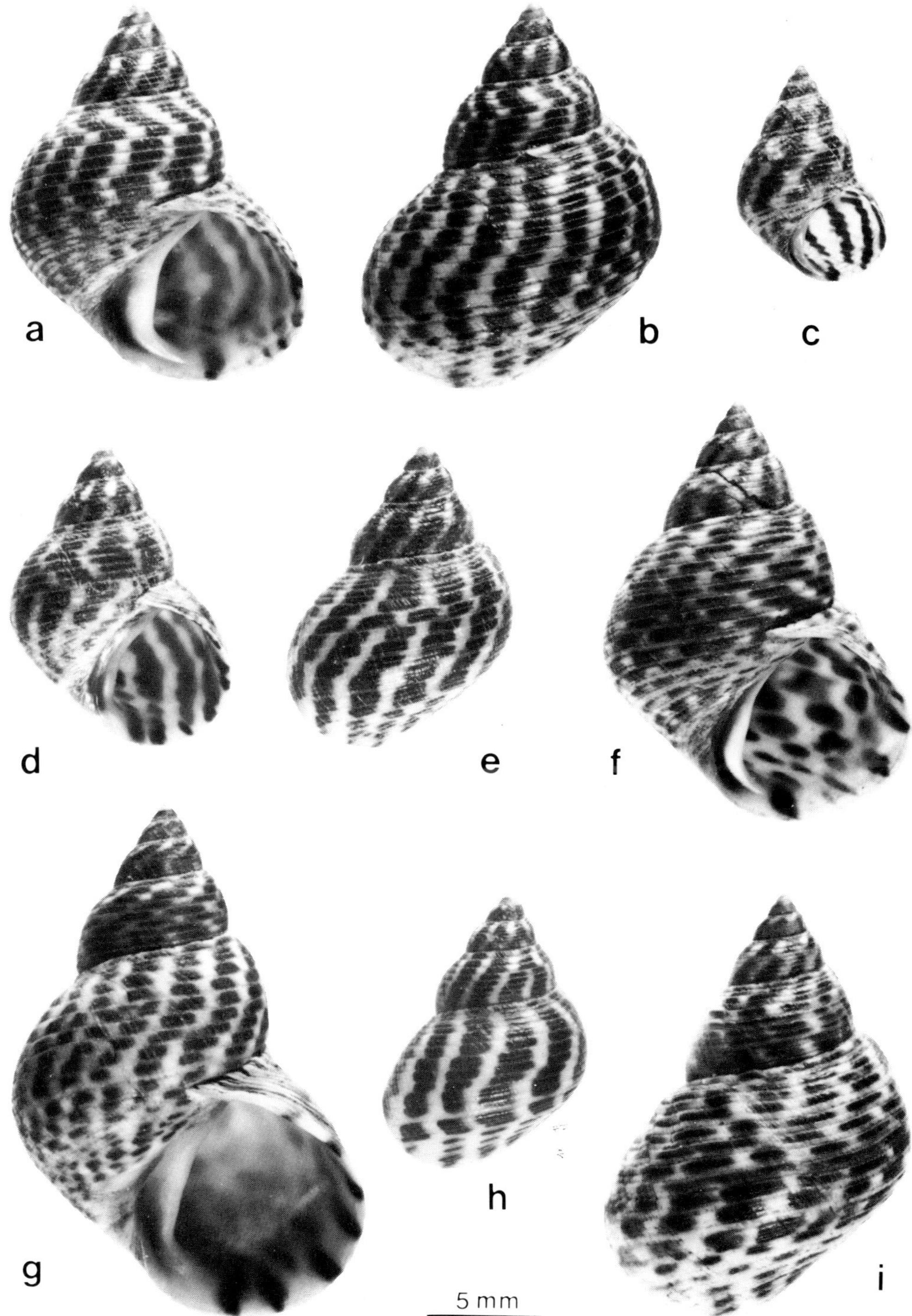


**Fig. 19** *Littoraria* (*Littoraria*) *vespacea:* **(a,b)** paratype, ♀, Santubong, Sarawak (DGR); **(c)** ♀, Port Kelang, Malaysia (DGR); **(d,e)** holotype, ♂, Santubong, Sarawak (BMNH 198345); **(f–i)** paratypes, Santubong, Sarawak (DGR); **(f,i)** ♀; **(g)** ♀; **(h)** ♂.

**Table 8** Dimensions of *Littoraria* (*Littoraria*) *vespacea*.

| Specimen | Locality | Sex | Primary grooves | H (mm) | B (mm) | LA (mm) | WA (mm) | C (mm) | P | S | SH |
|---|---|---|---|---|---|---|---|---|---|---|---|
| L. *vespacea* holotype BMNH 198345 | Santubong, Sarawak | ♂ | 8 | 12.8 | 10.2 | 7.9 | 6.1 | 1.9 | 1.25 | 0.77 | 1.62 |
| DGR | Santubong, Sarawak | ♂ | 8 | 14.7 | 10.5 | 9.0 | 7.0 | 2.3 | 1.40 | 0.78 | 1.63 |
| DGR | Santubong, Sarawak | ♂ | 8 | 11.2 | 8.1 | 6.6 | 5.1 | 1.3 | 1.38 | 0.77 | 1.70 |
| DGR | Santubong, Sarawak | ♀ | 8 | 17.9 | 13.7 | 11.2 | 9.0 | 2.0 | 1.31 | 0.80 | 1.60 |
| DGR | Santubong, Sarawak | ♀ | 8 | 15.7 | 12.1 | 9.2 | 7.4 | 1.8 | 1.30 | 0.80 | 1.71 |
| AMS C. 131737 | Sembawang Estuary, Singapore | | 8 | 23.8 | 17.9 | 13.2 | 11.4 | 3.0 | 1.33 | 0.86 | 1.80 |
| DGR | 10 km N. Port Kelang, Malaysia | ♂ | 8 | 9.5 | 6.8 | 5.5 | 4.5 | 1.2 | 1.40 | 0.82 | 1.73 |
| DGR | 10 km N. Port Kelang, Malaysia | ♀ | 8 | 8.7 | 6.9 | 5.5 | 4.5 | 1.2 | 1.26 | 0.82 | 1.58 |
| DGR, mean of 10 | Santubong, Sarawak | ♂ | | 10.52 | | | | | 1.332 | 0.800 | 1.657 |
| standard error | | | | 0.62 | | | | | 0.013 | 0.007 | 0.014 |
| DGR, mean of 10 | Santubong, Sarawak | ♀ | | 14.43 | | | | | 1.337 | 0.801 | 1.700 |
| standard error | | | | 0.76 | | | | | 0.007 | 0.009 | 0.025 |
| statistic t or U | | | | 3.980 | | | | | 52.5 | 51 | 67 |
| probability | | | | 0.001 | | | | | 0.883 | 0.970 | 0.218 |

excavated that inner lip of aperture stands up sharply; pillar straight or slightly convex, strongly narrowed towards base. Sexual dimorphism: males smaller.

*Dimensions:* Table 8.

*Sculpture* (Fig. 20a,b). Protoconch not seen. All whorls of teleoconch sculptured with spiral grooves. Primary grooves 8(9), almost equidistant, posterior 3 ribs usually a little wider. Secondary sculpture develops on whorl 5; posterior rib usually remains undivided, but the several ribs immediately below it become divided each by a narrow secondary groove, and later all other ribs may be thus divided at $\frac{1}{4}$–$\frac{1}{2}$ width. Tertiary sculpture and intercalated riblets may appear at end of whorl 6, on largest shells. Ribs total 23–35(40) on last whorl; all ribs remain flattened, but for the rounded posterior rib. Grooves narrow, less than $\frac{1}{5}$ rib width; secondary grooves impressed lines only. Microsculpture of faint spiral striae on ribs, sometimes absent; faint axial growth striae cover surface, stronger in primary grooves. Periostracum sometimes produced into small bristles on ribs.

*Colour.* Rather constant; ground colour whitish or cream yellow, with pattern of short black dashes on ribs. Dashes usually remain discrete, and are almost always entirely aligned to form axial or somewhat oblique stripes, numbering (9)12–15 on last whorl. Alignment may become less perfect in largest shells. On base, alignment is less pronounced; characteristically 2 or more secondary ribs in middle of base lack pigment, while the primary rib between shows normal black dashes. Exterior pattern visible within aperture, but clouded by white callus except at margins. Columella and parietal callus white, excavation of columella brown or dull purple.

ANIMAL. *Colour.* Pigmentation black; sides of foot darkly mottled; head black; tentacles black banded, broad unpigmented stripe each side of base, often meeting behind tentacle.

*Penis* (Fig. 21a–f). Length to 4.0 mm. Base simple, incorporating glandular disc. Filament large, tapering near tip, constricted at base. Sperm groove open. Base cream, ochraceous at base of glandular disc and of filament; glandular disc opaque cream; filament off white.

*Sperm.* Eupyrene sperm 279–303 μm. Nurse cells (Fig. 21g-i) 15–23 μm, oval to rounded; rods 1(2–3), small, central or asymmetrically placed, not projecting, oval or fusiform; yolk granules large.

*Pallial oviduct* (Fig. 21j–m). Length to 6.0 mm. Spiral section to 5.5 mm diam., $6\frac{1}{2}$ whorls; opaque albumen gland $\frac{1}{3}$ whorl, white; translucent albumen gland white; opaque capsule gland $1\frac{1}{2}$–2 whorls, pale pink or cream; translucent capsule gland red brown; spiral distinct externally; egg groove darkly pigmented. Straight section to 1.9 mm, grey to pale brown, brownish translucent capsule

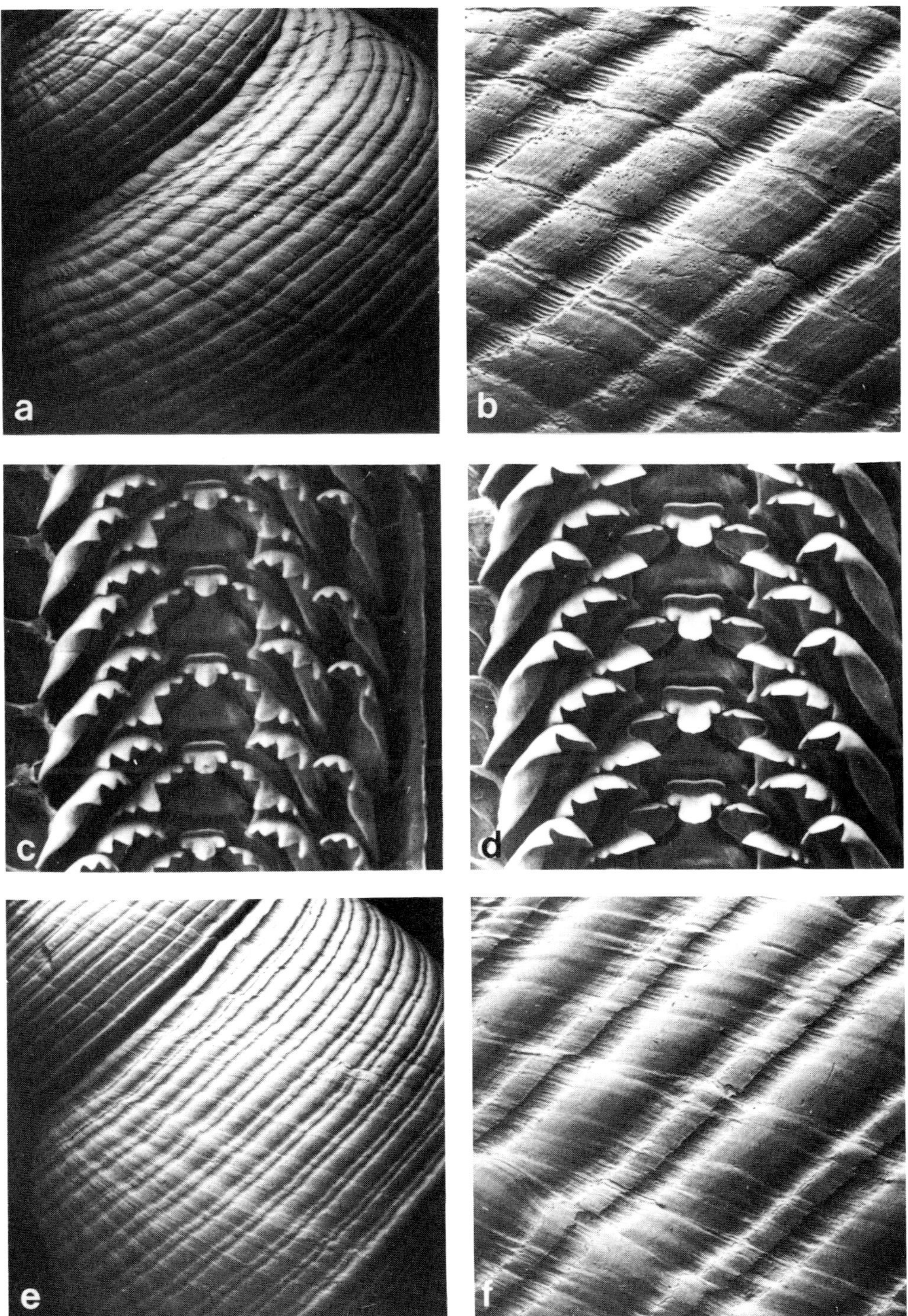


**Fig. 20** **(a–c)** *Littoraria* (*Littoraria*) *vespacea*, Santubong, Sarawak: **(a)** last whorl ( × 9); **(b)** detail ( × 38); **(c)** radula ( × 250). **(d–f)** *Littoraria* (*Littorinopsis*) *lutea*, Ubin I., Singapore: **(d)** radula ( × 200); **(e)** last whorl ( × 8); **(f)** detail ( × 33).

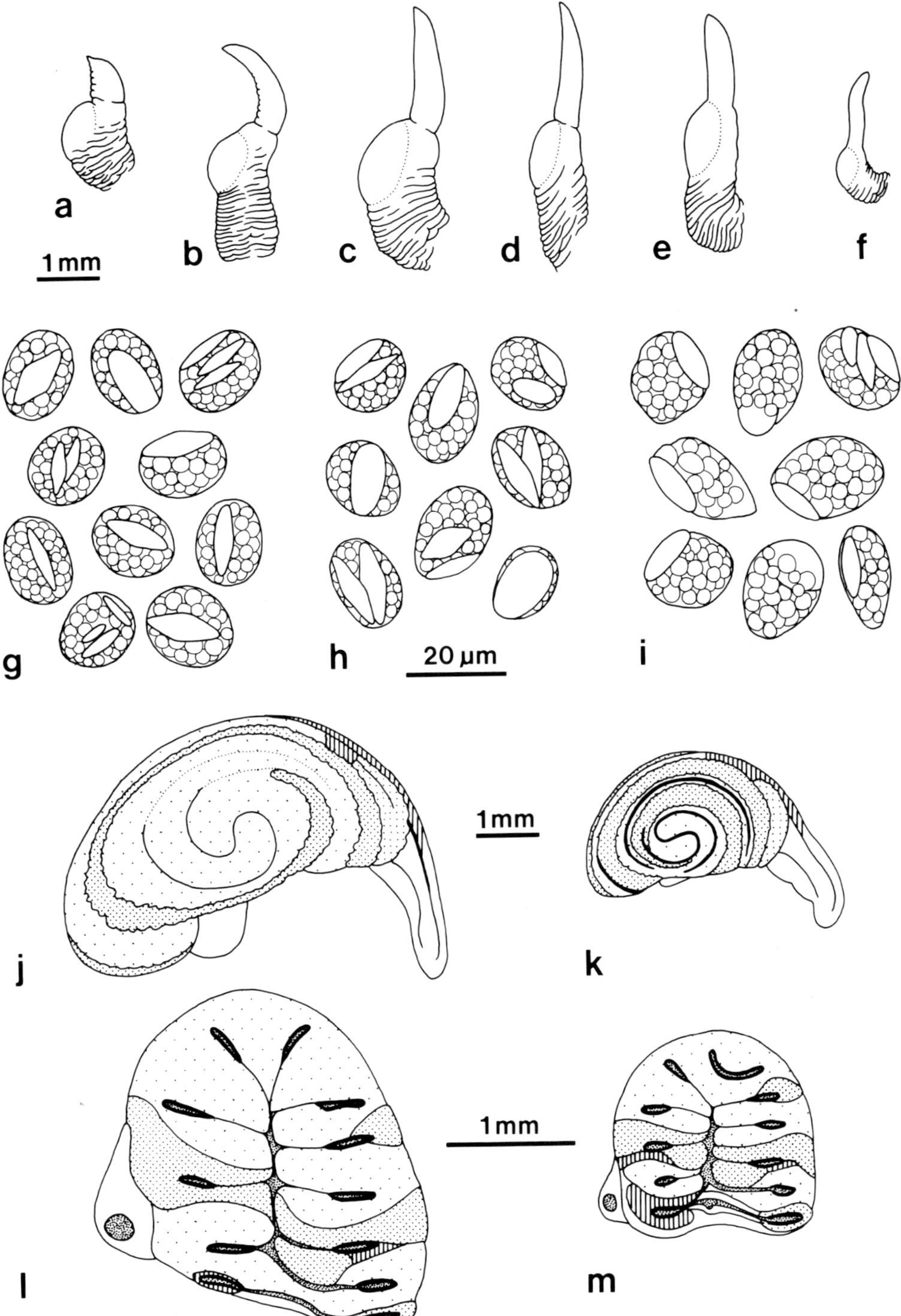


**Fig. 21** *Littoraria* (*Littoraria*) *vespacea:* **(a)** penis, contracted; **(b–f)** penes, relaxed; **(a–e)** Santubong, Sarawak; **(f)** Port Kelang, Malaysia; **(g–i)** sperm nurse cells; **(g,h)** Port Kelang, Malaysia; **(i)** Santubong, Sarawak; **(j–m)** pallial oviducts, with transverse sections, Santubong, Sarawak.

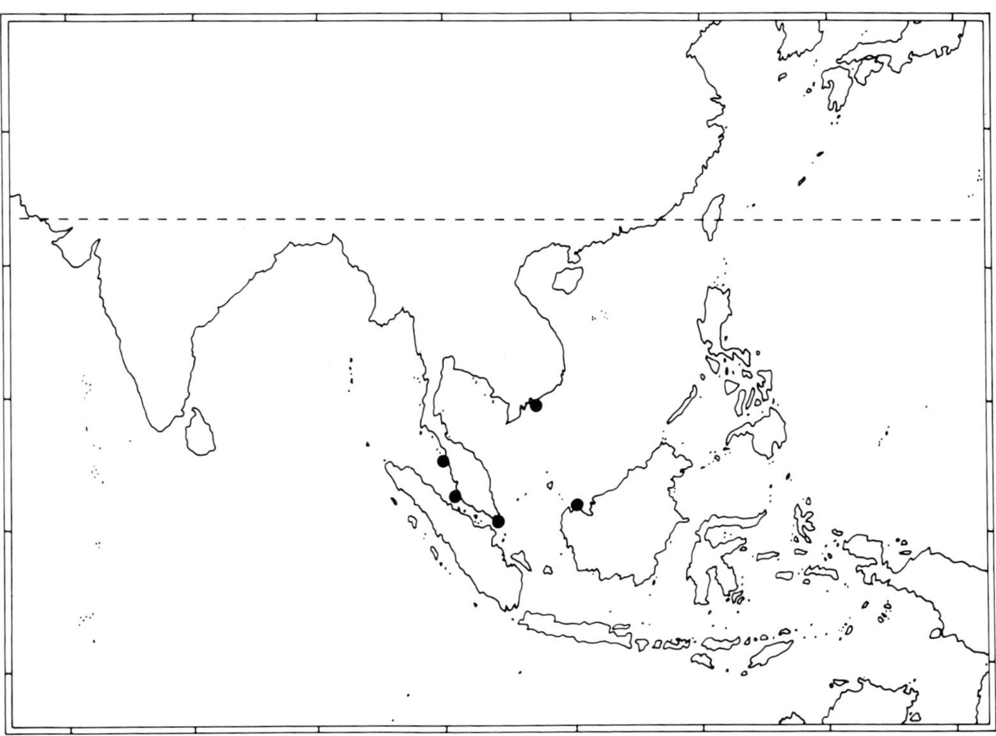

**Fig. 22** Distribution of *Littoraria* (*Littoraria*) *vespacea*.

gland visible posteriorly; no terminal papilla. Bursa to 1.4 mm, anterior. Development assumed oviparous.

*Radula* (Fig. 20c). Length to 19 mm; relative length 1.14–1.50. Saw-toothed type; central rachidian cusp square, edge slightly pointed; cusps of paired teeth almost equilaterally triangular; lateral with no gap anterior to main cusp.

*Pallial complex*. Hypobranchial gland 0.5–1.0 mm wide, lemon yellow when fresh, grey brown when preserved.

DISTRIBUTION. *Habitat*. On trunks in *Avicennia* and *Sonneratia* fringes, 0.8–1.9 m above ground; on roots in *Rhizophora* and *Bruguiera* forests, 0.3–0.8 m above ground; extending as far as landward zone; also in crevices on wooden pilings. A continental species.

*Range* (Fig. 22). West coast of peninsular Malaysia, Singapore, Sarawak and southern Vietnam.

*Records*. **Malaysia: Peninsula:** Batu Maung, Penang (DGR); 10 km N. of Port Kelang (DGR); Kuala Selangor (USNM, ANSP); **Sarawak:** Santubong (DGR); **Singapore:** Jelutong, Ubin I. (DGR); Sembawang Estuary (AMS); Kranji (BMNH, ANSP); **Vietnam:** Vung Tau (ANSP).

REMARKS. This species may have been hitherto overlooked as a result of its resemblance to *L. articulata*, *L. strigata* and *L. intermedia*. In addition, the distribution of *L. vespacea* is restricted and specimens are rare in museum collections. Despite the close similarity of the shell to *L.* (*Palustorina*) *articulata* and *L.* (*P.*) *strigata*, the anterior position of the bursa copulatrix and non-flagellate nurse cells preclude placement of *L. vespacea* in the same subgenus. Since the oviduct is of the oviparous type, the species is assigned to the subgenus *Littoraria*. *L. vespacea* does not appear to be closely related to any of the other Indo-Pacific species of the genus. The Eastern Pacific species *L. zebra*, *L. fasciata* and *L. varia* also show an oviduct of the oviparous type, with a bursa in the anterior position, as does the western Atlantic *L. irrorata*. It is interesting to note a resemblance of the shell of *L. vespacea* to those of young *L. zebra* and *L. fasciata*, in which the columella is also very deeply excavated and the microsculpture and colour pattern are similar, while the penis is reminiscent of that of *L. fasciata*.

The habitat of *L. vespacea* broadly overlaps that of *L. articulata* and *L. strigata*, but it does not occur in such large numbers and its distribution extends through the *Rhizophora* and *Bruguiera* zones to the outer edge of the landward fringe.

SIMILAR SPECIES. *L. vespacea* is clearly separated from *L. articulata* and *L. strigata* by the form of the penis, the non-flagellate sperm nurse cells, the anterior bursa copulatrix and the more numerous whorls of the pallial oviduct. The shells are similar, but those of *L. vespacea* are usually distinguished by their more inflated whorls, very deeply excavated columella and more numerous axial series of dark dashes in a more regular arrangement. *L. intermedia* should not be confused with *L. vespacea*, for in the former the spire is taller, the ground colour of the shell is grey or brown, the columella purple, the penis is bifurcate and the oviduct of the ovoviviparous type.

## Subgenus *LAMELLILITORINA*

*Littorina* (*Lamellilitorina*) Tryon, 1887: 230 [type species by subsequent designation (Wenz, 1938) *Littorina albicans* Metcalfe, 1852]

DIAGNOSIS. As for the only known member, *L. albicans*.

### ***Littoraria* (*Lamellilitorina*) *albicans*** (Metcalfe, 1852)

*Littorina albicans* Metcalfe, 1852: 73 [Borneo; lectotype (Rosewater, 1970) BMNH 1968355]; Reeve, 1857: *Littorina* pl. 9, figs 44a,b; Nevill, 1885: 150–151
*Litorina albicans*—Weinkauff, 1882: 81–82, pl. 11, figs 2,3
*Littorina* (*Lamellilitorina*) *albicans*—Tryon, 1887: 253, pl. 46, figs 25,26
*Littorina* (*Littorinopsis*) *albicans*—von Martens, 1897: 199
*Littorina* (*Littorinopsis*) *scabra scabra*—Rosewater, 1970: 456–461, pl. 352, figs 5,31 [in part; not Linnaeus, 1758]

DIAGNOSIS. Shell: thin; spite flat sided, last 2 whorls rounded; lip thickened and flared; varices frequent, 0–20; columella rather narrow, only slightly excavated; protoconch large, broadly conical, 5 whorls; primary grooves 20–23; ribs on body whorl numbering over 100, flattened; microsculpture of deep axial lines in primary grooves; colour polymorphic, yellow, pink or brown when young, sometimes with 9–12 narrow oblique stripes on last whorl, but adult shells usually faded entirely to white or lilac. Animal: penis bifurcate, small, filament extremely small; oviparous.

SHELL (Frontispiece, Fig. 23). *Shape*. Height 12–24 mm. Teleoconch 5.5–6.5 whorls. Shell thin, delicate, translucent. Spire outline: slightly convex, whorls 1–2 rounded, 3–5 rather flat with indistinct sutures, 5–6 rounded with impressed sutures. Peripheral keel slight in young shells, becoming obsolete on whorl 6. Adult lip thickened and flared; varices 0–20, typically numerous on last whorl. Columella rather narrow, flat or shallowly excavated; pillar almost straight. Sexual dimorphism: slight, males smaller, relatively lower spire and larger aperture.

*Dimensions:* Table 9.

*Sculpture* (Fig. 24a–c,e,f). Protoconch large, 610–660 $\mu$m in length, of 5 whorls, broadly conical. First half whorl of teleoconch smooth. Primary grooves (17)20–23(26), equally spaced. Grooves $\frac{1}{3}$–1 rib width just before secondary sculpture begins. Secondary sculpture may form after a major

**Fig. 23** *Littoraria* (*Lamellilitorina*) *albicans:* **(a–c)** Santubong, Sarawak (DGR); **(a)** ♀; **(b)** ♀; **(c)** ♂; **(d)** lectotype of *Littorina albicans* Metcalfe, Borneo (BMNH 1968355); **(e–i)** Santubong, Sarawak (DGR); **(e,f)** ♀; **(g,h)** ♂; **(i)** ♀.

**Table 9** Dimensions of *Littoraria* (*Lamellilitorina*) *albicans*.

| Specimen | Locality | Sex | Primary grooves | H (mm) | B (mm) | LA (mm) | WA (mm) | C (mm) | P | S | SH |
|---|---|---|---|---|---|---|---|---|---|---|---|
| *Littorina albicans* lectotype, BMNH 1968355 | Borneo | | 25 | 18.7 | 12.0 | 10.1 | 7.9 | 1.1 | 1.56 | 0.78 | 1.85 |
| DGR | Santubong, Sarawak | ♂ | 26 | 21.1 | 13.7 | 12.3 | 9.5 | 1.8 | 1.54 | 0.77 | 1.72 |
| DGR | Santubong, Sarawak | ♂ | 22 | 12.2 | 7.8 | 6.7 | 5.1 | 0.9 | 1.56 | 0.76 | 1.82 |
| DGR | Santubong, Sarawak | ♀ | 21 | 24.0 | 14.5 | 12.2 | 9.6 | 1.8 | 1.66 | 0.79 | 1.97 |
| DGR | Santubong, Sarawak | ♀ | 20 | 14.4 | 9.0 | 7.5 | 5.5 | 0.8 | 1.60 | 0.73 | 1.92 |
| DGR, mean of 10 | Santubong, Sarawak | ♂ | | 14.01 | | | | | 1.577 | 0.768 | 1.840 |
| standard error | | | | 0.84 | | | | | 0.020 | 0.009 | 0.022 |
| DGR, mean of 10 | Santubong, Sarawak | ♀ | | 17.16 | | | | | 1.624 | 0.770 | 1.909 |
| standard error | | | | 1.02 | | | | | 0.026 | 0.009 | 0.017 |
| statistic t or U | | | | 2.390 | | | | | 65 | 63 | 83 |
| probability | | | | 0.028 | | | | | 0.280 | 0.352 | 0.012 |

varix on last whorl, by intercalation of 1–5 riblets between each pair of primary ribs. Secondary riblets at first narrow, but towards end of last whorl all ribs are subequal, remaining low and flattened, and numbering over 100. Spiral sculpture becomes irregular and less distinct where it is crossed by numerous varices and growth lines. Microsculpture of faint axial growth striae over whole surface, developed into deep, regular axial lines in primary grooves; on whorls 4–5 the primary grooves are narrow and therefore appear punctate; spiral microsculpture absent.

*Colour* (Frontispiece). Polymorphic; yellow shells with faint pattern are most frequent, brown or pink shells not uncommon. Ground colour lemon yellow, cream, pale brown or orange pink, darkest towards apex. Alternate primary ribs often opaque white on penultimate whorl, all opaque by last whorl. Pattern of dark brown or lilac dashes on ribs, sometimes densely scattered over entire shell without conspicuous axial alignment; but usually first 4 whorls lack dark pigment, while last 2 are patterned with narrow, oblique stripes from suture to periphery, numbering 9–12 per whorl; pattern absent after first major varix. Yellow and pink shells fade to pure white with age, while in those which have been predominantly brown the dark pigment on the spire whorls fades to lilac and the body whorl and varices to white. Exterior pattern clearly visible within aperture of small translucent shells; aperture white in large individuals. Columella white; sometimes red brown in small pink or brown shells.

ANIMAL. *Colour*. Pigmentation pale to dark grey, correlated with shell colour, never unpigmented even in white shells; sides of foot faintly to darkly mottled; head darkest, red buccal mass clearly visible; tentacles red, with little dark pigment, no white stripes at either side of base, only groove between tentacle base and snout lacks pigment. Outer surface of mantle lemon yellow.

*Penis* (Fig. 25a–d). Rather small, to 3.1 mm. Base bifurcate; limb bearing glandular disc short. Filament very small, tip mucronate. Sperm groove open. Base and filament off white; glandular disc cream.

*Sperm*. Eupyrene sperm 152–170 $\mu$m. Nurse cells (Fig. 25e–g) 14–22 $\mu$m; oval, pyriform (with eupyrene sperm attached at blunt end) or rounded; rods often apparently absent, seldom clearly visible, but if so then single, parallel sided, blunt to rounded, sometimes projecting to form papillose tip; yolk granules closely packed, indistinct, polygonal, often elongate or lozenge shaped.

*Pallial oviduct* (Fig. 25h–k). Length to 6.4 mm. Spiral section to 2.9 mm diam., $6\frac{1}{2}$ whorls; opaque albumen gland $\frac{1}{3}$ whorl, white; translucent albumen gland pale fawn; opaque capsule gland $1\frac{1}{2}$–2 whorls, cream; translucent capsule gland red brown; spiral distinct externally; egg groove darkly pigmented. Straight section to 2.5 mm, fawn, reddish translucent capsule gland visible posteriorly; no terminal papilla. Bursa long, to 2.4 mm, anterior. Development assumed oviparous.

*Radula* (Fig. 24d). Length to 12 mm; relative length 0.54–0.79. Chisel-toothed type; cusps of

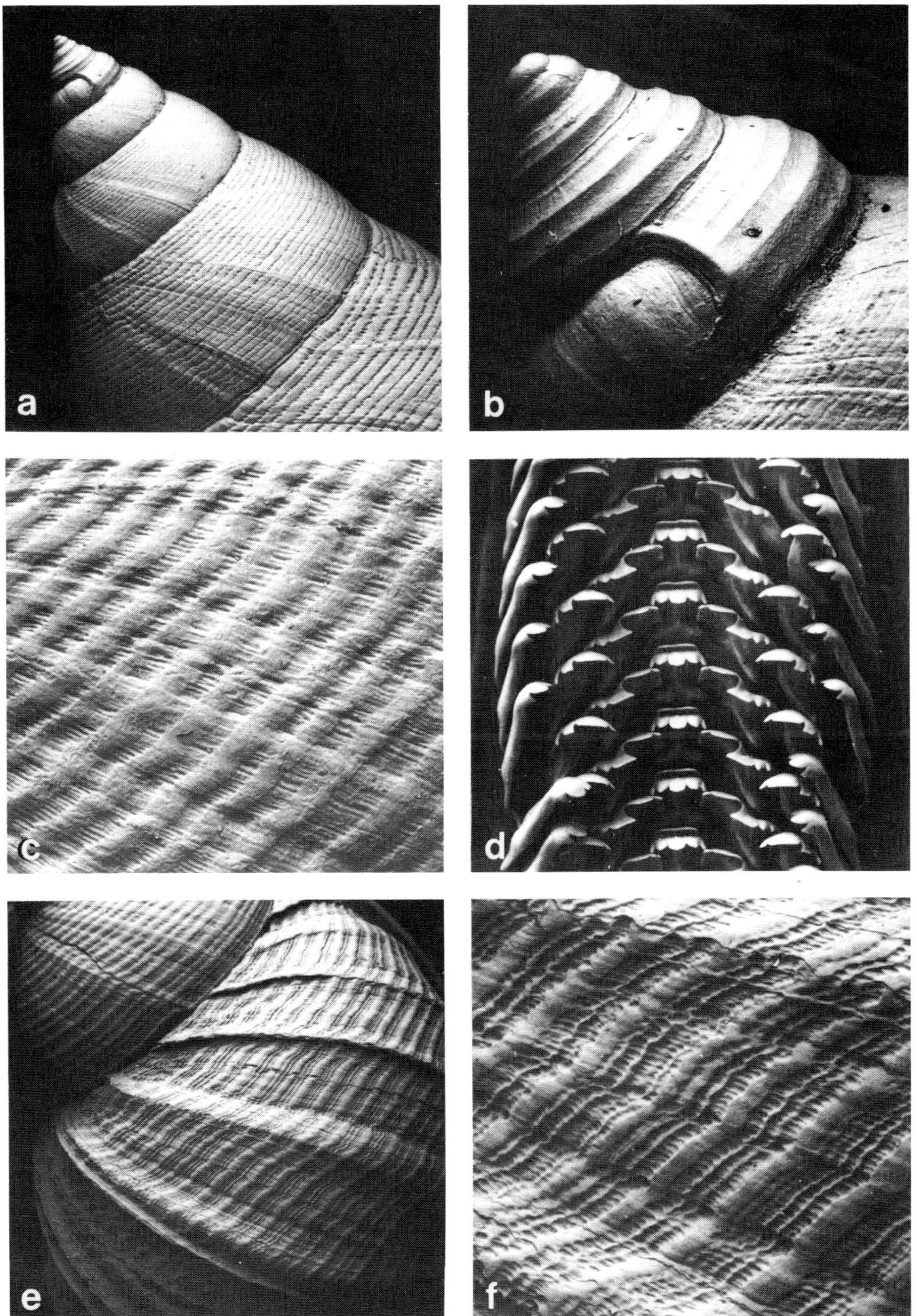


**Fig. 24** *Littoraria* (*Lamellilitorina*) *albicans*, Santubong, Sarawak: **(a)** spire ( × 20); **(b)** protoconch ( × 90); **(c)** detail of penultimate whorl ( × 30); **(d)** radula ( × 190); **(e)** last whorl ( × 7); **(f)** detail ( × 27).

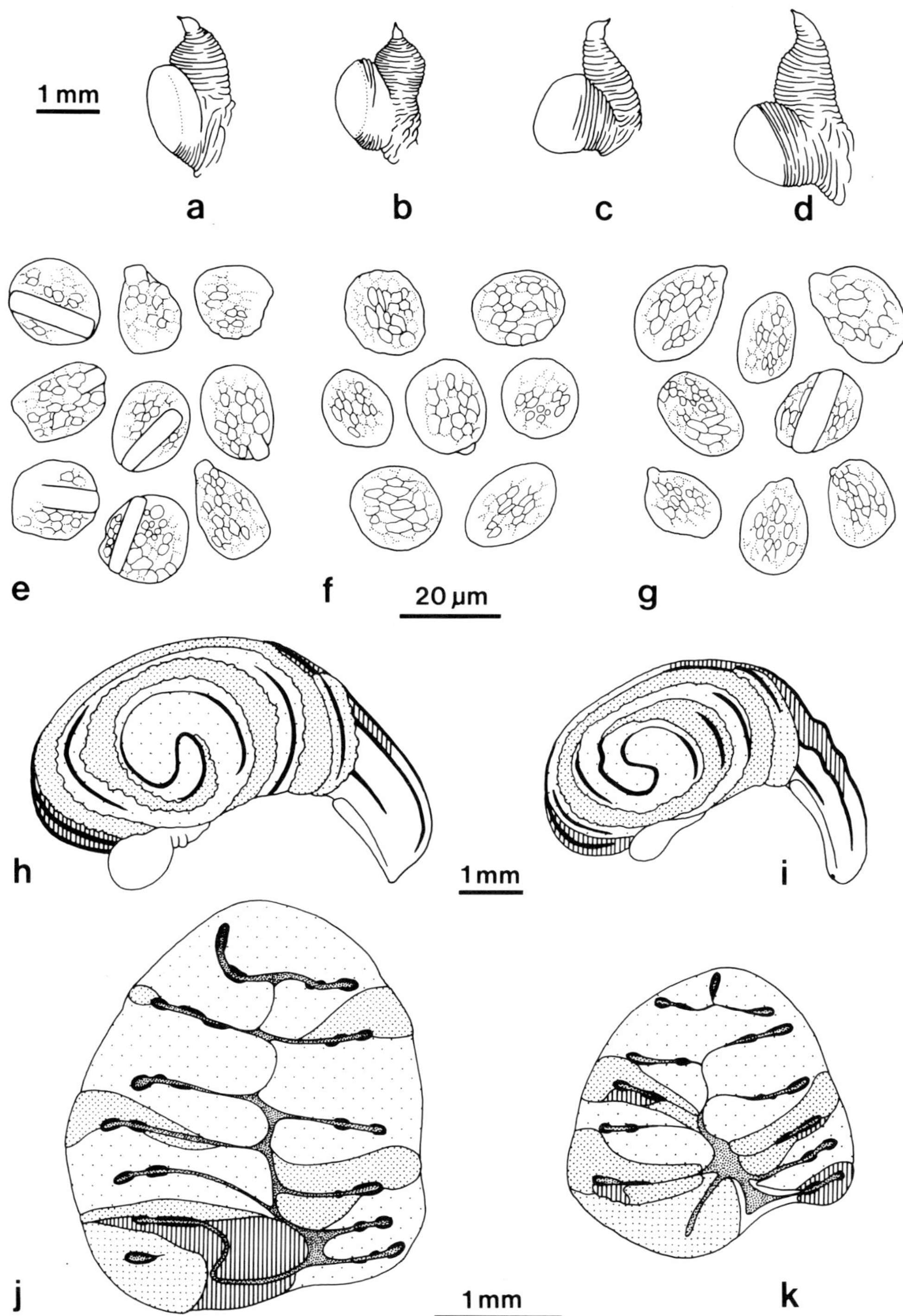


**Fig. 25** *Littoraria* (*Lamellilitorina*) *albicans*, Santubong, Sarawak: **(a–d)** penes, relaxed; **(e–g)** sperm nurse cells; **(h–k)** pallial oviducts, with transverse sections.

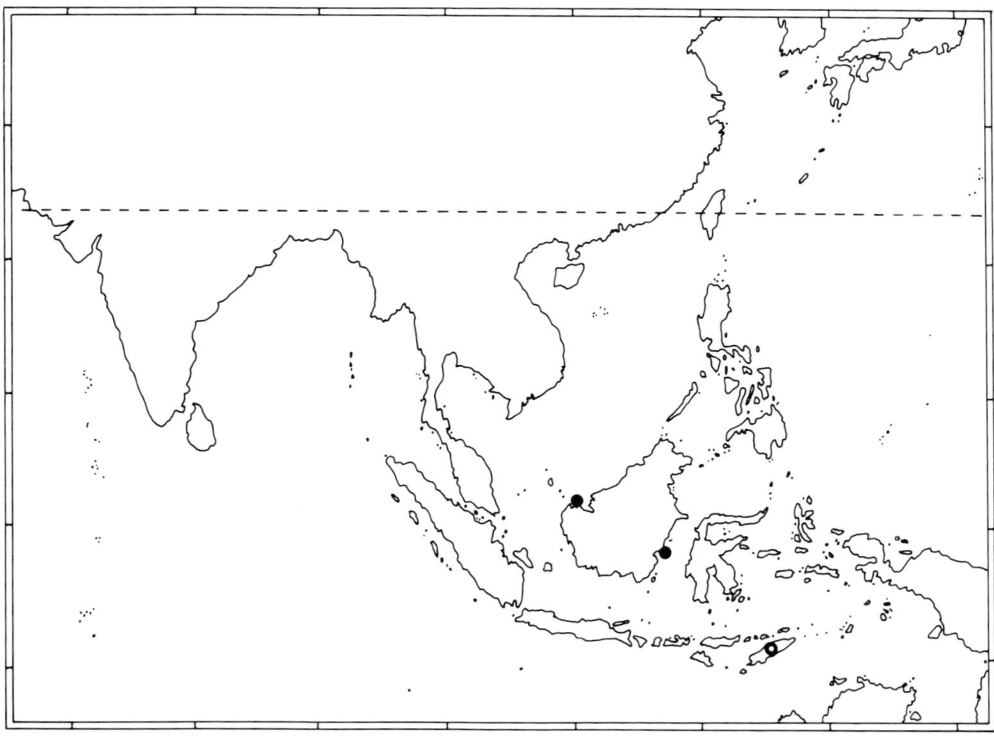

**Fig. 26** Distribution of *Littoraria (Lamellilitorina) albicans*.

rachidian of equal length, central one slightly wider; cusps of paired teeth obliquely triangular; lateral with gap anterior to main cusp.

DISTRIBUTION. *Habitat*. Leaves of *Avicennia* from near seaward edge to landward fringe, 0.5–3.5 m above ground; scarce on *Sonneratia* leaves; in landward fringe also on *Nypa* fronds.

*Range* (Fig. 26). Southern Borneo; perhaps extending to Timor (a single record, doubtful since species not recorded at any intervening localities in Indonesia).

*Records*. **Malaysia: Sarawak:** Santubong (DGR, BMNH, USNM, SM); Buntal (SM); **Indonesia**: **Kalimantan**: Balikpapan Harbour (AMS); **Lesser Sunda Is**: Timor (RNHL, record doubtful).

REMARKS. Although common in Borneo, at least in Sarawak, *L. albicans* is rare in museum collections and does not seem to have been recognized as a distinct species in the literature since 1897. The white, lamellose shells of large adults are so distinctive that Tryon (1887) placed the species in a new subgenus, *Lamellilitorina*. This monotypic subgenus is retained here in recognition of the large protoconch (unique in the genus), the unusual shell sculpture (large number of primary grooves, strong axial microsculpture), small penis, distinctive radula and remarkable coloration of the animal.

In other *Littoraria* species varices are formed during interruptions to growth corresponding with an annual reproductive season, and possibly also during otherwise unfavourable periods. It would seem difficult thus to account for the formation of as many as 20 varices in *L. albicans*, especially in an equatorial area without extremes of climate. However, it is possible that the longevity of this species is indeed much greater than others in the genus, for alone of species of *Littoraria* the shell fades to pure white, suggesting a most prolonged exposure to the sun.

SIMILAR SPECIES This is one of the most distinctive of *Littoraria* species and when large, with varices well developed, cannot be confused with any other. Small and immature shells are superficially close to *L. delicatula* and *L. conica*, but are immediately distinguished by their very numerous primary grooves, large protoconch and characteristic microsculpture.

## Subgenus *LITTORINOPSIS*

*Littorina* (*Littorinopsis*) Mörch, 1876: 135 [type species by original designation *Littorina subangulata* Lamarck, error for *Littorina angulifera* (Lamarck, 1822)]

DIAGNOSIS. Shell often of thin texture; a few varices may be present; microsculpture, if present, is of spiral striae, usually only in grooves; shell coloration often polymorphic. Penial base bifurcate, glandular disc borne on a limb separate from that bearing filament; penial sperm groove open or closed. Sperm nurse cells not flagellate. Bursa copulatrix opens at anterior end of straight section of pallial oviduct; capsule glands absent. Development ovoviviparous. Radula with 'hooded' rachidian tooth.

### *Littoraria* (*Littorinopsis*) *scabra* (Linnaeus, 1758)

Helix scabra Linnaeus, 1758: 770 [neotype here designated BMNH 198347; additional specimens from neotype lot AMS, USNM; Magnetic I., N. Qld., Australia]; Linnaeus, 1767: 1243; Chemnitz, 1795: 283, pl. 210, fig. 2074 [in part]

*Litorina scabra*—Philippi, 1847, vol. 2: 221–223 [in part]; von Martens, 1871: 39; Weinkauff, 1878: 37–39, pl. 4, figs 7,12 [in part]; von Martens, 1880: 284 [in part]

*Littorina scabra*—Reeve, 1857: *Littorina* pl. 5, figs 21b,c [in part]; Nevill, 1885: 145 [in part]; von Martens, 1887: 169 [in part]; Fischer, 1891: 170 [in part]; Hidalgo, 1904–5: 207 [in part]; Taylor, 1971*b*: 197, pl. 15, fig. 14; Nielsen, 1976: fig. 1c [in part; ecology]; Wilson & Gillett, 1979: 52, pl. 8, figs 5,5a; Kilburn & Rippey, 1982: 51, pl. 10, fig. 4

*Littorina* (*Melaraphe*) *scabra*—Tryon, 1887: 243–244, pl. 42, fig. 19 [in part]; Dautzenberg, 1929: 289 [in part]

*Littorina* (*Malaraphe*) *scabra*—Casto de Elera, 1896: 309 [in part]

*Littorina* (*Littorinopsis*) *scabra*—von Martens, 1897: 194; Schepman, 1909: 193–194 [in part]; Oostingh, 1927: 1–2 [in part]; Adam & Leloup, 1938: 75, pl. 1, fig. 2a, text fig. 25A [radula]; Cernohorsky, 1972: 56, pl. 12, fig. 7

*Littorinopsis scabra*—Hirase & Taki (undated): pl. 79, fig. 13; Kuroda & Habe, 1952: 64; Oyama & Takemura, 1961: fig. 11; Brandt, 1974: 53–54 [in part]

*Melaraphe* (*Littorinopsis*) *scabra*—Abe, 1942: fig. 17, no. 11 [in part; ecology and behaviour]

*Littoraria scabra*—Azuma, 1960: 10; Habe, 1964: 29, pl. 9, fig. 32 right [in part]; Habe & Kosuge, 1966: pl. 6, fig. 16; Higo, 1973: 46

*Littorina* (*Littorinopsis*) *scabra scabra*—Rosewater, 1970: 456–461, pl. 325, figs 2,4; pl. 352, figs 1,11,32; pl. 353, figs B,C [in part]; Rosewater, 1980: figs 1,4,7,8

*Buccinum lineatum* Gmelin, 1791: 3493 [lectotype figure (Rosewater, 1970) Knorr, 1768: pl. 14, fig. 4; not da Costa, 1778]

*Buccinum foliorum* Gmelin, 1791: 3493 [lectotype figure here designated Rumphius, 1705: pl. 29, fig. Y; type locality restricted to Amboina, Moluccas]

*Littorina novaehiberniae* Lesson, 1831: 348 [Port Praslin, New Ireland; lectotype (Rosewater, 1970) MHNG]

*Littorina angulifera*—Quoy & Gaimard, 1833: 474–476, pl. 33, figs 1–3 [not Lamarck, 1822]

*Phasianella angulifera*—Deshayes & Milne Edwards, 1843: 244–245 [in part; not Lamarck, 1822]

*Littorinopsis angulifera*—Cossmann, 1916: pl. 3, figs 2,3 [not Lamarck, 1822]

*Littorina* (*Littorinopsis*) *scabra angulifera*—Rosewater, 1981: 20–23, pl. 2, fig. F [in part; not Lamarck, 1822]

*Turbo striatus* Schumacher, 1838: 198 [lectotype figure here designated Chemnitz, 1795: pl. 210, fig. 2074; not da Costa, 1778]
? *Litorina scabra* var. *flammulata* Philippi, 1847, vol. 2: 222
*Littorina scabra* subvar. *flammulata*—Nevill, 1885: 145
*Litorina scabra* var. *ventricosa* Philippi, 1847, vol. 2: 222, pl. 5, fig. 3 [not fig. 8 as cited in text; lectotype here designated BMNH 198341; type locality restricted to Cagayan, province of Misamis, Mindanao, Philippines; not *Littorina ventricosa* Brown, 1838]
*Littorina scabra* subvar. *tenuis* Nevill, 1885: 146 [Arakan [Burma]; lectotype here designated ZSI, 31.0 mm]
*Littorina* (*Melaraphe*) *scabra* var. *intermedia*—Tryon, 1887: 244 [in part, not Philippi, 1846]
*Melarhaphe* (*Littorinopsis*) *scabra intermedia*—Hirase, 1934: 47, pl. 79, fig. 13 [not Philippi, 1846]

NOMENCLATURE. Since the monograph of the genus *Litorina* by Philippi (1847–48) the name '*Littorina scabra*' has been generally accepted as applicable to the most frequently collected and well known of the Indo-Pacific mangrove littorinids. However, to equate the species with the name requires some justification. The original description by Linnaeus (1758) reads '*testa subcarinata imperforata ovata acuminata striata ... fasciis fuscis dissectis, in inferiore anfractu linea elevata*', and is quite inadequate to define the species. No locality was given. Philippi (1847) argued that the colour pattern excluded the other carinate species recognized in his monograph, but his own concept of '*Litorina scabra*' embraced four species as here defined, and still others could equally well be described as subcarinate, with a pattern of dissected dark bands. Unfortunately, no Linnaean specimen is known to exist. Dance (1967) listed *Helix scabra* amongst the species described by Linnaeus which are not represented by specimens in the Linnaean shell collection, but noted that some of the missing specimens may be located amongst the residual unsorted material. An examination of this unauthenticated material revealed six unlabelled and unmarked specimens of this species, and four of *L. angulifera* (Lamarck), also with no data. From the comments of Dance on the containers used in the collection, these specimens are probably subsequent additions to the collection of Linnaeus and have no claim as type material. In edition ten of the *Systema Naturae* (Linnaeus, 1758), but not in edition twelve (Linnaeus, 1767), the abbreviation 'M.L.U.' appeared following the description of *Helix scabra*, implying that a specimen of this species was contained in the Museum Ludovicae Ulricae. However, *Helix scabra* was not mentioned in the printed (Linnaeus, 1764) nor manuscript version (Hanley, 1860) of Linnaeus' catalogue of this collection, nor was the species listed by Odhner (1953) in the only modern account.

The identification of the present species with *Helix scabra* rests largely upon the observations of Hanley (1855) on a shell from the Linnaean collection which has since been lost. Hanley accepted an annotation in Linnaeus' personal copies of the *Systema Naturae* as evidence of the author's possession of *Helix scabra* (evidence which was disputed by Dance, 1967) and proceeded to identify as the holotype 'the shell, which alone of his collection agrees with the description' (Hanley, 1855, p. 365). Hanley added: 'It is the *Littorina scabra* of modern conchologists, and was described and illustrated by Chemnitz (Conch. Cab. vol. xi, pl. 210, f. 2074, 2075) under the Linnaean name; that author having happily guessed at a species, which was neither defined by an adequate diagnosis nor illustrated by a reference to any engraving'. The monographs of the genus *Littorina* by Philippi (1847–48) and by Reeve (1857) can be taken as representative of the concept of the species '*Littorina scabra*' in Hanley's day. The varieties of '*Litorina scabra*' described by Philippi included four species as here recognized, chiefly *L. scabra s. s.* and *L. lutea*, but also *L. intermedia* and *L. pallescens*. Reeve illustrated *L. scabra s. s.*, *L. lutea* and *L. angulifera* (Reeve pl. 3, fig. 15a) as '*Littorina scabra*'. Of the two figures of Chemnitz referred to by Hanley, fig. 2074, although poorly drawn, is certainly the species here recognized as *L. scabra s. s.*, as indicated by the characteristic pattern, the concave spire outline and the description in the text of the wide, white columella. However, fig. 2075 is *L. angulifera*, as suggested by the coloration and outline, and confirmed by the locality 'Guinea'.

Rosewater (1970) designated the figure 2074 of Chemnitz as lectotype of *Helix scabra*, but such a designation is invalid since a lectotype must be either a specimen from a syntypic series or a figure quoted by the author of the species. In view of Hanley's comparison with the holotype, there would be some justification for the designation of the original shell of the Chemnitz figure as a neotype. There is in the Zoological Museum, Copenhagen, a specimen (inscribed 66/14; 38.3 mm) almost certainly from the collection of Spengler, which, according to T. Schiøtte (pers. comm.) and as suggested by Rosewater (1970) could be this figured specimen. This shell has been examined and neither pattern nor shape are closely similar to the published figure, although both are undoubtedly of the same species. Furthermore, in his text Chemnitz stated only that figure 2075 (of *L. angulifera*) was based on a shell belonging to Spengler; it is likely that the original of figure 2074 belonged to Chemnitz himself and is now lost.

In conclusion, none of the pieces of evidence presented above is of itself conclusive of the identity of the Linnaean species. However, restriction of the name '*scabra*' to the present species is consistent with all this evidence and does not conflict with accepted usage. To ensure future stability of nomenclature, a neotype is here designated, comprising both shell and preserved animal of a male specimen. No preserved specimens were available from Amboina, Moluccas, the type locality designated by Rosewater (1970) and the type locality of *Helix scabra* Linnaeus now becomes Magnetic Island, Queensland, Australia. This change in locality is unfortunate, but at Magnetic Island the ecology, reproduction, behaviour and anatomy have been studied in detail.

Of the synonyms of *L. scabra*, *Buccinum lineatum* and *B. foliorum* of Gmelin are based upon earlier figures by Knorr (1768) and Rumphius (1705) respectively. Schumacher (1838) proposed the new name *Turbo striatus* for the shells illustrated by Chemnitz as *Helix scabra*, objecting that the figures 2074 and 2075 depicted shells insufficiently depressed and carinate to be the Linnaean species. In fact, shells of *L. scabra* are often a little broader and the periphery more strongly keeled than in these figures. As discussed above, figures 2074 and 2075 represent *L. scabra s. s.* and *L. angulifera* respectively, and to minimize confusion the former is designated lectotype of *Turbo striatus*. Of the varieties of *Litorina scabra* described by Philippi (1847), var. *flammulata* was not illustrated, but is probably this species, since none of the others included by Philippi under the name '*scabra*' can be described as flammulate. The lectotype of *Litorina scabra* var. *ventricosa* Philippi has been selected on the basis of an original label by Cuming, bearing one of the localities listed by Philippi, and with the name inscribed by Philippi himself (p. 1). Types of *Littorina novaehiberniae* Lesson and of *Littorina scabra* subvar. *tenuis* Nevill are typical specimens of *L. scabra*.

Rosewater (1981) designated a lectotype of *L. angulifera* from the collection of Lamarck in the MHNG. Unfortunately the supposed syntypic series is a mixed collection of *L. scabra* and *L. angulifera*, and the selected lectotype belongs to the former species. The original description by Lamarck (1822) could apply to both species, but the type locality was given as the Antilles, so that Rosewater's designation of an Indo-Pacific specimen is inappropriate. In order to maintain the accepted usage of the name '*angulifera*', the lectotype of *Phasianella angulifera* Lamarck is here re-designated as the specimen MHNG 1096/92-11. Although smaller than the dimension given by Lamarck, this shell shows an angulated periphery, conforming to his description.

DIAGNOSIS. Shell: large (25–35 mm); spire outline concave near apex; strong peripheral keel; columella wide, excavated, white; primary grooves 9–11; ribs low, numbering 36–41 on last whorl; microsculpture of spiral striae in grooves; colour pale with more or less dense pattern of black or dark brown dashes, aligned at suture and periphery to form oblique axial stripes numbering 9–14 on last whorl. Animal: penis bifurcate, ochraceous, limb with glandular disc large, filament small; ovoviviparous.

SHELL (Fig. 27). *Shape*. Height (15)25–35(44) mm. Teleoconch (6.5)7–8(8.5) whorls. Shell of moderate thickness, solid. Spire outline at first concave, becoming convex after whorls 5–6; whorls rounded; sutures impressed, narrowly channelled; peripheral keel prominent, becoming indistinct at end of last whorl. Adult lip a little thickened, not flared; varices absent. Columella wide, shallowly excavated; pillar straight or gently convex at base. Sexual dimorphism: males smaller, relatively lower spire and larger aperture.

*Dimensions:* Table 10.

*Sculpture* (Fig. 28a–d). Protoconch normal. First 2–3 whorls of teleoconch smooth. Primary grooves 9–11(12), almost equally spaced, posterior rib usually narrowest, sometimes 1–2 ribs below sutural rib are a little wider than the rest; 2 posterior grooves wider and deeper. On whorl 7 all, or only widest, primary ribs become divided each by a secondary groove at anterior $\frac{1}{3}$–$\frac{1}{2}$ width; single

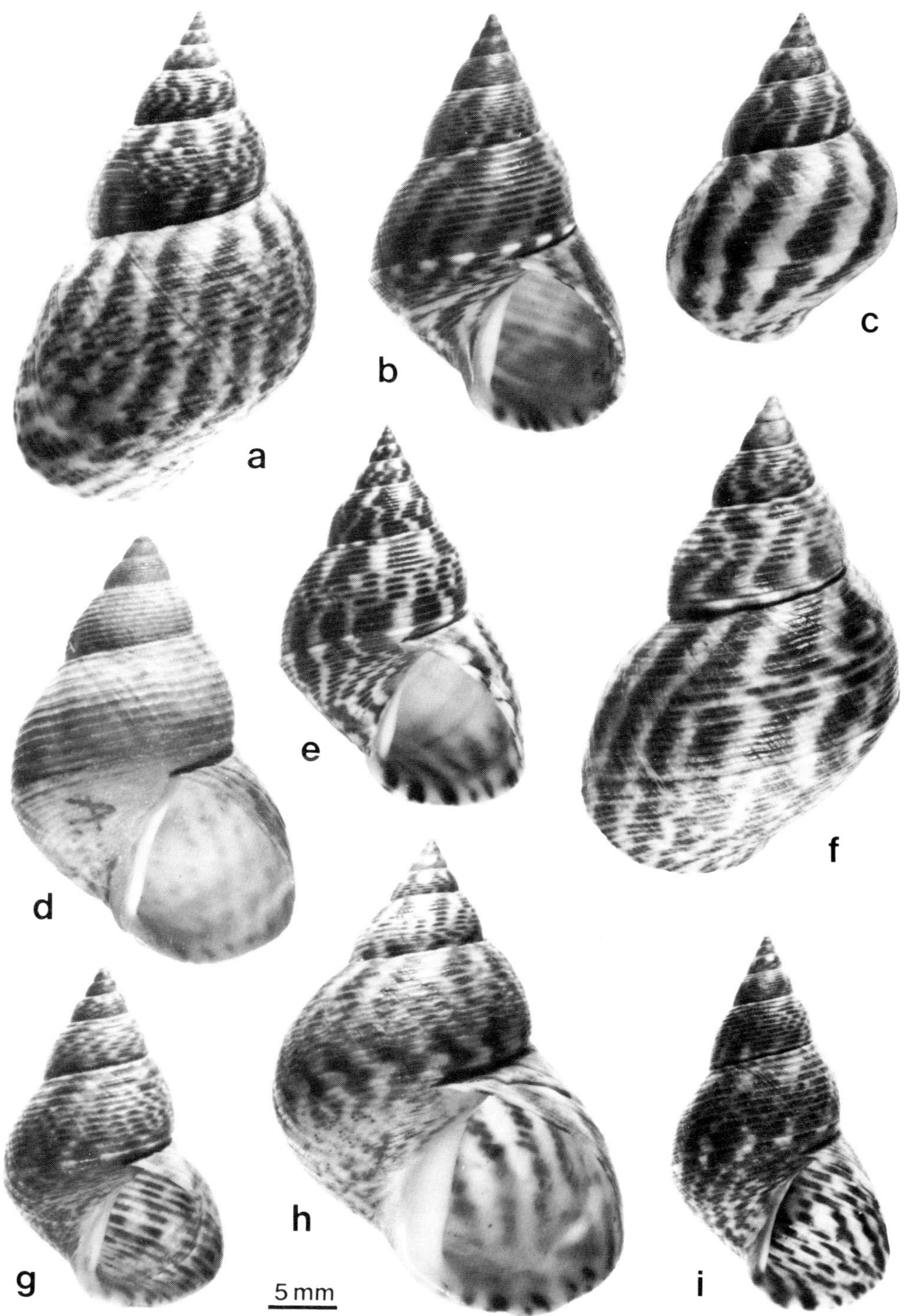


**Fig. 27** *Littoraria* (*Littorinopsis*) *scabra:* **(a)** ♂, Moa I., Qld. (DGR); **(b)** Namalungo, Mozambique (NM); **(c)** Durban, S. Africa (WAM); **(d)** lectotype of *Littorina scabra* subvar. *tenuis* Nevill, Arakan, Burma (ZSI); **(e)** neotype of *Helix scabra* Linnaeus, ♂, Magnetic I., Qld. (BMNH 198347); **(f)** lectotype of *Litorina scabra* var. *ventricosa* Philippi, Cagayan, Mindanao, Philippines (BMNH 198341); **(g)** lectotype of *Littorina novaehiberniae* Lesson, Port Praslin, New Ireland (MHNG); **(h)** Green I., Qld. (NMV); **(i)** Pialba, Qld. (QM).

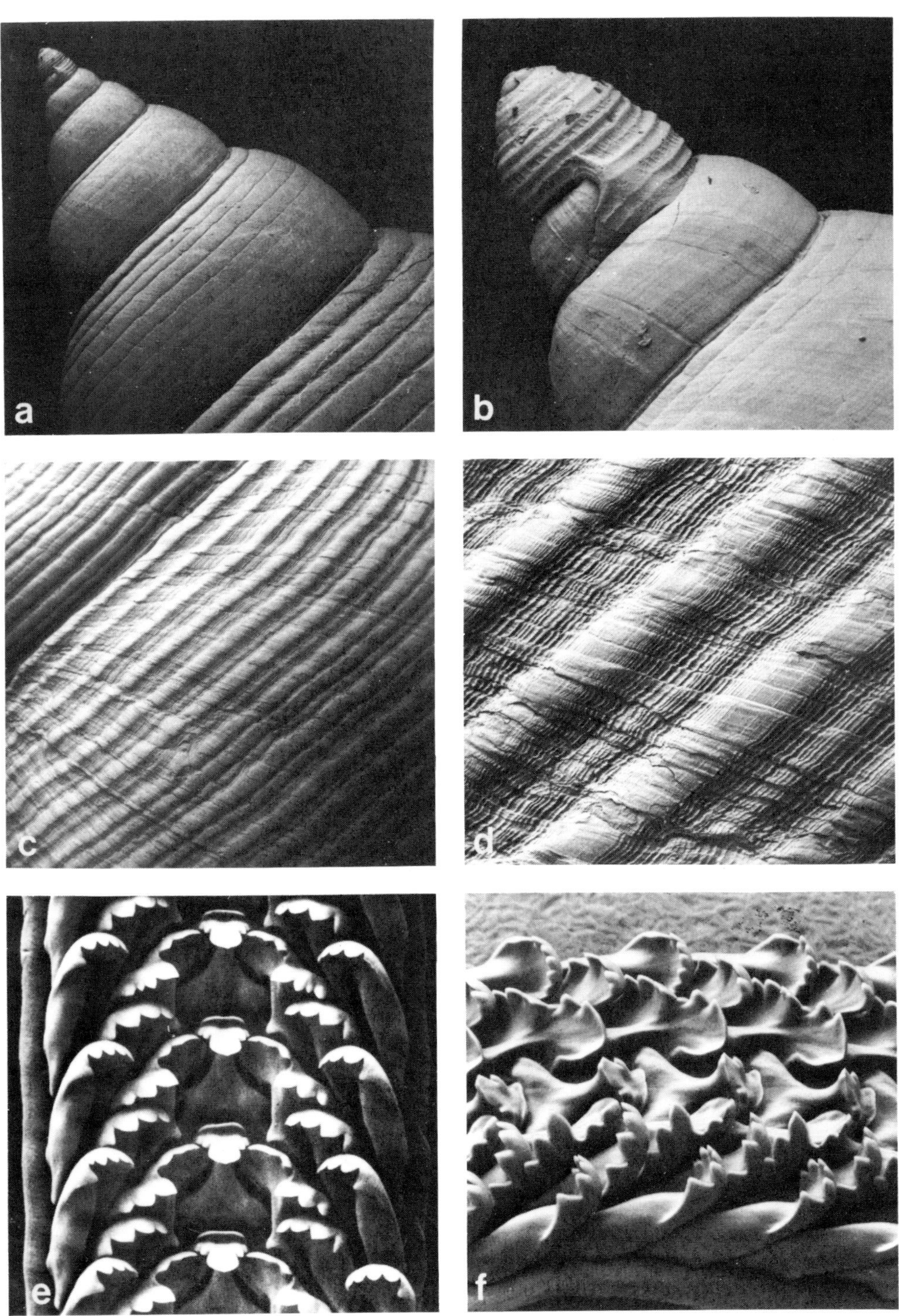


**Fig. 28** *Littoraria (Littorinopsis) scabra*, Magnetic I., Qld.: **(a)** spire ( × 19); **(b)** protoconch ( × 85); **(c)** last whorl ( × 8); **(d)** detail ( × 30); **(e,f)** radula ( × 140).

**Table 10** Dimensions of *Littoraria (Littorinopsis) scabra.*

| Specimen | Locality | Sex | Primary grooves | H (mm) | B (mm) | LA (mm) | WA (mm) | C (mm) | P | S | SH |
|---|---|---|---|---|---|---|---|---|---|---|---|
| *Helix scabra* neotype, BMNH 198347 | Magnetic I., Qld. | ♂ | 12 | 28.1 | 19.1 | 16.4 | 11.6 | 2.7 | 1.47 | 0.71 | 1.71 |
| *Littorina novaehiberniae* lectotype, MHNG | Port Praslin, New Ireland | | 10 | 24.0 | 16.7 | 13.3 | 10.6 | 2.2 | 1.44 | 0.80 | 1.80 |
| *Litorina scabra* var. *ventricosa* lectotype, BMNH 198341 | Cagayan, Mindanao, Philippines | | 10 | 34.6 | 24.3 | 19.4 | 15.6 | 4.5 | 1.42 | 0.80 | 1.78 |
| *Littorina scabra* subvar. *tenuis* lectotype, ZSI | Arakan, Burma | | 11 | 31.0 | 21.3 | 17.7 | 13.6 | 3.5 | 1.46 | 0.77 | 1.75 |
| DGR | Magnetic I., Qld. | ♀ | 11 | 40.6 | 25.8 | 21.8 | 15.7 | 3.9 | 1.57 | 0.72 | 1.86 |
| DGR | Moa I., Qld. | ♀ | 9 | 33.8 | 23.5 | 18.6 | 14.9 | 4.0 | 1.44 | 0.80 | 1.82 |
| DGR | Moa I., Qld. | ♂ | 10 | 32.2 | 20.9 | 17.8 | 12.8 | 3.2 | 1.54 | 0.72 | 1.81 |
| AMS C.43106B | Mauritius | | 11 | 36.0 | 23.6 | 18.9 | 13.7 | 3.2 | 1.53 | 0.72 | 1.90 |
| BMNH | Okinawa, Japan | ♂ | 10 | 19.2 | 12.8 | 11.0 | 8.4 | 2.1 | 1.50 | 0.76 | 1.75 |
| BMNH | Okinawa, Japan | ♀ | 9 | 23.9 | 15.8 | 12.8 | 10.5 | 2.6 | 1.51 | 0.82 | 1.87 |
| AMS C.131829 | Ilha do Ibo, Mozambique | ♂ | 9 | 21.7 | 14.0 | 11.7 | 8.9 | 2.4 | 1.55 | 0.76 | 1.85 |
| AMS C.131829 | Ilha do Ibo, Mozambique | ♀ | 11 | 25.7 | 17.4 | 14.0 | 11.2 | 2.8 | 1.48 | 0.80 | 1.84 |
| BMNH acc. 1838 | Trincomalee, Sri Lanka | | 11 | 30.8 | 24.5 | 20.5 | 15.8 | 4.7 | 1.26 | 0.77 | 1.50 |
| DGR, mean of 10 | Moa I., Qld. | ♂ | | 32.37 | | | | | 1.450 | 0.754 | 1.759 |
| standard error | | | | 0.51 | | | | | 0.019 | 0.009 | 0.014 |
| DGR, mean of 10 | Moa I., Qld. | ♀ | | 37.58 | | | | | 1.478 | 0.761 | 1.834 |
| standard error | | | | 0.52 | | | | | 0.013 | 0.010 | 0.011 |
| statistic t or U | | | | 7.158 | | | | | 63 | 57 | 93 |
| probability | | | | <.001 | | | | | 0.352 | 0.630 | <.001 |

narrow riblets may be intercalated in primary grooves on last whorl. Tertiary sculpture absent. Ribs remain low and flattened, except for peripheral rib which is raised and rounded, and posterior rib which is pushed up to form a sutural channel; intercalated riblets always remain small and inconspicuous, less than $\frac{1}{2}$ width of ribs; total number of ribs on last whorl 36–41, rarely up to 59 when riblets are well developed. Primary grooves $\frac{1}{2}$–1 rib width, wider posteriorly; secondary grooves usually remain narrower. Sculpture becomes less distinct at end of whorl 8. Microsculpture of faint, irregular axial growth striae over surface; strong, regular spiral striae in grooves.

*Colour*. Rather constant; ground colour whitish, cream or pale brown, with dense pattern of long black or dark brown dashes, mostly on ribs; dashes aligned at suture and periphery to form oblique axial stripes, numbering 9–14 on last whorl; alignment sometimes complete from suture to base, especially on spire whorls; peripheral keel usually with conspicuous white gaps between dashes. Occasional pale shells lack alignment and show uniform speckling of grey or pale brown. Aperture pale yellow, external pattern visible as black or brown lines and dashes, obscured within by white or grey callus in larger shells. Columella almost invariably white, occasionally stained brown or purple at edge of inner apertural lip.

ANIMAL. *Colour* (Fig. 3). Pigmentation black; sides of foot mottled; head darkest, often with pale central streak; tentacles banded, unpigmented stripe each side of base.

*Penis* (Figs 3, 29a–d). Length to 11.0 mm. Base thick, bifurcate; limb bearing glandular disc the larger of the two; glandular disc bulbous. Filament small, tapering. Sperm groove open. Base ochraceous cream, darkest near glandular disc; disc opaque dark cream; filament off white.

*Sperm* (Fig. 5a,b). Eupyrene sperm 154–168 μm. Nurse cells (Fig. 29e,f) 21–37 μm, fusiform

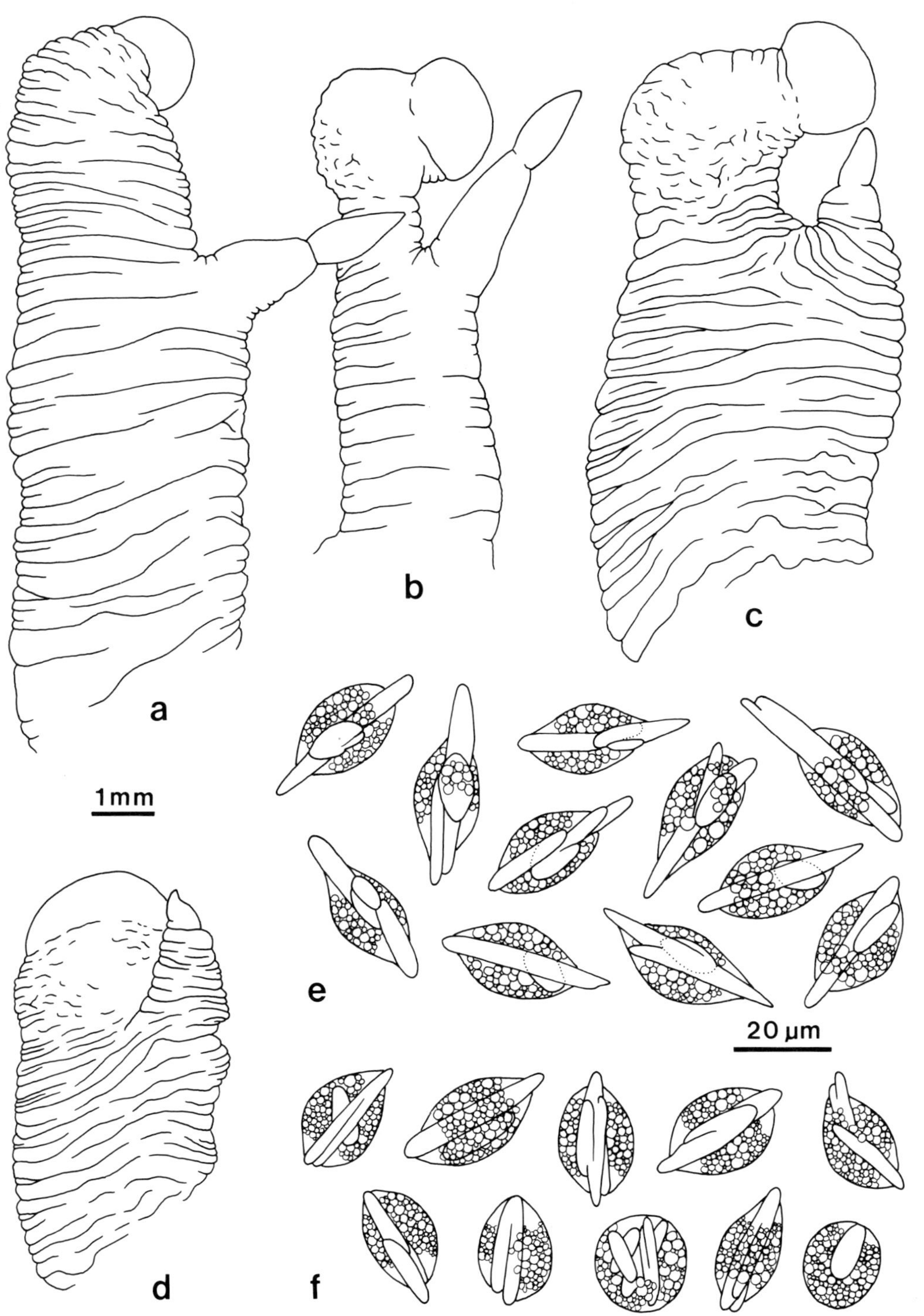


**Fig. 29** *Littoraria* (*Littorinopsis*) *scabra*, Magnetic I., Qld.: **(a–c)** penes, relaxed; **(d)** penis, contracted; **(e,f)** sperm nurse cells.

(sometimes oval); rods 2–3, central, projecting (rarely not so), usually one elongate and pointed, another small and oval; yolk granules small.

*Pallial oviduct* (Figs 9,10). Length to 7.0 mm. Spiral section to 3.2 mm diam., 3½ whorls; opaque albumen gland 1 whorl, cream to white; translucent albumen gland fawn to pale brown; capsule glands absent; spiral indistinct externally; egg groove only lightly pigmented. Straight section to 4.2 mm, fawn, terminating in a papilla. Bursa long, to 2.7 mm, anterior. Development ovoviviparous.

*Radula* (Fig. 28e,f). Length to 55 mm; relative length 0.91–1.54. Saw-toothed type; central rachidian cusp large, broad, edge slightly pointed, flanked by small cusp on each side; cusps of paired teeth almost equilaterally triangular; lateral with gap anterior to main cusp.

DISTRIBUTION. *Habitat*. Trunks and roots of *Rhizophora* trees near seaward edge of mangrove forests and on isolated trees, 0.5–2.0 m above the ground; sometimes on trunks in *Avicennia* fringe; occasionally on trunks of *Hibiscus* or on driftwood on sandy shores where mangroves are absent; rarely on sheltered rocks. An oceanic species.

*Range* (Fig. 30). Throughout tropical and subtropical Indo-Pacific, excluding only northern Indian Ocean, northern Bay of Bengal and mainland South East Asia. Rare in Hawaiian Islands where mangroves were only established this century. Only 1 specimen recorded from south-eastern Polynesia, where mangroves are absent, and 1 from the Red Sea.

*Records*. **South Africa:** Isipingo (BMNH); Durban Bay (NM, WAM); **Mozambique:** Inhaca I. (NM); Porto Belmore, NW. Nacala Bay (NM); Ilha do Ibo (AMS); **Tanzania:** Bagamoyo (BMNH); Chukwani, Zanzibar (ANSP); **Kenya:** Shimoni (BMNH); **Madagascar:** Tuléar (AMS, MCZ); Ile Ste. Marie (NMV); Nossi Bé (RNHL, ANSP); **Mauritius:** Le Chaland (AMS); Mouth of Camisard R. (ANSP); **Seychelles:** Cosmoledo I. (BMNH, USNM); Mahé I. (BMNH, ANSP); **Aldabra Atoll** (BMNH); **Chagos Arch.:** Diego Garcia (BMNH, USNM); **Saudi Arabia:** Obhor Greek, Jeddah (BMNH); **Maldive Is:** Gan, Addu Atoll (ANSP); Fadiffolu Atoll (BMNH, ANSP); **India:** Netravata R., Mangalore (USNM); Manali I., off Mandapam (MCZ); Madras (USNM); **Sri Lanka:** Trincomalee (BMNH); **Burma:** Arakan (ZSI); **Andaman Is:** Port Blair (BMNH); **Nicobar Is** (BMNH); **Cocos-Keeling Is** (BMNH, AMS, ANSP); **Thailand:** Ao Nam-Bor, Phuket I. (DGR); **Malaysia: Sabah:** Labuan I. (RNHL); Kudat Bay (ANSP); Po Bui I., Sandakan (USNM); **Singapore** (BMNH); **Indonesia: Sumatra:** Weh I. (RNHL); Sinabang, Simeulue I. (RNHL); Bai I., Batu Is (USNM); Sanding I., Mentawai Is (USNM); **Java:** Djakarta Bay (RNHL); Menscheneter I. (ANSP); Tjilaoet Eureum (RNHL); **Lesser Sunda Is:** Koeta Beach, Bali (MCZ); Komodo (USNM); Maumere, Flores (USNM); Kambang I., Timor (RNHL); **Kalimantan:** Maratua I. (RNHL); Sandanan, S. Natuna Is (NUS); **Sulawesi:** Manado (RNHL); Mamudju (RNHL); Makassar (RNHL); Gorontalo (RNHL); Tanah Djampea (RNHL); **Moluccas:** Beo, Talaud I. (RNHL); Ternate, Halmahera (WAM); Taliabu I. (RNHL); Ambon I. (WAM); Mitak I., Tanimbar Is (WAM); Tamberfane, Aru Is (WAM); Kisar I. (RNHL); **Irian Jaya:** Manokwari (RNHL); Japen I., Schouten Is (RNHL, AMS, ANSP); **Philippines:** Sanga Sanga I., Sulu Arch. (NMV, ANSP); Cagayan de Oro, Mindanao I. (BMNH); Banacon I., Bohol I. (AMS); Linapacan I. (USNM); Lubang I. (ANSP); Cuyo Is (USNM); Catanauan, S. Quezon (WAM); Balibatikan, Batangas (ANSP); Panukulan, Polillo I., E. Quezon (WAM); **Taiwan** (USNM); **Japan:** Toguchi, Kitanakagusuku, Okinawa, Ryukyu Is (BMNH): **Papua New Guinea:** Madang (AMS); Collingwood Bay (AMS); Kapa Kapa (AMS); Woodlark I. (AMS); Kuia I., Trobriand Is (AMS); **Bismarck Arch.:** Duke of York Is, New Britain (AMS); **Australia: W.A.:** Norwegian Bay, N. of Pt Cloates (WAM); Barrow I. (WAM, USNM); Broome (NMV); Swan Pt (NMV); Admiralty Gulf (WAM); **N.T.:** Darwin (DGR); Cape Don (AMS); Yirrkala, Arnhemland (USNM); **Qld.:** Thursday I., Torres Str. (DGR, QM); Quintell Beach (QM); Lizard I. (AMS); Green I. (NMV); Dunk I. (QM, NMV); Magnetic I. (DGR, AMS, USNM); Bowen (AMS); Proserpine (NMV); N. Keppel I. (NMV); Heron I. (AMS); Caloundra (QM); Stradbroke I. (QM, NMV, WAM); **New Caledonia:** Ile des Pins (AMS); Baie des Citrons, Nouméa (AMS); Poum (ANSP); **New Hebrides:** Aneityum and Tana (BMNH); Efate I. (AMS); Santo I. (AMS); Vanua Lava (BMNH); **Solomon Is:** Santa Cruz I. (AMS); Malaita (BMNH, AMS); Choiseul I. (BMNH); Shortland I. (ANSP); **Caroline Is:** Gorokotau I., Palau (ANSP); Yap I. (ANSP); Moen I., Truk Is (BPBM); Ponape (BMNH); Werua, Kapingamarangi (USNM); **Mariana Is:** Saipan (ANSP); Guam (AMS, USNM); **Marshall Is:** Eniwetok (USNM); Roi I., Kwajalein (BPBM); Enybor I., Jaluit (USNM); **Gilbert Is:** Apaiang (MCZ); **Ellice Is:** Vaitupu (USNM); Nukulailai (USNM);

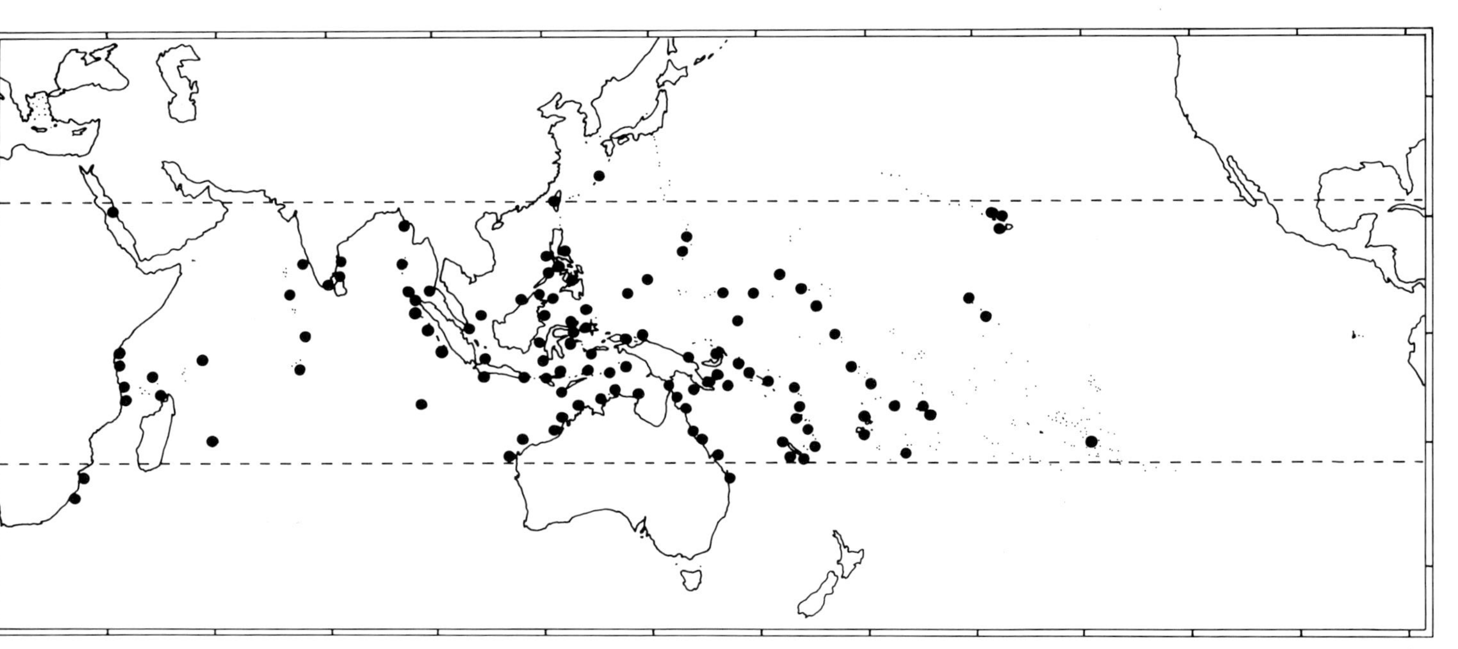

**Fig. 30** Distribution of *Littoraria* (*Littorinopsis*) *scabra*.

**Wallis Is:** between Luanna and Fungalei Is (USNM); **Fiji:** NW. Vanua Levu (USNM); Nananu-i-ra I., NNE. Viti Levu (MCZ); Lomololo, S. Viti Levu (WAM); **Tonga:** Velitoa, nr. Tongatapu (BPBM); **Samoa Is:** Salua Pata, Upolu I. (ANSP); Tafuna Cove, Tutuila (USNM); **Line Is:** Fanning I. (USNM); Palmyra I. (ANSP); **Hawaiian Is:** Coconut I., Kaneohe Bay, Oahu (DGR, BPBM); Molokai (BPBM); Puako, Hawaii (BPBM); **Tuamotu Is:** Tauere Atoll (USNM).

REMARKS. *L. scabra* has a very wide geographical distribution, but throughout the range the characters of both shell and animal are rather constant. In particular the penis shape is always diagnostic. In contrast to several other widely distributed *Littoraria* species, no geographical forms can be recognized. Rather, variation in shell characters is mainly associated with habitat differences. Shells reach large size in mangrove forests, but are usually smaller if they occur on rocks or in other unfavourable habitats. On *Rhizophora* trees shells are heavily pigmented, while in those from *Avicennia* trees (e.g. Fig. 27d) the pattern is paler and more diffuse, or rarely almost absent. Fig. 27c,h,i illustrates several unusual shape and colour varieties.

SIMILAR SPECIES. Some shells of *L. intermedia* and of *L. pallescens* may approach *L. scabra* in shape and coloration, but lack the white columella, strong spiral microsculpture and smooth apical whorls of this species. The white columella and lack of an intercalated secondary rib in each primary groove separate *L. scabra* from *L. lutea*. *L. angulifera* from the tropical Atlantic is quite distinct (Figs 4o, 49e, 99d–f), with 67–91 ribs on the last whorl, indistinct or absent microsculpture, slight peripheral keel, lilac columella and large penial filament. The penial form of *L. scabra* is diagnostic.

## *Littoraria* (*Littorinopsis*) *lutea* (Philippi, 1847)

*Litorina scabra* var. *lutea* Philippi, 1847, vol. 2: 222, *Litorina* pl. 5, fig. 11 [not fig. 6 as cited in text; fig. 11 here designated lectotype; type locality restricted to Masbate, Philippines]
*Littorina scabra* var. *lutea*—Nevill, 1885: 146 [in part]
*Littorina* (*Littorinopsis*) *scabra* var. *lutea*—Schepman, 1909: 194
*Litorina scabra* var. *punctata* Philippi, 1847, vol. 2: 222, *Litorina* pl. 5, fig. 5 [figure here designated lectotype; type locality restricted to Singapore; not *Littorina intermedia* var. *punctata* Philippi, 1846]
*Littorina scabra* var. *punctata*—Nevill, 1885: 145
? *Litorina scabra* var. *rubra* Philippi, 1847, vol. 2: 222 [Chemnitz (1795) fig. 2075 is not the lectotype figure; type locality restricted to Mindoro, Philippines; not *Litorina rubra* Anton, 1839]
*Littorina scabra*—Reeve, 1857: *Littorina* pl. 5, fig. 21a [in part; not Linnaeus, 1758]
*Litorina scabra*—Weinkauff, 1878: 37–39, pl. 4, figs 8–10 [in part; not Linnaeus, 1758]
*Littorina* (*Melaraphe*) *scabra*—Tryon, 1887: 243 [in part; not Linnaeus, 1758]
*Littorina* (*Malaraphe*) *scabra*—Casto de Elera, 1896: 309 [in part; not Linnaeus, 1758]
*Littorina* (*Littorinopsis*) *scabra scabra*—Rosewater, 1970: 456–461 [in part; not Linnaeus, 1758]
*Littorina arboricola* var. *philippiana*—Nevill, 1885: 148 [in part; not Reeve, 1857]
*Littorina scabra* var. *philippiana*—Hidalgo, 1904–5: 207 [in part; not Reeve, 1857]
*Littorina pallescens* ? var. *erronea* Nevill, 1885: 148 [Singapore; lectotype here designated ZSI, 28.9 mm]

NOMENCLATURE. This species has not hitherto been regarded as distinct, but on account of its similarity in shape to *L. scabra* has most often been considered a colour variety of that species. Some shell forms of *L. pallescens* resemble *L. lutea* closely, resulting in further confusion in the literature. In 1847 Philippi described seven varieties of *Litorina scabra*, based mainly upon differences in colour and pattern of the shell. In fact, the coloration of *L. scabra s. s.* is rather constant and four of Philippi's colour varieties are based wholly or in part upon the species here defined as *L. lutea*. Variety number 2, *Litorina scabra* var. *articulata*, is placed in the synonymy of *L. intermedia*, since the figured specimen is a large form of that species. However, the three paralectotypes are dark forms of *L. lutea* similar in appearance to the figure. No type specimen of variety number 3, var. *punctata*, is known to exist, but the figure is determined as *L. lutea* by reason of the spotted peripheral keel, colour pattern and shell shape. The name is preoccupied by *Littorina intermedia* var. *punctata* Philippi,

**Fig. 31** *Littoraria* (*Littorinopsis*) *lutea:* **(a–c)** Ubin I., Singapore (DGR); **(a,b)** ♂; **(c)** ♀; **(d)** Balikpapan, Kalimantan (AMS); **(e,f)** ♀, Ubin I., Singapore (DGR); **(g)** locality unknown (BMNH); **(h)** lectotype of *Littorina pallescens* ? var. *erronea* Nevill, Singapore (ZSI); **(i)** lectotype figure of *Litorina scabra* var. *lutea* Philippi, Masbate, Philippines (Philippi, 1847, *Litorina* pl. 5, fig. 11).

1846. The figure of variety number 5, var. *lutea*, is here designated lectotype of the species, in the absence of a type specimen. The combination of yellow colour, purple columella and keeled outline leave no doubt as to the identity of the figure. Variety number 6, var. *rubra*, is not figured, but Philippi mentioned fig. 2075 of Chemnitz (1795), remarking that in the latter the keel was not sufficiently developed, and following the reference with a query. This figure of Chemnitz is clearly *L. angulifera*, from the shell shape and colour, as confirmed by the locality 'Guinea'. Philippi himself did not regard the Chemnitz figure as typical, so that it can hardly be designated as the lectotype of var. *rubra*, as proposed by Rosewater (1970). Furthermore, the localities Mindoro, Philippines, and Canton given by Philippi militate against the variety being a form of *L. angulifera*. Philippi's concept of *L. scabra* included, besides the Linnaean species, only *L. lutea* and a few forms closely resembling it, while *L. angulifera* was correctly recognized. Since pink shells of *L. lutea* are known, *Litorina scabra* var. *rubra* is tentatively included in the synonymy of this species.

The name *Litorina scabra* var. *concolor*, attributed to Weinkauff (1878), is listed in synonymies by Tryon (1887) and by Rosewater (1970). However, in Weinkauff's original text '*concolor*' is clearly the initial adjective in a Latin description and is not intended as a varietal name. The lectotype of *Littorina pallescens ?* var. *erronea* Nevill is a typical shell of *L. lutea*.

DIAGNOSIS. Shell: strong peripheral keel; columella wide, excavated, purple; primary grooves 8–11; secondary sculpture of single intercalated riblet in each primary groove; ribs low, numbering 33–50 on last whorl; microsculpture faint; colour polymorphic, usually yellow or pink, speckled with brown, keel spotted. Animal: penis bifurcate, filament large, glandular disc black; ovoviviparous.

SHELL (Fig. 31). *Shape*. Height 18–35 mm. Teleoconch 7–8 whorls. Shell of moderate thickness, solid. Spire outline gently convex, whorls rounded; suture impressed. Peripheral keel marked by a strong rib. Adult lip a little thickened, not flared; varices absent. Columella wide, excavated; pillar straight or slightly convex. Sexual dimorphism: males smaller, relatively lower spire and larger aperture.

*Dimensions:* Table 11.

*Sculpture* (Fig. 20e,f). Protoconch normal. First 1–2 whorls of teleoconch smooth. Primary grooves 8–11(12), equally spaced, posterior 2–3 grooves wider and deeper than others, posterior 3 ribs

**Table 11** Dimensions of *Littoraria* (*Littorinopsis*) *lutea*.

| Specimen | Locality | Sex | Primary grooves | H (mm) | B (mm) | LA (mm) | WA (mm) | C (mm) | P | S | SH |
|---|---|---|---|---|---|---|---|---|---|---|---|
| *Litorina scabra* var. *lutea* lectotype figure (Philippi, 1847) | [Masbate, Philippines] | | | 27.5 | 17.3 | 11.8 | 9.5 | | 1.59 | 0.81 | 2.33 |
| *Littorina pallescens* var. *erronea* lectotype, ZSI | Singapore | | 9 | 28.9 | 17.3 | 14.5 | 11.3 | 2.8 | 1.67 | 0.78 | 1.99 |
| BMNH 1969381 | Unknown | | 9 | 34.8 | 19.9 | 17.0 | 12.2 | 3.5 | 1.75 | 0.72 | 2.05 |
| BMNH acc. 1829 | Masbate, Philippines | | 8 | 23.2 | 14.7 | 12.5 | 9.9 | 2.6 | 1.58 | 0.79 | 1.86 |
| AMS C.131831 | Balikpapan, Kalimantan | ♀ | 10 | 22.2 | 14.6 | 11.6 | 9.0 | 2.2 | 1.52 | 0.78 | 1.91 |
| AMS C.131831 | Balikpapan, Kalimantan | ♂ | 8 | 22.5 | 14.5 | 12.7 | 10.0 | 2.7 | 1.55 | 0.79 | 1.77 |
| DGR | Ubin I., Singapore | ♂ | 10 | 23.4 | 15.2 | 13.5 | 10.0 | 2.6 | 1.54 | 0.74 | 1.73 |
| DGR | Ubin I., Singapore | ♀ | 8 | 26.3 | 17.7 | 14.6 | 11.1 | 2.7 | 1.49 | 0.76 | 1.80 |
| DGR, mean of 10 | Ubin I., Singapore | ♂ | | 22.81 | | | | | 1.584 | 0.745 | 1.764 |
| standard error | | | | 0.36 | | | | | 0.019 | 0.007 | 0.021 |
| DGR, mean of 10 | Ubin I., Singapore | ♀ | | 26.01 | | | | | 1.565 | 0.768 | 1.859 |
| standard error | | | | 0.70 | | | | | 0.021 | 0.009 | 0.012 |
| statistic t or U | | | | 4.060 | | | | | 55 | 72 | 89 |
| probability | | | | 0.001 | | | | | 0.740 | 0.106 | 0.002 |

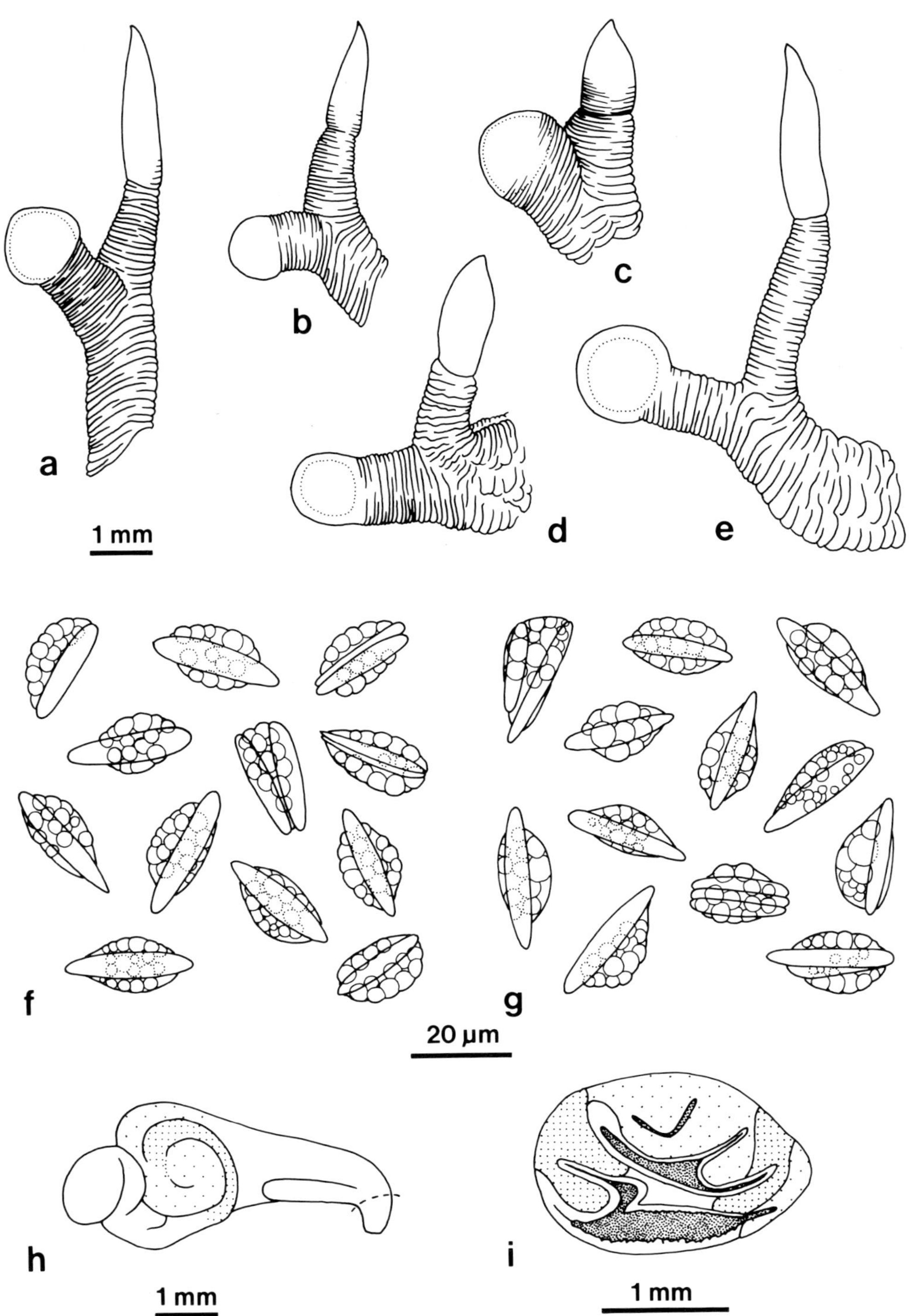


**Fig. 32** *Littoraria* (*Littorinopsis*) *lutea*, Ubin I., Singapore: **(a,b,d,e)** penes, relaxed; **(c)** penis **(a)** contracted; **(f,g)** sperm nurse cells; **(h,i)** pallial oviduct, with transverse section.

correspondingly narrower. Secondary sculpture develops on whorl 6, with intercalation of a single narrow riblet in each primary groove; primary ribs usually remain undivided, only rarely do 1 or 2 develop a secondary groove at about $\frac{1}{2}$ width. Narrow tertiary riblets may appear on whorl 8. On last whorl all primary ribs are flattened, but for the prominent rib at periphery; primary ribs of equal width, about twice as wide as secondary ribs intercalated between; total number of ribs 33–50. Grooves on last whorl less than $\frac{1}{3}$ maximum rib width; on whorls 6 and 7 posterior 2–3 grooves may equal width of intervening ribs. Microsculpture of irregular axial growth striae over surface; faint spiral striae sometimes visible in wider grooves.

*Colour.* Polymorphic; usually yellow, sometimes pink, with variable development of brown pattern. Ground colour pale yellow, whitish or orange pink. Pattern of dark brown or black dashes and flecks on ribs; darkest shells are densely patterned, with rough alignment at suture and periphery of spire whorls; pale shells may lack pigment entirely, but usually show a characteristic faint speckling, often developed only on last whorl; in most shells the keel is conspicuously spotted. Pattern visible within aperture only in darker shells, clouded by a cream callus. Columella and parietal callus dark purple or pink, pillar whitish.

ANIMAL. *Colour.* Pigmentation grey to black, correlated with shell colour; sides of foot sparsely mottled; head darkest; tentacles banded, unpigmented stripe each side of base.

*Penis* (Fig. 32a–e). Length to 7.7 mm. Base bifurcate. Filament fairly large, tapering at tip. Sperm groove open. Base pale ochre, filament white, penial glandular disc black to dark grey.

*Sperm.* Eupyrene sperm 152–164 $\mu$m. Nurse cells (Fig. 32f,g) 19–27 $\mu$m, oval to fusiform; rods 1–2(3), usually central, projecting, fusiform with rather pointed ends; yolk granules large.

*Pallial oviduct* (Fig. 32h,i). Length to 4.4 mm. Spiral section to 2.2 mm diam., $3\frac{1}{2}$ whorls; opaque albumen gland $\frac{1}{2}$ whorl, cream; translucent albumen gland pale brown; capsule glands absent; spiral indistinct externally; egg groove not darkly pigmented. Straight section to 2.5 mm, pale brown, terminating in a papilla. Bursa to 1.6 mm, anterior. Development ovoviviparous.

*Radula* (Fig. 20d). Length to 22 mm; relative length 0.88–0.90. Intermediate between saw- and chisel-toothed types; central rachidian cusp square, edge rounded; cusps of paired teeth obliquely triangular; lateral with gap anterior to main cusp.

DISTRIBUTION. *Habitat.* Trunks and occasionally leaves of trees in *Avicennia* fringe, to 1.7 m above ground; occasionally on roots and trunks of outermost trees of *Rhizophora* zone. A species of relatively clear water, and therefore 'oceanic' conditions, although range does not extend to the oceanic islands of the western Pacific.

*Range* (Fig. 33). Philippines, Indonesia and Singapore.

*Records.* **Singapore:** Jelutong, Ubin I. (DGR); Tanjong Penuru (ANSP); **Malaysia: Sabah:** Labuan I. (BMNH); Tanjong Aru, Kota Kinabalu (ANSP); Po Bui I., Sandakan (USNM); **Indonesia: Java:** Djakarta Bay (RNHL); Madura (RNHL); **Lesser Sunda Is:** Bali (AMS); Koeta Beach, Bali (MCZ); Bolang, SW. Timor (RNHL); **Kalimantan:** Balikpapan (AMS); **Sulawesi:** Buka I., Gulf of Tomini (USNM); Makassar (RNHL); **Moluccas:** Ambon I. (RNHL); Maikoor R., Aru Is (USNM); **Irian Jaya:** Seroei, Japen I. (RNHL); **Philippines:** Puerto Princessa, Palawan I. (USNM); Dapitan, Mindanao I. (MCZ); Iloilo, Panay I. (BMNH); Hadayan I., Bohol I. (ANSP); Coron I. (USNM); Manoquid, Sorsogon, Luzon I. (AMS); Baruyan R., Calapan, Mindoro I. (MCZ); Manila, Luzon I. (ANSP, MCZ); Bacoor Bay, Luzon I. (USNM).

REMARKS. This species is rare in museum collections. Live material has been seen from Singapore and preserved specimens from Balikpapan, Borneo; from these collections the black penial disc appears to be characteristic of the species. Shell shape varies little and the regular intercalation of secondary ribs in the primary grooves is diagnostic. At Singapore, shells from *Rhizophora* trees were almost always dark brown and those from adjacent *Avicennia* trees yellow with pale patterning. No pink shells were seen at this locality, but do occur in the Philippines. *L. lutea* is apparently somewhat more tolerant of continental conditions than the typically oceanic species *L. scabra*.

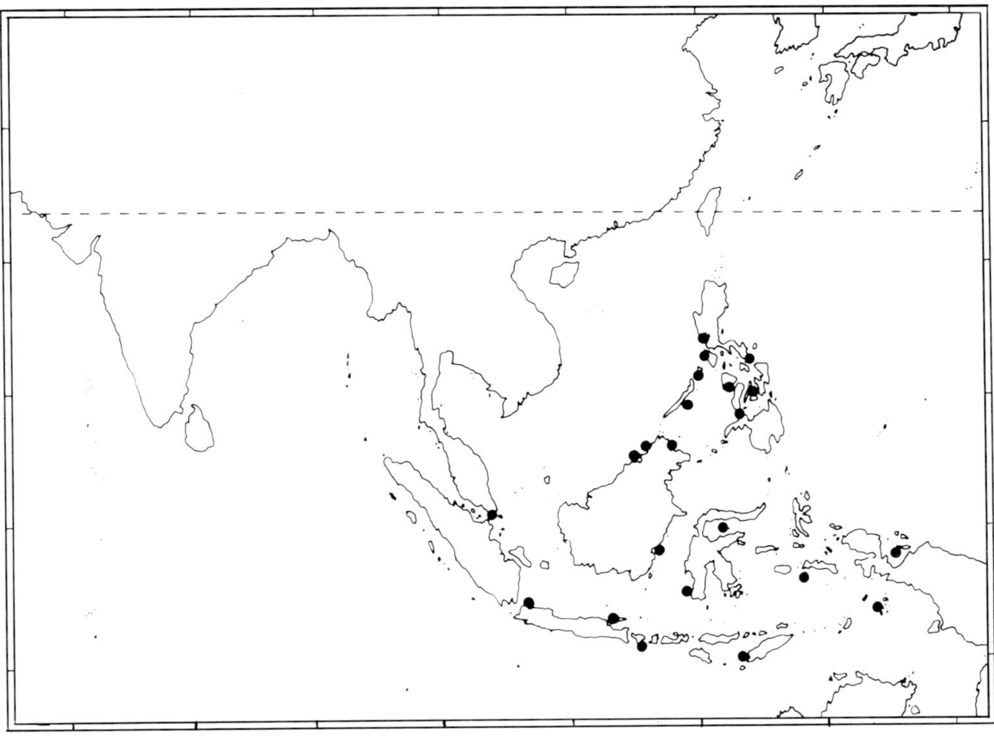

**Fig. 33** Distribution of *Littoraria* (*Littorinopsis*) *lutea*.

SIMILAR SPECIES. Shells of *L. lutea* are similar in shape to those of *L. scabra*, but in the former the regular intercalation of secondary riblets in the primary grooves is diagnostic, spiral microsculpture in the grooves is faint or absent, and the columella is purple. Penial form, the shape of the sperm nurse cells and radular characters also distinguish these two species. The shape of the penis of *L. lutea* is somewhat similar to that of *L. philippiana*, although the glandular disc is black and the sperm groove open. The shell of *L. philippiana* is superficially similar, but secondary sculpture between the primary ribs is irregular and crowded, and the peripheral rib is not conspicuously spotted. Large forms of *L. intermedia* could be confused with *L. lutea*, but penial characters and the form of the secondary sculpture at once separate these two. Penial and sperm characters distinguish *L. pallescens*, and most shells lack secondary sculpture, but in the form of that species described by Philippi as *Littorina sieboldii* (Fig. 34g) the shell is very close to that of *L. lutea*. In the '*sieboldii*' form of *L. pallescens* a transverse profile of the ribs and grooves is saw-toothed, secondary sculpture consists of incised lines in the V-shaped grooves, and secondary ribs do not become prominent, while the columella is white and the peripheral keel not marked with a conspicuous row of dashes or spots. In contrast, in *L. lutea* the primary ribs are rounded and the grooves wider, the intercalated secondary riblets are distinctive, and the columella always purple.

## *Littoraria* (*Littorinopsis*) *pallescens* (Philippi, 1846)

*Littorina pallescens* Philippi, 1846: 142 [Cagayan, province of Misamis, Mindanao I., [Philippines]; lectotype (Rosewater, 1970) BMNH 1968277]; Reeve, 1857: *Littorina* pl. 9, fig. 43; Nevill, 1885: 148 [in part]; Hidalgo, 1904–5: 207

*Litorina pallescens*—Philippi, 1847, vol. 3: 10, *Litorina* pl. 6, fig. 4 [as *Litorina filosa* in text, corrected in index]; von Martens, 1871: 39; Weinkauff, 1882: 58–59, pl. 7, figs 14,15
*Littorina (Malaraphe) pallescens*—Casto de Elera, 1896: 310
*Littorina filosa* var. *pallescens* 'Reeve, *non* Philippi'—Hidalgo, 1904–5: 206
*Littorina sieboldii* Philippi, 1846: 142 [Japan; holotype BMNH 1968278]; Reeve, 1857: *Littorina* pl. 5, figs 23a,b
*Litorina sieboldii*—Philippi, 1847, vol. 3: 9–10, *Litorina* pl. 6, fig. 3; Weinkauff, 1882: 56, pl. 7, figs 6,7
*Littorina (Malaraphe) sieboldii*—Casto de Elera, 1896: 310
*Littorinopsis sieboldii*—Kuroda & Habe, 1952: 64
*Littoraria sieboldii*—Higo, 1973: 46
? *Litorina scabra* var. *suturalis* Philippi, 1847, vol. 2: 222, *Litorina* pl. 5, fig. 10 [not fig. 7 as cited in text; fig. 10 here designated lectotype]
*Littorina arboricola* Reeve, 1857: *Littorina*, pl. 6, figs 27,27a [Singapore; lectotype (Rosewater, 1970) BMNH 1968321]; Nevill, 1885: 147
*Littorina (Melaraphe) scabra* var. *filosa*—Tryon, 1887: 244, pl. 42, figs 25,26,30 [in part; not Sowerby, 1832]
*Littorina (Littorinopsis) scabra* var. *filosa*—Adam & Leloup, 1938: 75–77, pl. 1, figs 2b,c; text fig. 25B [radula]
*Littorina (Malaraphe) scabra*—Casto de Elera, 1896: 309 [in part; not Linnaeus, 1758]
*Melaraphe (Littorinopsis) scabra*—Abe, 1942: 391–435, fig. 17, nos 1–10 [in part; not Linnaeus, 1758; ecology and behaviour]
*Littorina (Littorinopsis) scabra scabra*—Rosewater, 1970: 456–461, pl. 325, fig. 1; pl. 352, figs 2, 12,16,24,25 [in part; not Linnaeus, 1758]
*Littorinopsis scabra*—Brandt, 1974: 53–54 [in part; not Linnaeus, 1758]
*Littorina scabra—Nielsen*, 1976: fig. 1B [in part; not Linnaeus, 1758; ecology and behaviour]
*Littorina scabra* var. *philippiana*—Hidalgo, 1904–5: 207 [in part; not Reeve, 1857]
*Littorinopsis intermedia*—Oyama & Takemura, 1961: figs 12,13 [not Philippi, 1846]; Brandt, 1974: 54, pl. 4, fig. 62 [in part; not Philippi, 1846]

NOMENCLATURE. In the literature *L. pallescens* has been confused with other colourful, strongly ribbed and carinate forms under the name '*filosa*'. The species is common in collections, but the name '*pallescens*' has seldom been employed this century. There is no doubt that the syntypic series is the collection from which the species was described, since Cuming's original label, bearing Philippi's determination, accompanies the specimens. The lectotype may perhaps be the specimen figured by Philippi in 1847, although neither the type locality, nor material from the Cuming Collection, was mentioned in the text. No specimen of *Litorina scabra* var. *suturalis* Philippi, 1847 has been traced; tentative placement of this variety in the synonymy of *L. pallescens* is based upon the colour pattern and wide, white columella, which seem to distinguish the figure from the similar *L. lutea*.

DIAGNOSIS. Shell: columella wide, excavated, inner lip of aperture callused and rounded; primary grooves 9–10; secondary sculpture usually absent; ribs rounded or flattened, numbering 21–26 on last whorl; grooves wide or narrow; microsculpture faint; colour polymorphic; if present, dark pigment may form 6–10 axial stripes and a spiral band on base. Animal: penis bifurcate, filament small, glandular disc large, with thin, margined edge; ovoviviparous.

SHELL (Frontispiece, Figs 34,35). *Shape*. Height (9)15–25(31) mm. Teleoconch 6.5–8.5 whorls. Shell thick, solid. Spire outline gently convex; whorls rounded; sutures impressed. Peripheral keel evident on young shells, often indistinct on last whorl. Adult lip thickened, a little flared. Varices uncommon, usually strong growth lines only, sometimes up to 3 varices. Columella wide, excavated, pillar straight; inner lip of aperture callused and rounded in adult shells. Sexual dimorphism: males a little smaller (not quite significant in sample), relatively lower spire and larger aperture, aperture more elongate.

*Dimensions:* Table 12.

**Fig. 34** *Littoraria* (*Littorinopsis*) *pallescens:* **(a)** lectotype of *Littorina pallescens* Philippi, Cagayan, Mindanao, Philippines (BMNH 1968277); **(b)** ♀, Gove, N. T. (DGR); **(c)** ♀, Ubin I., Singapore (DGR); **(d)** Polillo I., Philippines (WAM); **(e)** Sowek, Irian Jaya (AMS); **(f)** Hinchinbrook I., Qld. (DGR); **(g)** lectotype of *Littorina sieboldii* Philippi, Japan (BMNH 1968278); **(h)** Thursday I., Qld. (QM); **(i)** ♂, Phuket I., Thailand (DGR).

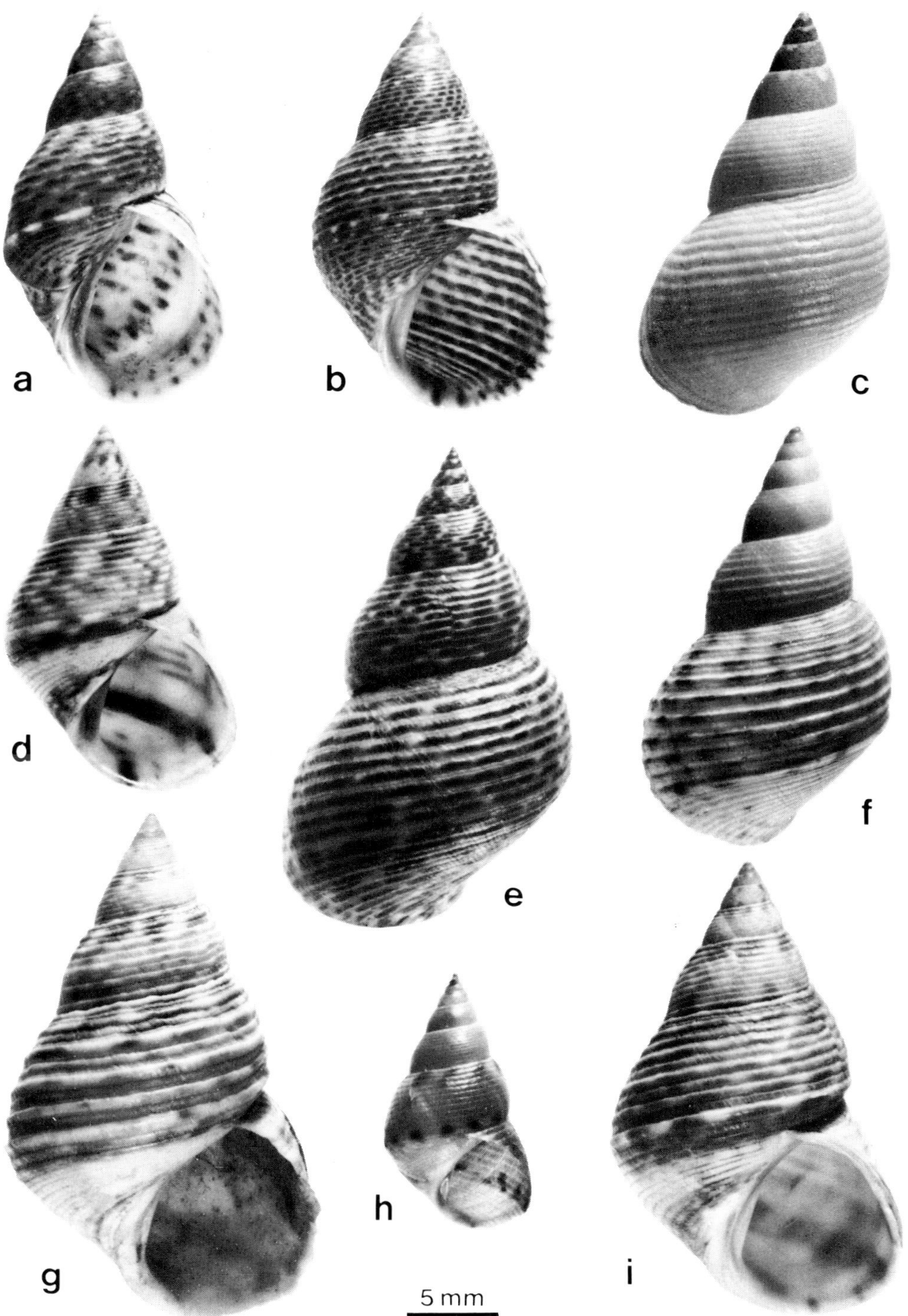


**Fig. 35** *Littoraria* (*Littorinopsis*) *pallescens:* **(a)** Kokenau, Irian Jaya (AMS); **(b)** Guadalcanal, Solomon Is (AMS); **(c)** ♀, Balikpapan, Kalimantan (AMS); **(d)** ♂, Ubin I., Singapore (DGR); **(e)** ♀, Gove, N. T. (DGR); **(f)** Kokenau, Irian Jaya (AMS); **(g)** lectotype of *Littorina arboricola* Reeve, Singapore (BMNH 1968321); **(h)** Lunga Bay, Mozambique (NM); **(i)** Sebarok, Singapore (WAM).

**Table 12** Dimensions of *Littoraria* (*Littorinopsis*) *pallescens*.

| Specimen | Locality | Sex | Primary grooves | H (mm) | B (mm) | LA (mm) | WA (mm) | C (mm) | P | S | SH |
|---|---|---|---|---|---|---|---|---|---|---|---|
| *Littorina pallescens* lectotype, BMNH 1968277 | Cagayan, Mindanao, Philippines | | 9 | 22.1 | 15.3 | 12.4 | 9.9 | 3.1 | 1.44 | 0.80 | 1.78 |
| *Littorina sieboldii* holotype, BMNH 1968278 | Japan | | 10 | 29.0 | 18.7 | 15.3 | 11.8 | 3.3 | 1.55 | 0.77 | 1.90 |
| *Littorina arboricola* lectotype, BMNH 1968321 | Singapore | | 8 | 26.8 | 17.4 | 13.2 | 10.6 | 2.5 | 1.54 | 0.80 | 2.03 |
| DGR | Penang, Malaysia | ♂ | 10 | 22.1 | 14.8 | 12.5 | 9.8 | 2.7 | 1.49 | 0.78 | 1.77 |
| DGR | Penang, Malaysia | ♀ | 9 | 29.0 | 18.9 | 14.2 | 11.7 | 2.7 | 1.53 | 0.82 | 2.04 |
| DGR | Kanchanadit, Thailand | ♀ | 9 | 31.3 | 18.8 | 15.6 | 11.9 | 3.0 | 1.71 | 0.76 | 2.01 |
| DGR | Phuket I., Thailand | | 9 | 10.9 | 7.2 | 5.7 | 4.5 | 0.8 | 1.51 | 0.79 | 1.91 |
| DGR | Ubin I., Singapore | ♂ | 9 | 21.6 | 14.3 | 12.8 | 9.7 | 3.0 | 1.51 | 0.76 | 1.69 |
| BMNH 1894.10.31 | Solomon Is | | 10 | 21.0 | 14.3 | 12.1 | 9.6 | 2.5 | 1.47 | 0.79 | 1.74 |
| DGR | Gove, N. T. | ♂ | 10 | 21.5 | 13.2 | 11.9 | 9.3 | 2.5 | 1.63 | 0.78 | 1.81 |
| DGR | Gove, N. T. | ♀ | 10 | 25.3 | 16.2 | 12.6 | 10.0 | 2.5 | 1.56 | 0.79 | 2.01 |
| DGR, mean of 10 | Ubin I., Singapore | ♂ | | 20.19 | | | | | 1.513 | 0.771 | 1.679 |
| standard error | | | | 0.36 | | | | | 0.015 | 0.005 | 0.019 |
| DGR, mean of 10 | Ubin I., Singapore | ♀ | | 21.91 | | | | | 1.519 | 0.808 | 1.843 |
| standard error | | | | 0.81 | | | | | 0.015 | 0.006 | 0.025 |
| statistic t or U | | | | 1.947 | | | | | 55 | 95 | 97 |
| probability | | | | 0.067 | | | | | 0.740 | <.001 | <.001 |

*Sculpture* (Figs 36, 53a). Protoconch normal. First 0–3 whorls of teleoconch smooth. Primary grooves (8)9–10(11), usually equally spaced and posterior 1–3 grooves deeper and wider, these sometimes also more closely spaced, so that posterior 1–2 ribs are narrowest. Secondary sculpture usually absent; rarely 1–2 ribs divide at anterior $\frac{1}{3}$–$\frac{1}{2}$ width on whorls 7–8; sometimes single minute riblets may appear by intercalation in the wider primary grooves of last whorl. Development of ribs and width of intervening grooves is variable: primary ribs may remain low and flattened on last whorl, of similar width, sometimes narrower posteriorly, often peripheral rib a little more prominent, primary grooves $\frac{1}{5}$–1 rib width or impressed lines only; frequently alternate, or even all, primary ribs become enlarged as prominent cords, primary grooves 1–3 times cord width; primary ribs between major cords may become obsolete on last whorl, leaving spaces between cords of 3–5 times their width. Total number of primary ribs on last whorl 21–26. Very rarely all primary ribs may divide and riblets are intercalated between, to produce as many as 43 ribs on last whorl, of which primary ribs remain most prominent. Microsculpture of axial growth striae covering surface, stronger in grooves; regular spiral striae may be visible in wide grooves.

*Colour* (Frontispiece). Polymorphic; various shades of brown, yellow and sometimes pink, often striped or banded. Ground colour cream, pale brown, chrome yellow or dark orange pink. Most shells show black or dark brown dashes and flecks on ribs, sometimes densely covering surface, usually variously aligned to form 6–10 oblique axial stripes near suture or from suture to periphery; often with a continuous dark spiral band just below periphery, sometimes another around columella, with pale or unpigmented zones at periphery and on middle of base; the dark pigment fades to lilac grey. Predominantly yellow or cream shells may lack pigment, may show faint pattern of pale orange, or may possess dark pigment only in stripes at suture and in a subperipheral band; apical 4–5 whorls of such shells often tinged pink. Predominantly orange pink shells show a similar range of patterns, but sometimes develop conspicuous dashes on peripheral keel only. External pattern and

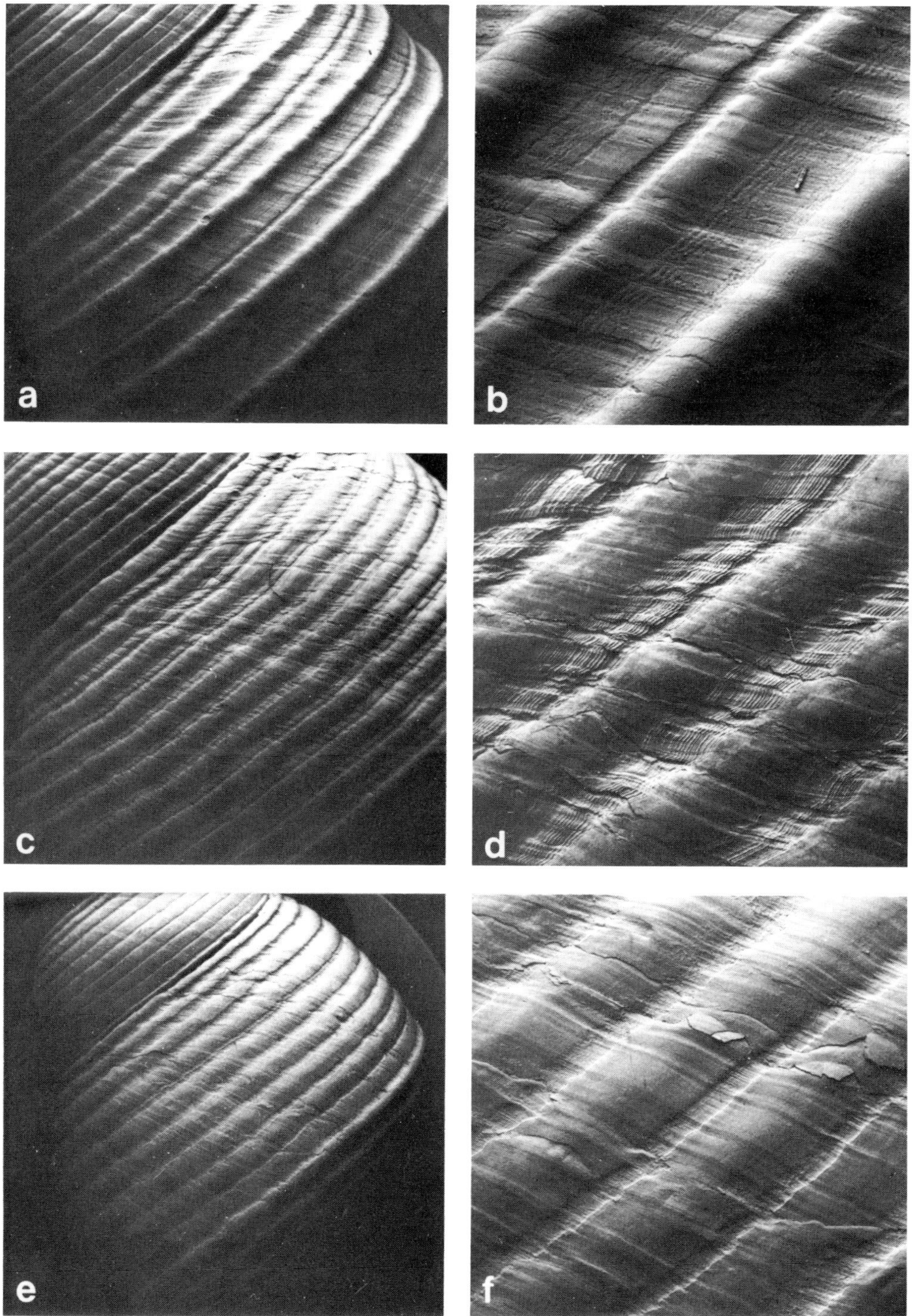


**Fig. 36** *Littoraria* (*Littorinopsis*) *pallescens:* **(a,b)** Solomon Is; **(a)** last whorl ( × 7); **(b)** detail ( × 32); **(c,d)** Gove, N.T.; **(c)** last whorl ( × 7); **(d)** detail ( × 32); **(e,f)** Phuket I., Thailand; **(e)** last whorl ( × 7); **(f)** detail ( × 28).

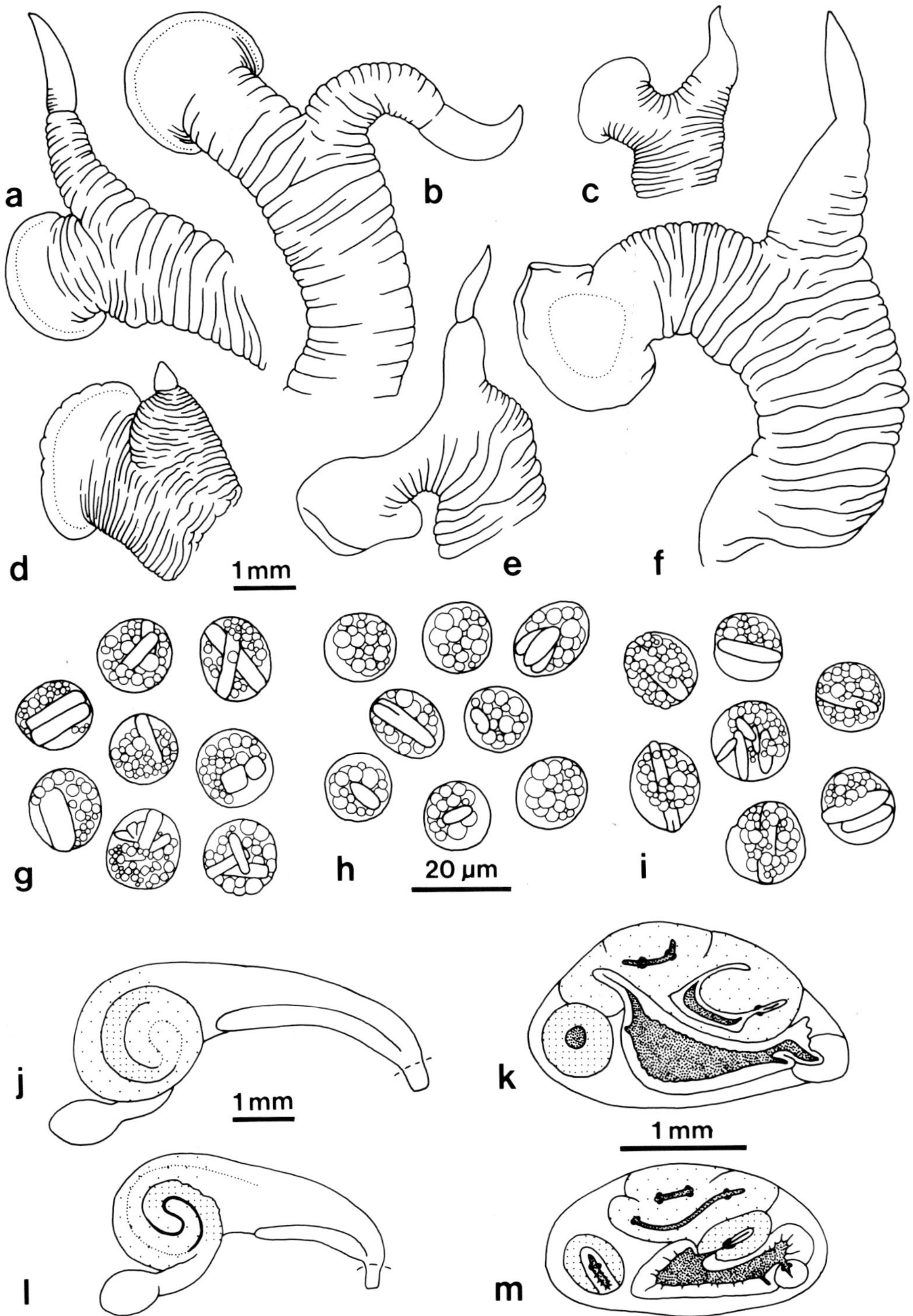


**Fig. 37** *Littoraria* (*Littorinopsis*) *pallescens:* **(a–c,e,f)** penes, relaxed; **(d)** penis, contracted; **(a,b,d)** Ubin I., Singapore; (c) Aldabra; **(e)** Santa Isabel, Solomon Is; **(f)** Gove, N. T.; **(g–i)** sperm nurse cells; **(g)** Ubin I., Singapore; **(h)** Phuket I., Thailand; **(i)** Gove, N. T.; **(j–m)** pallial oviducts, with transverse sections; **(j,k)** Gove, N. T.; **(l,m)** Phuket I., Thailand.

colour visible within aperture, clouded by whitish callus. Columella purple in patterned shells, pillar and inner lip of aperture whitish; entire columella white in shells lacking dark pigment.

ANIMAL. *Colour.* Pigmentation pale grey to black, correlated with shell colour; sides of foot mottled; head darkest, sometimes with pale central streak; tentacles banded, unpigmented stripe each side of base.

*Penis* (Fig. 37a–f). Length to 8.6 mm. Base bifurcate. Glandular disc large, with thin, margined edge. Filament small, tapering. Sperm groove open. Base and filament ochre to cream, glandular disc opaque cream or ochre, sometimes brown.

*Sperm.* Eupyrene sperm 180–185 μm. Nurse cells (Fig. 37g–i) 14–22 μm, round (sometimes somewhat oval); rods 1–2(3), small, variable in shape, elongate or oval, not projecting, sometimes absent; yolk granules fairly large.

*Pallial oviduct* (Fig. 37j–m). Length to 6.5 mm. Spiral section to 2.0 mm diam., $3\frac{1}{2}$ whorls; opaque albumen gland $\frac{3}{4}$–1 whorl, white to cream; translucent albumen gland off white, sometimes reddish; capsule glands absent; spiral indistinct externally unless egg groove is pigmented. Straight section to 5.0 mm, off white, terminating in a papilla. Bursa long, anterior, to 3.8 mm. Development ovoviviparous.

*Radula* (Fig. 53b). Length to 23 mm; relative length 0.85–1.10. Intermediate between saw- and chisel-toothed types; central rachidian cusp square, edge rounded; cusps of paired teeth obliquely triangular; lateral with gap anterior to main cusp.

DISTRIBUTION. *Habitat.* Typically on leaves in *Rhizophora* forest, less frequently on trunks and roots, 1.0–4.5 m above ground, most common at seaward edge, but extending into *Bruguiera* zone; also on leaves and sometimes trunks in *Avicennia* fringe, 0.7–2.0 m above ground; occasionally on leaves of *Sonneratia* and *Aegialitis*. An oceanic species.

*Range* (Fig. 38). Throughout the tropical Indo-Pacific, from central East Africa to Sri Lanka, Ryukyu Islands, Marshall Islands and Samoa. Three old and doubtful records from the Hawaiian Islands. This species is largely restricted to oceanic situations, occurring only rarely on the continental shoes of the Gulf of Siam, north-western Borneo, Hong Kong, eastern and north-western Australia (although common in Arnhemland).

*Records.* **Mozambique:** Lunga Bay to Memba Bay (NM); **Tanzania:** Chukwani, Zanzibar (ANSP); Tanga (MCZ); **Kenya:** Port Reitz, Mombasa (BMNH); **Madagascar:** Nossi Bé (ANSP); **Aldabra Atoll** (BMNH); **Mauritius:** Mouth of Camisard R. (ANSP); **Sri Lanka** (MCZ); **Andaman Is** (BMNH, NMW); **Nicobar Is** (BMNH); **Cocos-Keeling Is** (BMNH, ANSP); **Thailand:** Ao Nam-Bor, Phuket I. (DGR); Tanga I., Butang Group (USNM); Surat Thani (DGR); Bang Pakong R. (ANSP); Chantaburi (BMNH); Koh Kut (USNM); **Vietnam:** Chilins, Vung Tau district (ANSP); **Malaysia: Peninsula:** Penang (DGR); Pisang I. (NUS); **Sabah:** Tanjong Aru, Kota Kinabalu (ANSP); Kudat Bay (ANSP); Sandakan (ANSP, USNM); **Sarawak:** Santubong (SM); **Brunei:** Bandar Seri Begawan (RNHL); **Singapore:** Jelutong, Ubin I. (DGR); Sebarok (WAM); **Indonesia: Java:** Djakarta Bay (RNHL); Menscheneter I. (ANSP); **Lesser Sunda Is:** Koeta Beach, Bali (MCZ); Larantuka, Flores (RNHL); Maumere, Flores (USNM); Bolang, Timor (RNHL); **Kalimantan:** Lemukutan I. (RNHL); Balikpapan (AMS); **Sulawesi:** Manado (MCZ); Makassar (RNHL); Wowoni I. (MCZ); **Moluccas:** Dago Bay, Sangihe I. (MCZ); Kahatola I., S. Loloda Is (MCZ); Obi I. (BMNH); Ambon I. (BMNH; RNHL); Maikoor R., Aru Is (USNM); **Irian Jaya:** Sorong (AMS); Doreh Bay (RNHL); Japen I., Schouten Is (ANSP, RNHL); Kokenau (AMS); **Philippines:** Sanga Sanga I., Sulu Arch. (RNHL, ANSP); Tapiantana I. (USNM); Taganaan, Surigao, Mindanao (MCZ); Iloilo, Panay (USNM); Linapacan I. (USNM); Lopez Bay, S. Quezon (WAM); Matabunkay Cove, Batangas (ANSP); Prieto Diaz, Sorsogon (ANSP); Panukalan, NW. Polillo I. (WAM); **Hong Kong:** Shin Han (BMNH); Rocky Harbour and Port Shelter (ANSP); **Taiwan** (USNM, MCZ, ANSP); **Japan:** Fukuchi R., Okinawa I., Ryukyu Is (BMNH); **Papua New Guinea:** Madang (AMS); Popondetta (AMS); Losuia, Kiriwina Is, Trobriand Is (AMS); East Cape (AMS); **Australia: W.A.:** Broome (NMV); **N.T.:** Grose I. (AMS); Port Essington (AMS); Gove Peninsula, Arnhemland (DGR); Groote Eylandt (AMS); **Qld.:** Thursday I., Torres Str. (QM); Lizard I. (AMS); Machans Beach, Cairns (AMS); Dunk I. (NMV); Magnetic I. (DGR); Moreton Bay (MCZ); **New Caledonia:** San Gabriel, nr. Yaté (USNM); **New Hebrides:**

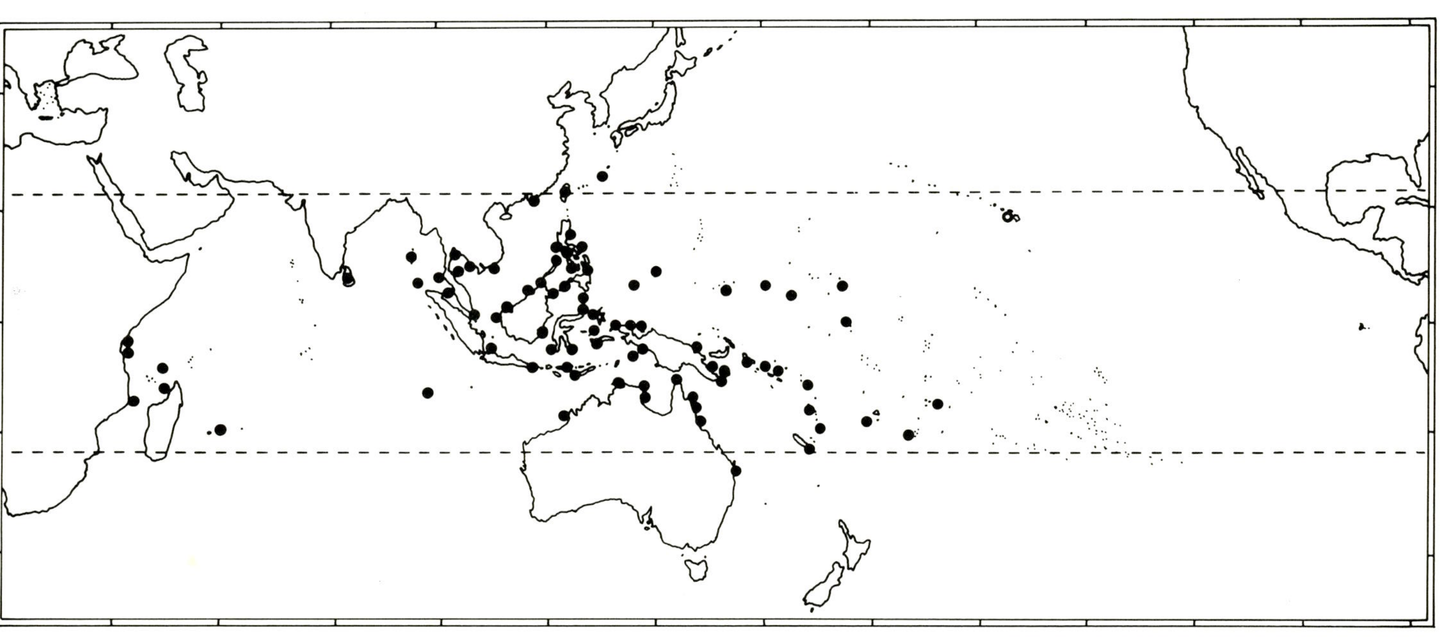

**Fig. 38** Distribution of *Littoraria* (*Littorinopsis*) *pallescens*.

Eromanga (AMS); Port Stanley, Malekula (BMNH); **Solomon Is:** Santa Cruz I., Vanikoro Is (USNM); Auki, Malaita (BMNH); Santa Isabel (BMNH); Shortland I. (ANSP); **Caroline Is:** Babelthuap I., Palau Is (USNM); Yap I. (T. Smalley); Ponape (BMNH); Tol Is, Truk Atoll (BPBM); Kusaie I. (USNM, BPBM); **Marshall Is:** Bikarej I., Arno Atoll (USNM); **Gilbert Is:** Apaiang (MCZ); **Fiji:** Lautoka, NW. Viti Levu (MCZ); Nadi Bay, Viti Levu (AMS); **Tonga:** Velitoa (BPBM); **Samoa Is:** Tafuna Cove, Tutuila (USNM); **Hawaiian Is** (RNHL, USNM, BPBM, all doubtful records).

REMARKS. Shells of *L. pallescens* are sometimes closely similar to those of *L. philippiana* and *L. lutea*; these three species show parallel colour polymorphism and somewhat similar shell shape, sculpture and columellar characters. Nevertheless, there can be no doubt that these are distinct species, for although the geographical range of *L. pallescens* overlaps that of each of the others, the diagnostic penial differences remain constant. Furthermore, details of shell sculpture and pattern can alone be used to identify all but a very few specimens. However, even when *L. philippiana* and *L. lutea* are separated from *L. pallescens*, there remains a wide variety of shell types, so that as here defined *L. pallescens* is the most variable of all the Indo-Pacific *Littoraria* species. Despite the range of sculpture, size and colour, there seems no reason to subdivide the species. Anatomical characters are constant and in particular the form of the penis at once separates *L. pallescens* from all other species in the genus. The existence of geographical forms accounts for much of the variation in shell characters, but even within local populations shells are variable in shape and sculpture, besides being colour polymorphic. Some of the variation in size and colour can be attributed to environmental effects.

The principal geographical forms are illustrated in Figs 34, 35 and 36. The typical form of the species occurs in the western Pacific islands, southern Japan, Indonesia, eastern Australia and the Indian Ocean; shells are rather small, secondary sculpture absent or limited to a few secondary grooves, yellow and orange pink shells are common, while in pigmented shells the pattern is of discrete dashes aligned near the sutures and prominent at the periphery. Sometimes, especially in pure yellow shells, the primary ribs become alternately prominent and obsolete (Figs 34a, 36a,b), as in the lectotype. In south-western Thailand, the Andaman and Nicobar Islands, pigmented shells frequently show a pattern of continuous spiral bands, especially on the base; primary grooves are narrow and the ribs flattened (Figs 34i, 36e,f). The form typical of Malaysia and Singapore was named *Littorina arboricola* by Reeve; shells are large and generally yellow, with axial stripes near the sutures and a spiral band on the base (Figs 34c, 35d,g,i). In a form from northern Australia, the Moluccas and parts of New Guinea, the shells are large and solid, the ribs equal and separated by wide grooves, the periphery of the last whorl rounded and the colour pattern similar to that of *L. intermedia* (Figs 34b,e, 35e, 36c,d). In only two rare forms is secondary sculpture regularly present. The first is the form described by Philippi as *Littorina sieboldii*, in which the shell is large, the spire rather elongate and the ribs sharp; the primary grooves are at first narrow, but on the last whorl each contains a secondary groove or an indistinct rib (Figs 34g, 35c). This form is known only from Balikpapan, Borneo (AMS) and Japen Island, Irian Jaya (RNHL), besides the rather doubtful type locality of 'Japan'. The second is a very curious form seen in a few collections from Irian Jaya (AMS, RNHL, ANSP), Sulawesi (RNHL) and Negros Island, Philippines (USNM), in which shells are thin and sculpture is indistinguishable from that of *L. philippiana*, with 2–3 secondary ribs between each primary rib. While female shells of this form are of normal shape (Fig. 35f), the males develop an extremely elongate, patulous aperture, usually after the first varix (Fig. 35a). Males of other forms develop a relatively larger aperture than females, but dimorphism is not so extreme. None of these varieties is sufficiently distinct to justify taxonomic recognition and numerous intermediates exist between them.

As in some other species, variation of shells within local populations suggests that the environment may influence shell size and colour. Specimens from small mangrove bushes in full sunlight and at high tidal levels are small, reaching a maximum size of only 9–11 mm in one collection from Phuket Island, Thailand. On *Avicennia*, *Sonneratia* and *Aegialitis* trees, shell pigmentation is in general paler and yellow shells more frequent than on *Rhizophora* trees, where brown colour morphs predominate. Differences between the proportions of colour morphs on adjacent trees of these two groups are often striking.

SIMILAR SPECIES. The small penial filament and marginate glandular disc distinguish *L. pallescens* from all other species, and the round sperm nurse cells are found elsewhere only in *L. subvittata* and *L. philippiana*. Shells of *L. pallescens* are almost always separable from those of *L. philippiana* by their lack of secondary sculpture, by the rounded inner apertural lip and often by the pattern of more discrete dashes confined to the ribs. From *L. lutea* this species can be distinguished by the narrower primary grooves on the early whorls and by the absence of the single intercalated secondary rib in each primary groove on the last whorl. The form found in northern Australia, the Moluccas and parts of New Guinea may resemble *L. intermedia* in shell characters, but grooves tend to be wider and the shell broader; penial form clearly separates these two species. The only other *Littoraria* with such a diversity of colour forms is *L. filosa*, which is easily recognized by its narrow columella and thin shell.

## *Littoraria (Littorinopsis) philippiana* (Reeve, 1857)

*Littorina philippiana* Reeve, 1857: *Littorina* pl. 5, figs 22a,b ['Philippine Islands', in error, type locality here corrected to E. Queensland, Australia; holotype BMNH 1968307]
*Litorina philippiana*—Weinkauff, 1882: 54–55, pl. 7, figs 2,3
*Littorina arboricola* var. *philippiana*—Nevill, 1885: 148 [in part]
*Litorina philippina* von Martens, 1900: 584 [unjustified emendation of *Littorina philippiana* Reeve, 1857]
*Littorina (Melaraphe) scabra* var. *filosa*—Tryon, 1887: 244, pl. 42, fig. 27 [in part; not Sowerby, 1832]
*Littorina (Littorinopsis) scabra scabra*—Rosewater, 1970: 456–461, pl. 352, fig. 13 [in part; not Linnaeus, 1758]
*Littorina scabra*—Wilson & Gillett, 1979: 52, pl. 8, fig. 5b [in part; not Linnaeus, 1758]

NOMENCLATURE. It is fitting that a littorinid should bear the name of Philippi, in recognition of his pioneering and accurate contribution to the study of the group. Although there is no doubt that the lectotype designated by Rosewater (1970) is the specimen figured by Reeve, there has clearly been a confusion of localities and labels. Reeve gave the type locality as the Philippine Islands, but the shell certainly belongs to a species endemic to Australia and south-eastern New Guinea. Accompanying the figured specimens are two shells of the form of *L. filosa* found in north-western Australia and the Lesser Sunda Islands, together with labels inscribed 'Kangaroo Is., S.A.' (presumably South Australia, where no *Littoraria* species occur). Since these two shells do not correspond with the description given by Reeve and this locality is not mentioned, they cannot be part of a syntypic series. The figured specimen therefore becomes the holotype.

DIAGNOSIS. Shell: columella wide, excavated, purple; primary grooves 8–9; secondary and tertiary sculpture well developed; primary ribs are raised as 13–18 narrow cords on last whorl, with 1–3 riblets and irregular striations in each interspace; microsculpture of irregular spiral striae in grooves; colour polymorphic, brown predominating, pattern diffuse. Animal: penis bifurcate, limb bearing glandular disc is long, disc small, filament long, sperm groove closed as a duct; ovoviviparous.

SHELL (Frontispiece, Fig. 39). *Shape*. Height (18)23–33(38) mm. Teleoconch 7.5–8.5(9) whorls. Shell of moderate thickness, solid. Spire gently convex; whorls rounded; sutures impressed. Peripheral keel prominent in young shells, often becoming obsolete on last whorl. Adult lip somewhat thickened, a little flared. Varices 0–3(7). Columella wide, excavated, pillar straight, narrowed at base. Sexual dimorphism: males with relatively lower spire and larger aperture.

*Dimensions:* Table 13.

*Sculpture* (Fig. 40a–d). Protoconch normal. First 0.5–1 whorl of teleoconch smooth. Primary grooves (7)8–9(11), equally spaced anteriorly, posterior 2–3 more closely spaced, deeper and broader. Secondary grooves appear on whorls 6–7, dividing each primary rib at anterior $\frac{1}{4}$–$\frac{1}{2}$ width; secondary riblets may also form by intercalation, especially in the wider posterior grooves. Higher

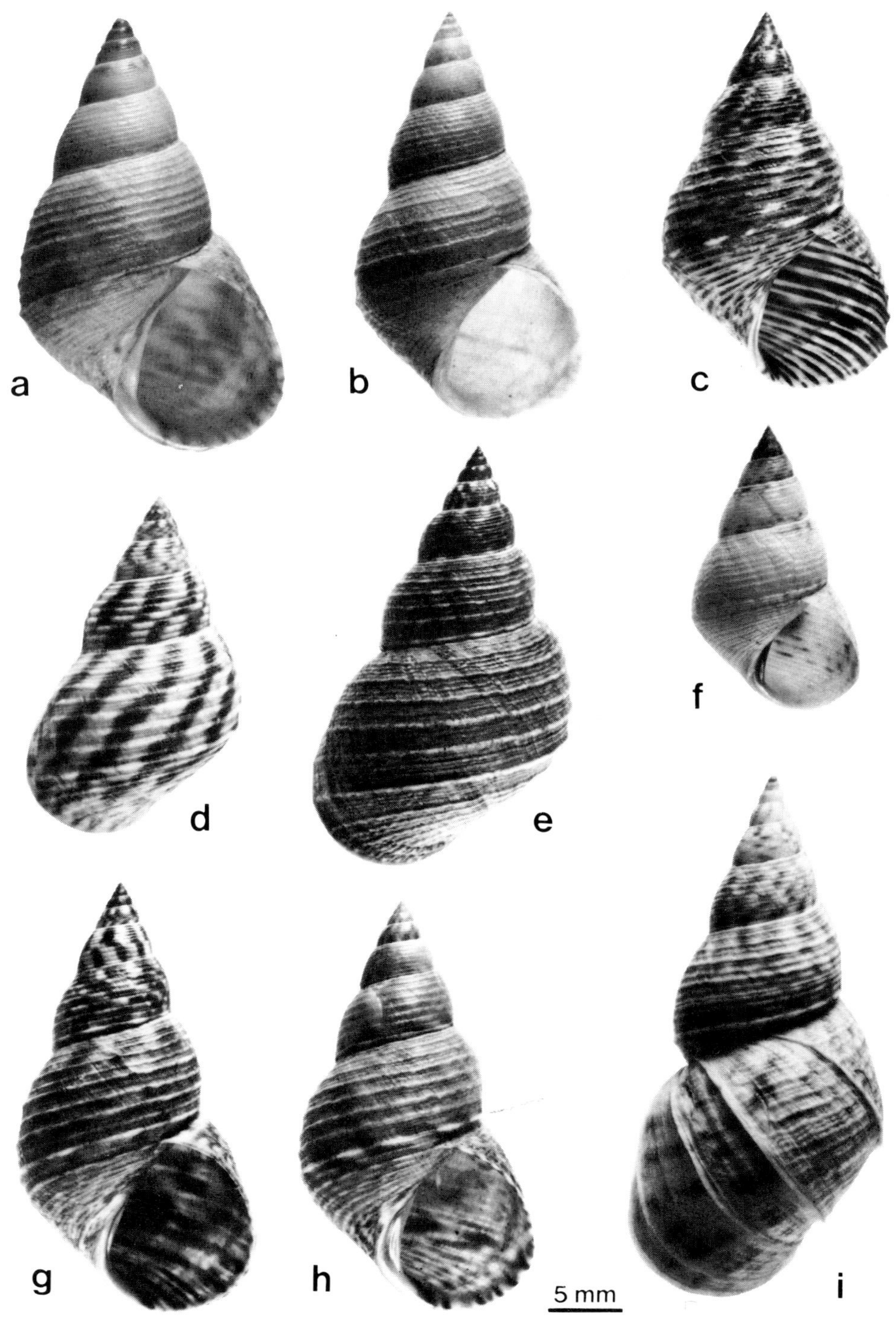


**Fig. 39** *Littoraria* (*Littorinopsis*) *philippiana:* **(a)** holotype of *Littorina philippiana* Reeve (BMNH 1968307); **(b)** Elliott R., Qld. (QM); **(c)** ♀, Orpheus I., Qld. (DGR); **(d,e)** Magnetic I., Qld. (DGR); **(d)** ♀; **(e)** ♀; **(f)** ♀, Moa I., Qld. (DGR); **(g,h)** Magnetic I., Qld. (DGR); **(g)** ♂; **(h)** ♀; **(i)** ♀, Bohle R., Townsville, Qld. (DGR).

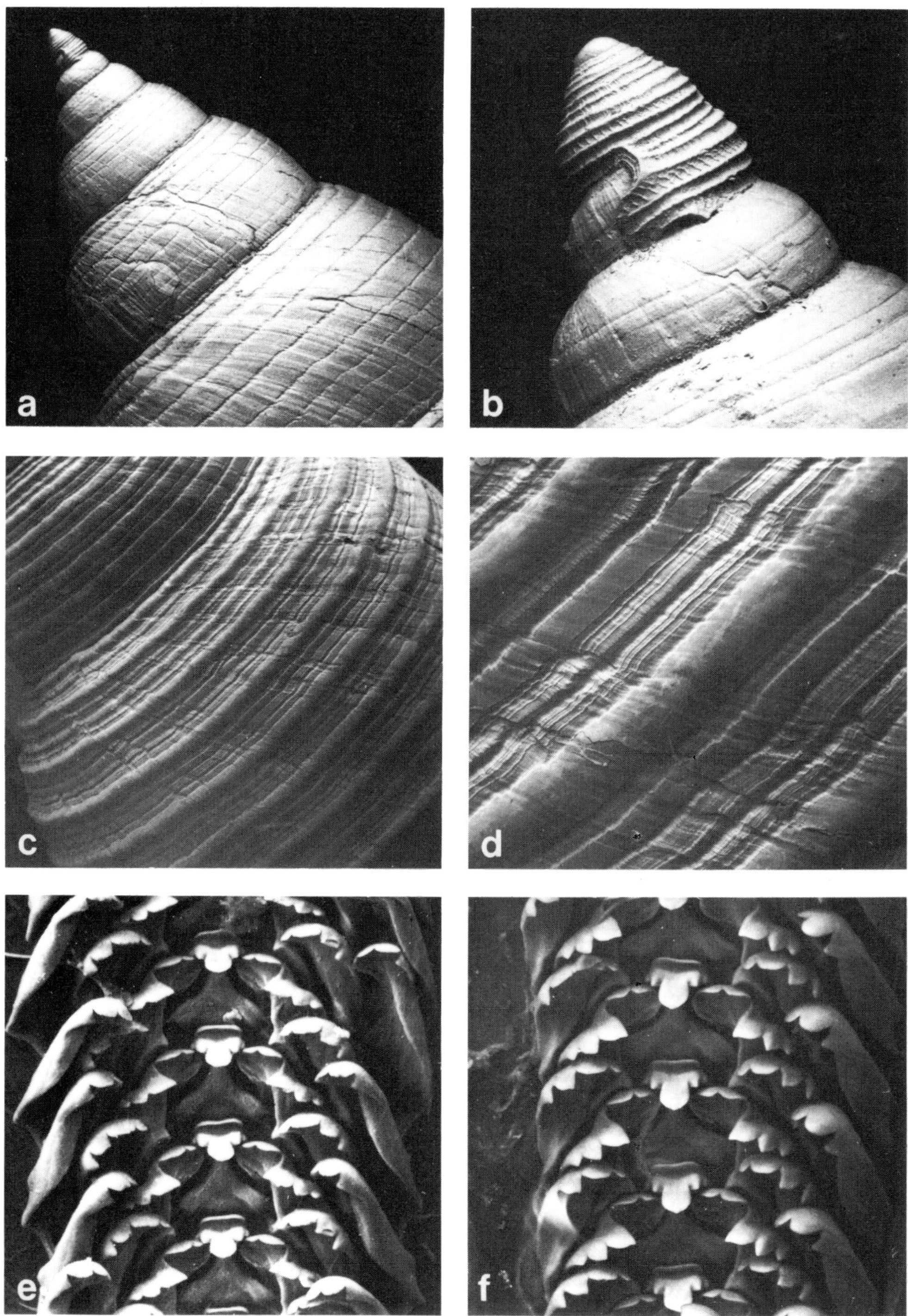


**Fig. 40** *Littoraria* (*Littorinopsis*) *philippiana:* **(a,b)** Magnetic I., Qld.; **(a)** spire ( × 17); **(b)** protoconch ( × 80); **(c,d)** Orpheus I., Qld.; **(c)** last whorl ( × 7); **(d)** detail ( × 28); **(e)** radula, Lizard I., Qld. ( × 225); **(f)** radula, Magnetic I., Qld. ( × 235).

**Table 13** Dimensions of *Littoraria (Littorinopsis) philippiana*.

| Specimen | Locality | Sex | Primary grooves | H (mm) | B (mm) | LA (mm) | WA (mm) | C (mm) | P | S | SH |
|---|---|---|---|---|---|---|---|---|---|---|---|
| *Littorina philippiana* holotype, BMNH 1968307 | [eastern Qld.] | | 7 | 29.8 | 18.4 | 15.5 | 12.1 | 2.8 | 1.62 | 0.78 | 1.92 |
| DGR | Thursday I., Qld. | ♀ | 8 | 35.8 | 21.9 | 18.0 | 14.4 | 2.9 | 1.63 | 0.80 | 1.99 |
| DGR | Thursday I., Qld. | ♀ | 9 | 36.1 | 19.2 | 17.5 | 12.7 | 3.1 | 1.88 | 0.73 | 2.06 |
| DGR | Moa I., Qld. | ♂ | 8 | 17.9 | 11.4 | 9.5 | 7.2 | 1.5 | 1.57 | 0.76 | 1.88 |
| DGR | Moa I., Qld. | ♀ | 8 | 19.6 | 12.4 | 9.7 | 8.0 | 1.6 | 1.58 | 0.82 | 2.02 |
| DGR | Bohle R., Qld. | ♂ | 8 | 26.8 | 15.5 | 13.5 | 10.5 | 2.1 | 1.73 | 0.78 | 1.99 |
| DGR | Bohle R., Qld. | ♂ | 8 | 23.2 | 14.4 | 12.7 | 10.6 | 2.6 | 1.61 | 0.83 | 1.83 |
| DGR | Bohle R., Qld. | ♀ | 9 | 33.4 | 19.0 | 15.9 | 12.5 | 2.3 | 1.76 | 0.79 | 2.10 |
| DGR | Bohle R., Qld. | ♀ | 10 | 30.2 | 18.0 | 14.9 | 12.3 | 2.3 | 1.68 | 0.83 | 2.03 |
| DGR, mean of 10 | Magnetic I., Qld. | ♂ | | 27.44 | | | | | 1.684 | 0.794 | 1.936 |
| standard error | | | | 1.28 | | | | | 0.041 | 0.009 | 0.021 |
| DGR, mean of 10 | Magnetic I., Qld. | ♀ | | 28.75 | | | | | 1.664 | 0.803 | 2.014 |
| standard error | | | | 0.47 | | | | | 0.020 | 0.005 | 0.021 |
| statistic t or U | | | | 0.963 | | | | | 50 | 56.5 | 82 |
| probability | | | | 0.348 | | | | | 1 | 0.657 | 0.014 |

orders of sculpture appear on whorls 7 and 8, as fine grooves and riblets in the broad spaces between major ribs. All primary ribs become raised and rounded on penultimate whorl, and on last whorl develop into prominent, rather narrow cords, numbering 13–18; peripheral rib is most prominent, those near suture least so; secondary riblets remain narrow, but may become rounded. Grooves remain small; on last whorl the spaces between primary ribs are 4–6 times cord width, each containing 1–3 riblets and numerous small, irregularly spaced grooves and striations. Microsculpture of axial growth striae, strongest between cords, and of fine, irregularly spaced spiral striae, only between cords; in some specimens the intersection of axial and spiral sculpture produces a beaded appearance between the ribs.

*Colour* (Frontispiece). Polymorphic; dark brown shells predominate, sometimes pink, rarely yellowish. Ground colour pale brown, cream, orange pink or yellow, with pattern of diffuse dashes and flecks of dark brown or black, fading to grey. On early whorls dark dashes are aligned at sutures to form short axial stripes; on last two whorls pattern is more diffuse, although indistinct, oblique, axial stripes numbering 10–11 per whorl are often visible. Most shells are densely patterned; palest pink or yellow shells show a faint orange or brown mottling or a marbled pattern, often with darkest pigment dashes at sutures and periphery. External pattern visible within aperture as lines and dashes of purple brown, clouded by grey or lilac callus. Columella purple or lilac.

ANIMAL. *Colour*. Pigmentation dark grey to black; sides of foot darkly mottled; head darkest, sometimes pale patch in centre; tentacles banded, unpigmented stripe each side of base.

*Penis* (Fig. 41a–d). Length to 10.0 (15.0) mm. Base bifurcate, limb bearing glandular disc long, disc small. Filament long, tapering towards tip. Sperm groove closed as a duct. Base and filament white to fawn; glandular disc opaque white to pale brown.

*Sperm* (Fig. 5d). Eupyrene sperm 160–187 $\mu$m. Nurse cells (Fig. 41e–g) 14–24 $\mu$m, rounded (sometimes somewhat oval); rods often absent, but if present 1–2, central, not projecting, parallel sided, blunt; yolk granules large.

*Pallial oviduct* (Fig. 41h,i). Length to 5.8 mm. Spiral section to 2.3 mm diam., $3\frac{1}{2}$ whorls; opaque albumen gland $\frac{3}{4}$ whorl, white to cream; translucent albumen gland white to fawn; capsule glands absent; spiral indistinct externally; egg groove not pigmented. Straight section to 3.9 mm, white to fawn, terminating in a papilla. Bursa to 2.0 mm, anterior. Reproduction ovoviviparous.

*Radula* (Fig. 40e,f). Length to 21 mm; relative length 0.63–0.79. Saw-toothed type; central rachidian cusp square, edge slightly pointed; cusps of paired teeth almost equilaterally triangular; lateral with gap anterior to main cusp.

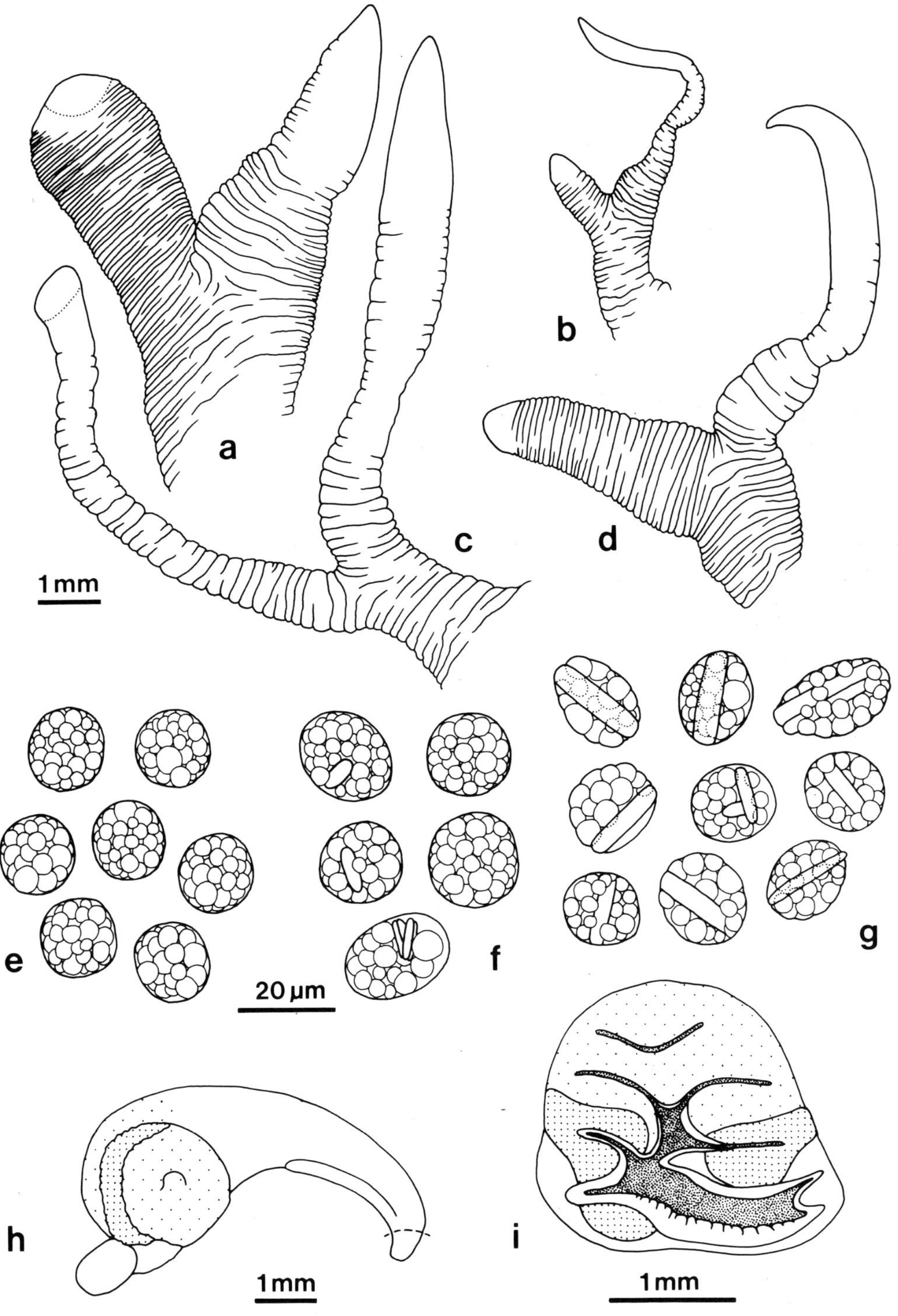


**Fig. 41** *Littoraria* (*Littorinopsis*) *philippiana*, Magnetic I., Qld.: **(a)** penis, partially relaxed; **(b–d)** penes, relaxed; **(e–g)** sperm nurse cells; **(h,i)** pallial oviduct, with transverse section.

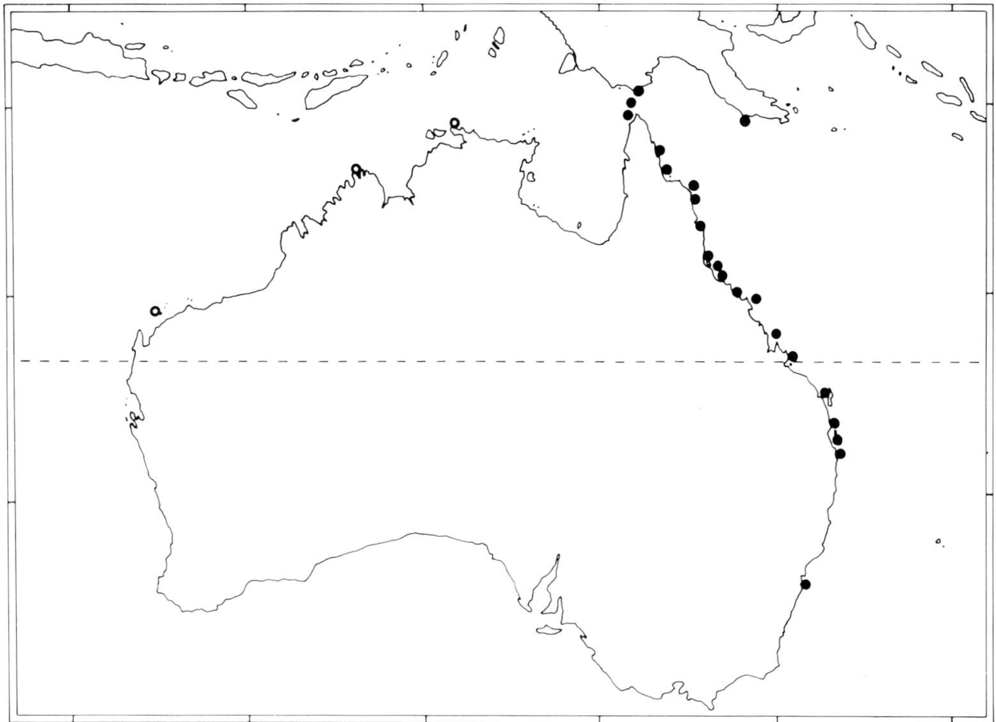

**Fig. 42** Distribution of *Littoraria* (*Littorinopsis*) *philippiana*.

DISTRIBUTION. *Habitat.* Characteristic of *Rhizophora* forests, usually on trunks, sometimes on leaves, 0.5–5.2 m above the ground; found also on trunks and leaves in the *Avicennia* fringe. A continental species.

*Range* (Fig. 42). Eastern Australia from Botany Bay, New South Wales to Torres Strait and south-eastern Papua New Guinea; rare in New South Wales. Three records from the Northern Territory and Western Australia are doubtful, either old collections or shells which could belong to the similar *L. pallescens*. Nineteenth century material from the Solomon Islands, New Hebrides, New Caledonia and Philippines is almost certainly incorrectly localized.

*Records.* **Australia: N.S.W.:** Bayview, Pittwater (AMS); Kurnell Pen. (DGR); Tweed Heads (NMV); **Qld.:** Dunwich, Stradbroke I. (WAM, QM); Caloundra (QM); Pialba (QM, AMS); Wathumba Creek, Fraser I. (AMS); Yeppoon (AMS, NMV); Percy I. (BMNH); Cannonvale Beach, Proserpine (AMS); Hayman I. (AMS); Bowen (AMS, WAM); Magnetic I. (DGR); Orpheus I., Palm Is (DGR); Missionary Bay, Hinchinbrook I. (DGR); Dunk I. (QM, NMV); Machans Beach, Cairns (NMV); Port Douglas (AMS); Low Isles (AMS); Lizard I. (AMS); Claremont I. (QM); Quintell Beach (QM); Thursday I., Torres Str. (DGR, QM, WAM); Moa I., Torres Str. (DGR); Saibai I., Torres Str. (AMS); **N.T.:** Port Essington (BMNH, doubtful record); **W.A.:** Vansittart Bay (AMS, doubtful record); Barrow I. (WAM, doubtful record); **Papua New Guinea:** Kapa Kapa (AMS); Marshall Lagoon (AMS).

REMARKS. The shell of *L. philippiana* is rather constant in sculpture, although coloration is variable. Specimens from *Rhizophora* forests are invariably darkly pigmented, while in those from *Avicennia* trees the pattern is usually paler and the pink or yellow ground colour visible. Of the anatomical characters the closed penial sperm duct is noteworthy, a feature shared only with *L. subvittata* and *L. intermedia* in the same subgenus. Sperm nurse cells are also similar in these species and the three are probably closely related. Shells of *L. pallescens* are sometimes extremely similar to those of *L. philippiana*, and sperm nurse cells are the same, although penial form differs in the two species.

Where their ranges overlap in eastern Australia they seldom occur together, since *L. pallescens* is an oceanic species, whilst *L. philippiana* occurs in continental situations. Only on high islands close to the North Queensland coast are both found together (Magnetic I., Hinchinbrook I., Dunk I., Lizard I., Thursday I.). However, despite their similarities there is no doubt that the two species are distinct, since at localities where both occur the diagnostic penial characters remain constant and shells can usually be distinguished.

SIMILAR SPECIES. As mentioned above, although penial form differs in the two species *L. philippiana* and *L. pallescens*, their shells can in some cases be closely similar. *L. philippiana* can usually be distinguished by its larger size, the sharp inner lip of the aperture adjacent ot the columella, and most important by the presence of numerous secondary and tertiary grooves in the spaces between the primary ribs. In addition there are differences in colour pattern; although both species are polymorphic, the pattern of *L. philippiana* is rather diffuse, whilst in *L. pallescens* the spiral dashes tend to be discrete and confined to the ribs. Neither pure yellow shells nor continuous spiral bands of pigment are found in *L. philippiana* and the columella is never white. Only a very few shells cannot be separated using shell characters alone. Of other similar species, *L. filosa* can bear a close resemblance in shell sculpture, but the columella is narrow and not excavated, and on the apical whorls the spacing of the primary grooves is unequal. *L. luteola* and *L. ardouiniana* can be superficially similar, but both lack the wide and excavated columella of *L. philippiana*. Sculptural details and penial form readily distinguish this species from *L. subvittata*, *L. intermedia* and *L. lutea*.

## ***Littoraria (Littorinopsis) intermedia*** (Philippi, 1846)

*Littorina intermedia* Philippi, 1846: 141 [in part; lectotype here designated BMNH 198343; type locality here restricted to Tahiti]; Reeve, 1858: *Littorina* pl. 18, fig. 101; Issel, 1869: 192; Nevill, 1885: 146 [in part]; Sowerby, 1892: 37; Dautzenberg, 1923: 49; Turton, 1932: 131; Fischer, 1970: 99

*Litorina intermedia*—Philippi, 1847, vol. 2: 223; Krauss, 1848: 103; von Martens, 1871: 39; von Martens, 1880: 284 [all in part]

*Littorina (Melaraphe) scabra* var. *intermedia*—Tryon, 1887: 244, pl. 42, figs 21,22,24 [in part]

*Littorina (Malaraphe) intermedia*—Casto de Elera, 1896: 309–310 [in part]

*Littorina (Melaraphe) intermedia*—Dautzenberg & Fisher, 1905: 146; Dautzenberg, 1929: 288–289 [both in part]

*Littorinopsis intermedia*—Kuroda & Habe, 1952: 64

*Littoraria intermedia*—Higo, 1973: 46

*Littorina intermedia* var. *punctata* Philippi, 1846: 141 [objective synonym of *Littorina intermedia* Philippi, 1846; not a secondary homonym of *Turbo punctatus* Gmelin, 1791 = *Nodilittorina punctata*]

*Litorina intermedia* var. *punctata*—Philippi, 1847, vol. 2: 223, *Litorina* pl. 5, fig. 6 [not fig. 11 as cited in text]

*Litorina scabra* var. *articulata* Philippi, 1847, vol. 2: 222, *Litorina* pl. 5, fig. 4 [lectotype (Rosewater, 1970) BMNH 1968354; type locality here corrected Jimaimilan [Himamaylan], Negros I., Philippines; not *Littorina intermedia* var. *articulata* Philippi, 1846]

*Litorina ambigua* Philippi, 1848, vol. 3: 62–63, *Litorina* pl. 7, fig. 6 [Sandwich Is [=Hawaiian Is]; lectotype (Rosewater, 1970) BMNH 1968314]

*Littorina ambigua*—Reeve, 1857: *Littorina* pl. 12, fig. 64; Hidalgo, 1904–5: 206

*Littorina newcombi* Reeve, 1857: *Littorina* pl. 6, figs 28a,b [Sandwich Is [=Hawaiian Is]; lectotype (Rosewater, 1970) BMNH 1968308]; Pease, 1868: 128; Nevill, 1885: 144; Sowerby, 1892: 37

*Litorina newcombi*—von Martens, 1871: 39

*Littorina scabra* var. *newcombi*—Dautzenberg, 1923: 49

*Littorina (Melaraphe) scabra* var. *newcombi*—Dautzenberg, 1929: 289

*Littorina fraseri* Reeve, 1857: *Littorina* pl. 10, fig. 47 [Lagos, West Africa, in error; corrected to Indo-Pacific; lectotype (Rosewater, 1981) BMNH 1968312]

*Littorina* (*Melaraphe*) *punctata* var. *fraseri*—Tryon, 1887: 248, pl. 44, figs 62,63
*Littorina pintado*—Pease, 1868: 127; Kay, 1979: 72 [both in part; not Wood, 1828]
*Litorina pintado*—von Martens, 1871: 39 [in part; not Wood, 1828]
*Littorina* (*Melaraphe*) *pintado*—Tryon, 1887: 250 [in part; not Wood, 1828]
*Littorina* (*Littoraria*) *pintado pintado*—Rosewater, 1970: 447–449, pl. 346, figs 3,4 [in part; not Wood, 1828]
*Litorina scabra*—Weinkauff, 1878: 37–39, pl. 4, figs 16–18 [not Linnaeus, 1758]
*Littorina* (*Melaraphe*) *scabra*—Tryon, 1887: 243, pl. 42, fig. 20 [in part; not Linnaeus, 1758]
*Littorina* (*Malaraphe*) *scabra*—Casto de Elera, 1896: 309 [in part;.not Linnaeus, 1758]
*Littorina scabra*—Risbec, 1942 [anatomy]; Whipple, 1965: figs 3c,4, pl. 25, figs 3a–c; Struhsaker, 1966 [reproduction and development]; Salvat & Rives, 1975: 263, no. 38; Nielsen, 1975: fig. 1a; Kay, 1979: 73, fig. 24g [all not Linnaeus, 1758]
*Littoraria scabra*—Habe, 1964: 29, pl. 9, fig. 32 left [in part; not Linnaeus, 1758]
*Littorina* (*Littorinopsis*) *scabra scabra*—Rosewater, 1970: 456–461, pl. 352, figs 6,10,14,15,19 [in part; not Linnaeus, 1758]
*Littorina* (*Littorinopsis*) *scabra*—Tadjalli-Pour, 1974: 58, pl. 20, fig. 3 [not Linnaeus, 1758]
*Littorinopsis scabra*—Brandt, 1974: 53–54, pl. 4, fig. 60 [in part; not Linnaeus, 1758]
*Littorina* (*Melaraphe*) *planaxis*—Tryon, 1887: 248, pl. 43, fig. 56 [in part; not Philippi, 1847]
*Littorina scabra* var. *philippiana*—Hidalgo, 1904–5: 206 [in part; not Reeve, 1857]
*Littorina* (*Melaraphe*) *scabra* var. *rhodea* Biggs, 1958: 272 [Bundar Abbas, S. Persia; holotype BMNH 1958.6.13.23]
*Littorina* (*Littoraria*) *cingulifera*—Rosewater, 1981: 18–20 [in part; not Dunker, 1845]

NOMENCLATURE. The identity of this species has been the subject of much confusion in the literature, as indicated by the lengthy synonymy. This has arisen firstly because the name '*intermedia*' has been applied to a group comprising at least three species, *L. articulata*, *L. strigata* and *L. intermedia s. s.*, and probably also including *L. subvittata*. Secondly, the species is so variable that several forms have been named and subsequently often incorrectly placed in the synonymies of other species.

Difficulties begin with the original diagnosis (Philippi, 1846) of *Littorina intermedia*, which was not sufficiently precise to define the species, and was followed by diagnoses of the principal colour varieties *punctata*, *articulata* and *strigata*. The varieties *punctata* and *strigata* were figured by Philippi in 1847, although the figures of his plate 5 were incorrectly cited in the text. From the accompanying descriptions it is evident that Philippi's plate 5, fig. 6 illustrates var. *punctata* and plate 5, figs 7,8,9 var. *strigata*. From these figures and the type localities given, each variety is recognizable as the type of a species here considered distinct, and it must be decided to which the name '*intermedia*' should be applied. Following the diagnosis of *Littorina intermedia*, Philippi (1846) gave the following dimensions: '*Altit.* 10, *diam.* $7\frac{1}{2}$, *altit. aperturae* $5\frac{1}{2}$ *lin.*' (22.0, 16.5, 12.1 mm respectively). Since neither *L. articulata* nor *L. strigata* ever attain such large size, the epithet '*intermedia*' is thereby restricted to the variety *punctata*. All the taxa described by Philippi (1846), and their locality records, were based on specimens in the Cuming Collection, now in the BMNH. In 1846, Philippi gave no localities for the variety *punctata*. However, of the localities listed for *Littorina intermedia*, the original material upon which the records were based has now been identified for the localities Natal, Elizabeth Island and Tahiti. All are the species illustrated by Philippi (1847) as *Litorina intermedia* var. *punctata*, as, from the known distribution of the species, was the record from the Red Sea. The majority of the specimens that Philippi had before him were therefore of the variety *punctata*, and in 1847 he listed all these localities for this variety. Only two other localities were given beneath the diagnosis of *Littorina intermedia*, which are the type localities of *L. articulata* and *L. strigata*. It seems that Philippi himself considered the variety *punctata* to be the typical colour form of his species, and on each occasion listed it as the first of the three varieties.

When Rosewater (1970) examined the collections of the BMNH, he distinguished only the type material of *Littorina intermedia* var. *strigata*, from which he designated the lectotype of *Littorina intermedia*. From the evidence presented above, this designation is inappropriate. The original specimen of Philippi's figure of *Litorina intermedia* var. *punctata* has now been identified, on the basis of correspondence of pattern and dimensions. The original label bears the locality 'Tahiete' inscribed by Cuming and '*intermedia* Ph.' in the hand of Philippi himself (p. 1). Tahiti is one of the localities

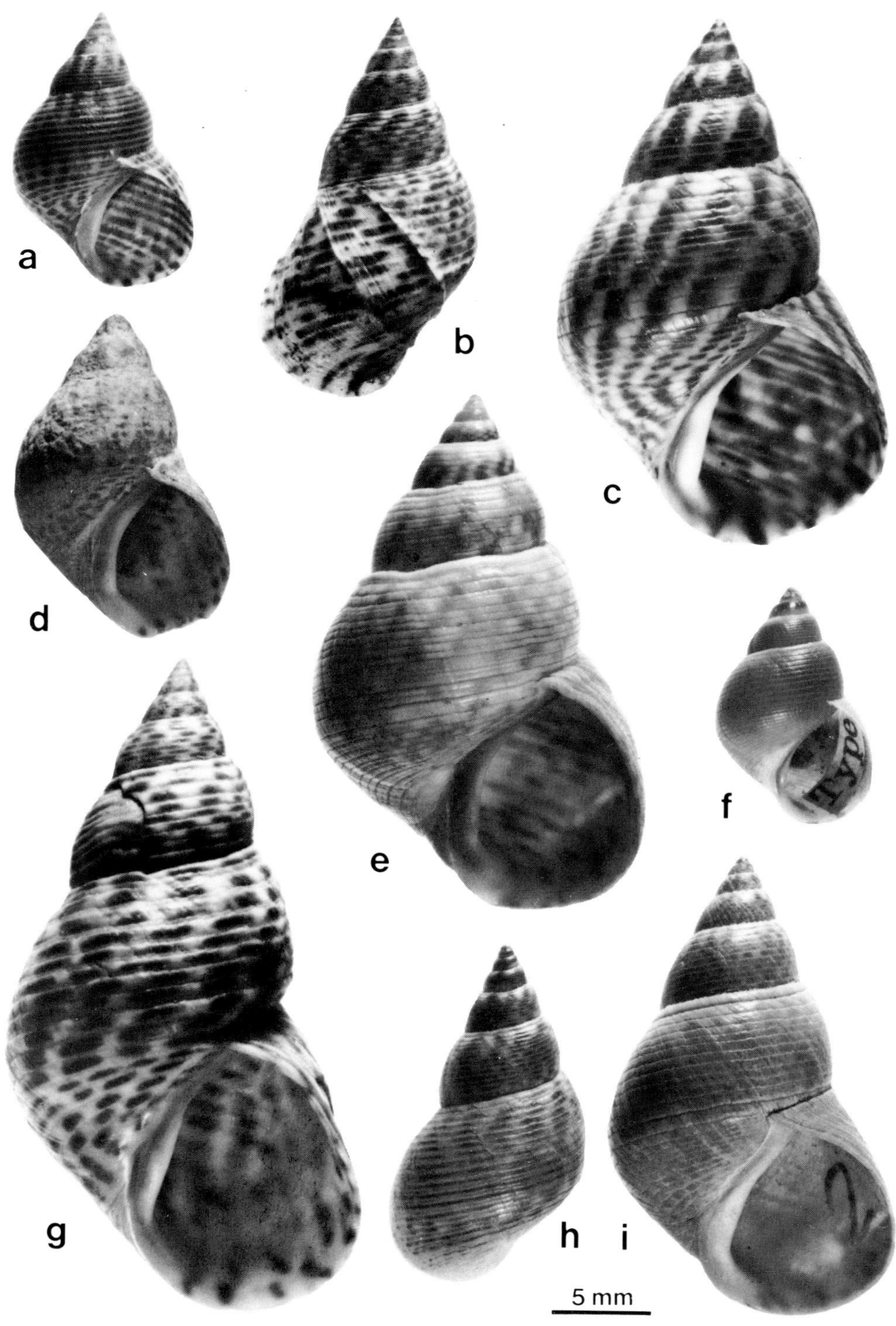


**Fig. 43** *Littoraria* (*Littorinopsis*) *intermedia:* **(a)** lectotype of *Littorina fraseri* Reeve, locality unknown (BMNH 1968312); **(b)** ♂, Sungei Merbok estuary, Malaysia (DGR); **(c)** Pearl Harbour, Hawaiian Is (AMS); **(d)** lectotype of *Litorina ambigua* Philippi, Hawaiian Is (BMNH 1968314); **(e)** lectotype of *Littorina newcombi* Reeve, Hawaiian Is (BMNH 1968308); **(f)** holotype of *Littorina* (*Melaraphe*) *scabra* var. *rhodea* Biggs, Bundar Abbas, Iran (BMNH 1958.6.13.23); **(g)** lectotype of *Litorina scabra* var. *articulata* Philippi, Himamaylan, Negros I., Philippines (BMNH 1968354); **(h)** Aqaba, Jordan (DGR); **(i)** lectotype of *Littorina intermedia* Philippi, Tahiti (BMNH 198343).

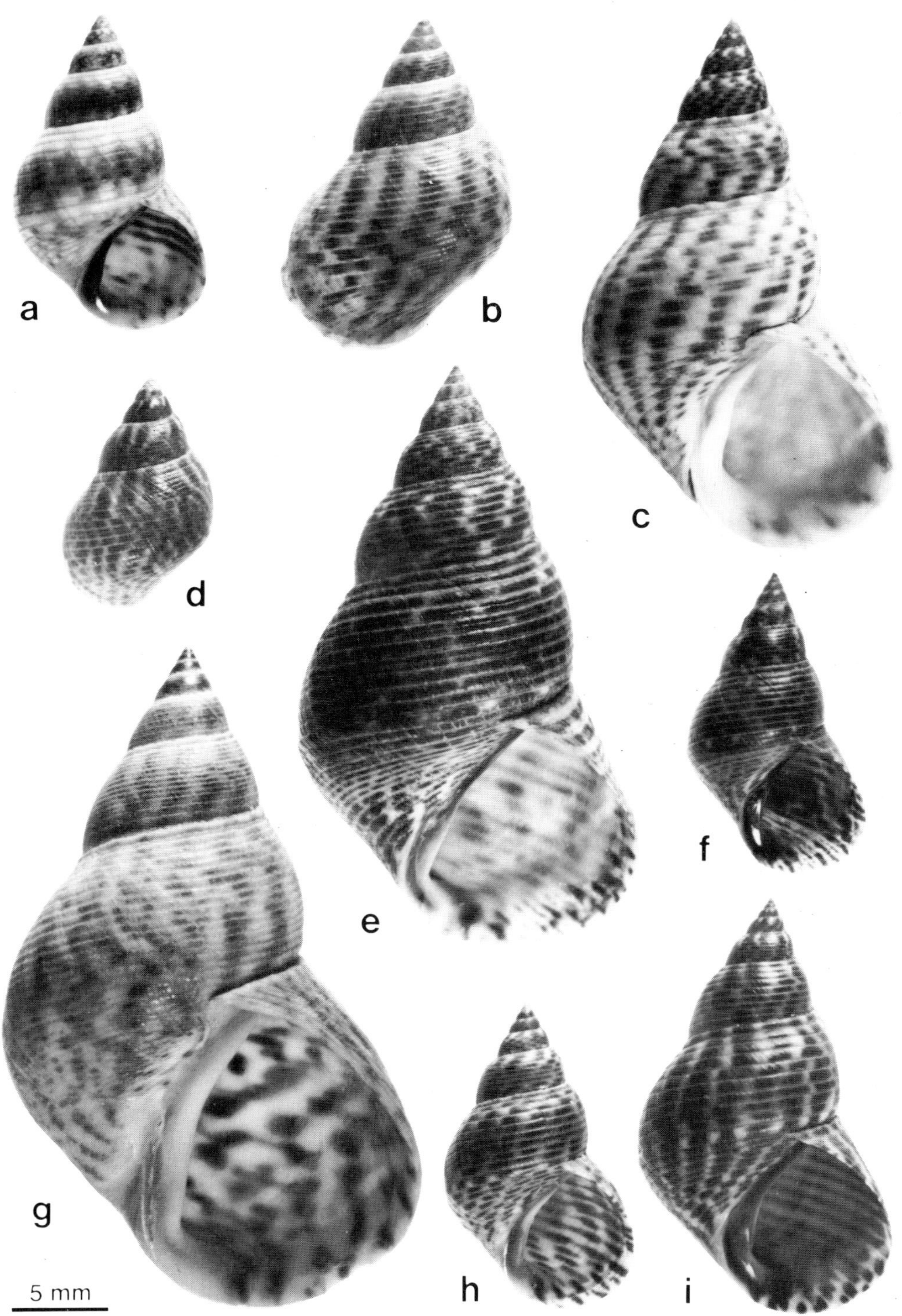


**Fig. 44** *Littoraria* (*Littorinopsis*) *intermedia:* **(a)** Okinawa, Japan (BMNH); **(b)** Ile Ste Marie, Madagascar (NMV); **(c)** ♀, Gove, N.T. (DGR); **(d)** Mangareva I., Gambier Is (WAM); **(e)** ♀, Penang, Malaysia (DGR); **(f)** ♂, Phuket I., Thailand (DGR); **(g)** Trincomalee, Sri Lanka (BMNH); **(h)** Polillo I., Philippines (WAM); **(i)** ♀, Magnetic I., Qld. (DGR).

given by Philippi for *Littorina intermedia* (1846) and for *Litorina intermedia* var. *punctata* (1847), although the origin and owner of the figured specimen were not explicitly stated. This specimen is here designated lectotype of *Littorina intermedia* Philippi, 1846, with type locality Tahiti. *Littorina intermedia* var. *punctata* becomes an objective synonym of this species. This redesignation of a lectotype for *L. intermedia* stabilizes the concept of the species as it has most frequently been used in the literature.

Of the several junior synonyms of *L. intermedia*, *Litorina scabra* var. *articulata* Philippi, 1847 was based upon an unusually large specimen of *L. intermedia*. The lectotype selected by Rosewater (1970) is probably the specimen figured by Philippi, but the original label does not bear the type locality given by Rosewater, which is corrected accordingly. The three paralectotypes are *L. lutea* (Philippi, 1847), to which the lectotype does indeed bear a close resemblance. *Litorina ambigua* Philippi, 1848 has been placed in the synonymy of *L. pintado* by several authors. Although the lectotype is an eroded shell, there is no doubt that it is a specimen of the Hawaiian form of *L. intermedia*. *Littorina newcombi* Reeve, 1857 is also the large Hawaiian form. Reeve (1857) gave the locality of *Littorina fraseri* as 'Lagos, West Africa', which misled Tryon (1887) to list it as a variety of *Littorina punctata* (Gmelin) and Rosewater (1981) to place the species in the synonymy of *L. cingulifera*. The type specimens are typical *L. intermedia*, however, and the type locality is therefore presumably incorrect. *Littorina (Melaraphe) scabra* var. *rhodea* Biggs, 1958 is a reddish shell lacking the usual colour pattern, and similar small, pallid shells are common in the NW. Indian Ocean. The name '*Litorina scabra* var. *minor*', attributed to Weinkauff (1878) in synonymies given by Tryon (1887) and Rosewater (1970), has arisen from a misreading of the original text, in which '*minor*' is used merely as the initial adjective in a Latin description and not as a varietal name.

Diagnosis. Shell: periphery rounded; columella wide, excavated, purple; primary grooves 8–10; secondary sculpture on last whorl only, not usually well developed; ribs low, except posterior rib which is often prominent, total number of ribs on last whorl 17–32; grooves narrow; microsculpture indistinct; colour variable, but shells often show dense pattern of discrete black or brown dashes, aligned to form 13–20 axial series on last whorl. Animal: penis bifurcate, base ochraceous, filament white and large, tip mucronate, sperm groove closed as a duct; ovoviviparous.

Shell (Figs 43, 44). *Shape.* Height (10)14–26(32) mm. Teleoconch 6.5–7.5(8) whorls. Shell usually thick, solid. Spire outline gently convex; whorls rounded; sutures impressed. Peripheral keel evident in young shells, becoming obsolete on last whorl. Adult lip slightly thickened, sometimes a little flared. Varices absent, but strong growth ridges may remain behind lip. Columella wide, excavated; pillar slightly convex, sharply pinched at base. Sexual dimorphism: males smaller, relatively lower spire and larger aperture.

*Dimensions:* Table 14.

*Sculpture* (Fig. 45a–e). Protoconch normal. All whorls of teleoconch sculptured by spiral grooves. Primary grooves (7)8–10(11), almost equally spaced, but all or one of the 3 posterior ribs usually wider than others; posterior rib almost always most prominent since posterior groove is deepest. Some or rarely all primary ribs may be divided by secondary grooves at $\frac{1}{2}$ width on whorls (6)7–8, although secondary sculpture often absent. Tertiary grooves on largest shells only. Ribs remain low and flattened, except posterior rib which is rounded, prominent, undivided and, on last whorl, often appears pushed up over suture; peripheral rib sometimes a little wider and more prominent than others; total number of primary ribs on last whorl 17–21, up to 32 ribs if secondary sculpture well developed. Intercalation of single, narrow secondary riblets occurs infrequently in the wider primary grooves of last whorl. Primary grooves remain narrow, usually less than $\frac{1}{5}$ rib width on last whorl, but posterior groove and that at periphery may reach $\frac{1}{2}$ rib width; secondary grooves are incised lines only. Microsculpture of faint axial growth striae over surface; spiral microsculpture usually absent, but faint spiral striae occasionally visible on ribs of spire whorls and in wider grooves of last whorl.

*Colour.* Variable over entire geographical range, but usually rather constant at each locality. Ground colour usually grey, sometimes pale brown, cream, whitish or rarely orange pink. Pattern of discrete black or dark brown dashes, closely spaced on ribs, often roughly aligned at suture or

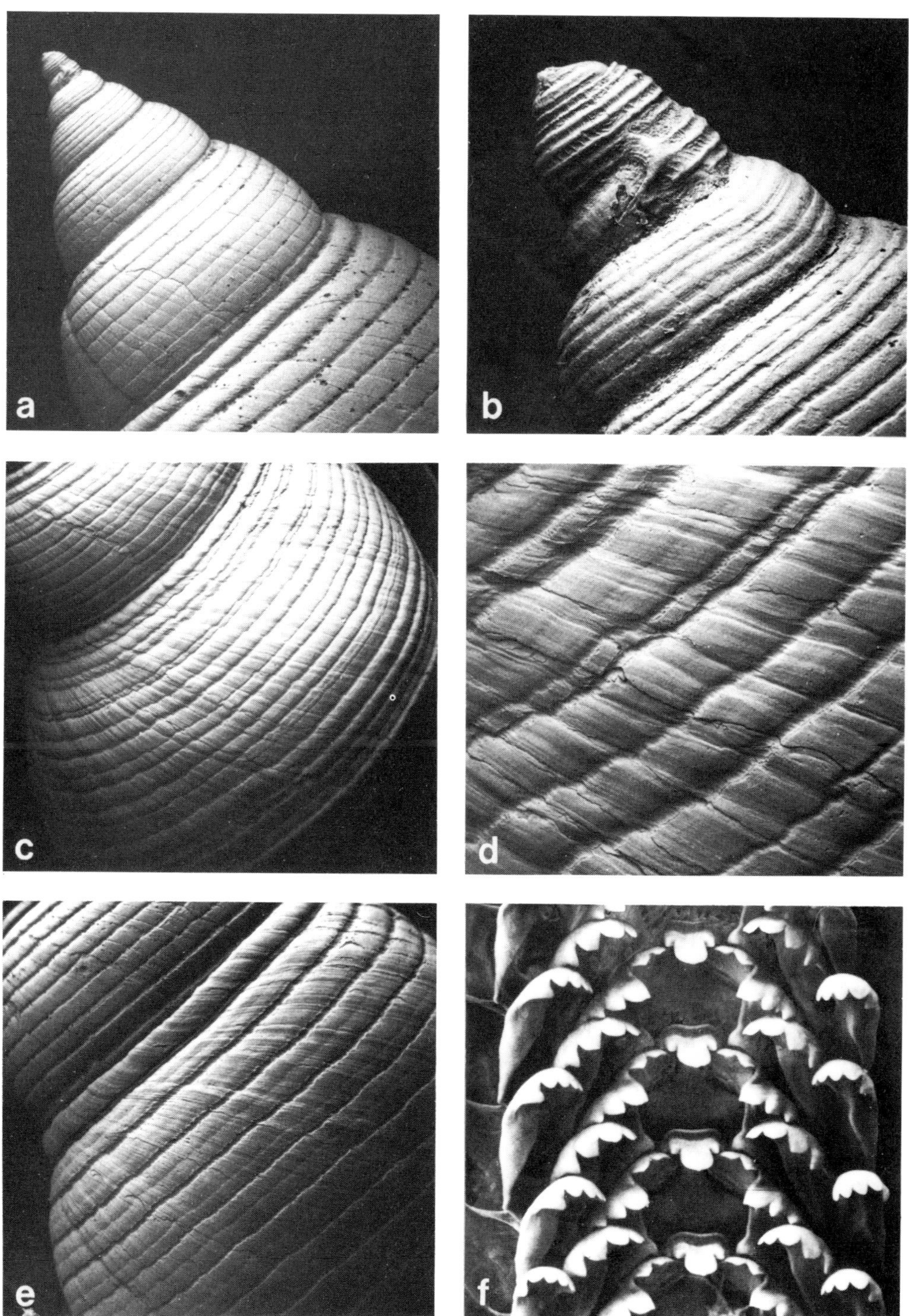


**Fig. 45** *Littoraria* (*Littorinopsis*) *intermedia:* **(a,b)** Magnetic I., Qld.; **(a)** spire ( × 19); **(b)** protoconch ( × 90); **(c,d)** Phuket I., Thailand; **(c)** last whorl ( × 7); **(d)** detail ( × 29); **(e,f)** Magnetic I., Qld.; **(e)** last whorl ( × 9); **(f)** radula ( × 210).

**Table 14** Dimensions of *Littoraria* (*Littorinopsis*) *intermedia*.

| Specimen | Locality | Sex | Primary grooves | H (mm) | B (mm) | LA (mm) | WA (mm) | C (mm) | P | S | SH |
|---|---|---|---|---|---|---|---|---|---|---|---|
| *Littorina intermedia* lectotype, BMNH 198343 | Tahiti | | 10 | 22.7 | 15.3 | 12.8 | 10.2 | 2.6 | 1.48 | 0.80 | 1.77 |
| *Litorina scabra* var. *articulata* lectotype, BMNH 1968354 | Himamaylan, Negros I., Philippines | | 8 | 31.9 | 19.0 | 16.1 | 12.2 | 3.0 | 1.68 | 0.76 | 1.98 |
| *Litorina ambigua* lectotype, BMNH 1968314 | Hawaiian Is | | 8 | 15.8 | 11.7 | 9.1 | 7.6 | 1.8 | 1.35 | 0.84 | 1.74 |
| *Littorina newcombi* lectotype, BMNH 1968308 | Hawaiian Is | | 8 | 25.4 | 17.2 | 13.4 | 10.4 | 2.7 | 1.48 | 0.78 | 1.90 |
| *Littorina fraseri* lectotype, BMNH 1968312 | [Indo-Pacific] | | 10 | 13.5 | 9.8 | 7.8 | 6.0 | 1.6 | 1.38 | 0.77 | 1.73 |
| *Littorina scabra* var. *rhodea* holotype, BMNH 1958.6.13.23 | Bundar Abbas, Iran | | 10 | 11.6 | 8.4 | 6.6 | 4.7 | 0.6 | 1.38 | 0.71 | 1.76 |
| DGR | Magnetic I., Qld. | ♂ | 9 | 18.8 | 12.5 | 10.2 | 8.2 | 2.2 | 1.50 | 0.80 | 1.84 |
| DGR | Magnetic I., Qld. | ♀ | 9 | 21.4 | 14.0 | 10.7 | 9.2 | 2.2 | 1.53 | 0.86 | 2.00 |
| DGR | Penang, Malaysia | ♂ | 10 | 22.2 | 14.5 | 13.0 | 10.0 | 2.8 | 1.53 | 0.77 | 1.71 |
| DGR | Penang, Malaysia | ♀ | 10 | 26.6 | 16.5 | 13.7 | 10.4 | 1.9 | 1.61 | 0.76 | 1.94 |
| BMNH 1969357 | Trincomalee, Sri Lanka | | 8 | 31.0 | 21.2 | 17.4 | 13.8 | 3.0 | 1.46 | 0.79 | 1.78 |
| BMNH 1969357 | Trincomalee, Sri Lanka | | 8 | 23.0 | 16.5 | 13.9 | 10.9 | 3.3 | 1.39 | 0.78 | 1.65 |
| DGR, mean of 10 | Magnetic I., Qld. | ♂ | | 17.37 | | | | | 1.501 | 0.786 | 1.778 |
| standard error | | | | 0.33 | | | | | 0.015 | 0.006 | 0.021 |
| DGR, mean of 10 | Magnetic I., Qld. | ♀ | | 19.67 | | | | | 1.516 | 0.803 | 1.881 |
| standard error | | | | 0.45 | | | | | 0.011 | 0.009 | 0.022 |
| statistic t or U | | | | 4.116 | | | | | 57 | 65.5 | 87 |
| probability | | | | <.001 | | | | | 0.630 | 0.264 | 0.004 |

over entire whorl to form axial series, numbering 13–20 on last whorl; alignment usually especially prominent at suture of spire whorls, where ground colour is palest. Degree of pigmentation variable, although grooves remain pale; some shells almost entirely black with pale patches at suture and periphery; palest shells only faintly mottled with light brown; sometimes with spiral paler zones at suture, periphery and around columella. Aperture with black lines and streaks, corresponding to external ribs, clouded with greyish callus. Columella and parietal callus dark purple or pink; columellar pillar usually paler pink or whitish.

ANIMAL. *Colour*. Pigmentation black; sides of foot densely mottled; head darkest, sometimes with pale central patch; tentacles banded, unpigmented stripe each side of base.

*Penis* (Fig. 46a–h). Length to 8.2 mm. Base bifurcate; glandular disc relatively small. Filament large, broad, tip mucronate. Sperm groove closed as a duct. Base ochraceous, darkest at base of filament and near disc, filament white, glandular disc dark ochre to cream.

*Sperm*. Eupyrene sperm 133–146 μm. Nurse cells (Fig. 46i–k) 18–24(27) μm, slightly irregularly oval to reniform; rods often indistinct, 1–4, narrow, pointed, central, usually not projecting or only minutely so (occasionally strongly); yolk granules large, indistinct.

*Pallial oviduct* (Fig. 46l–o). Length to 4.1(5.7) mm. Spiral section to 1.9(2.7) mm. diam., $3\frac{1}{2}$ whorls; opaque albumen gland $\frac{1}{2}$–$\frac{3}{4}$ whorl, white; translucent albumen gland white to fawn; capsule glands

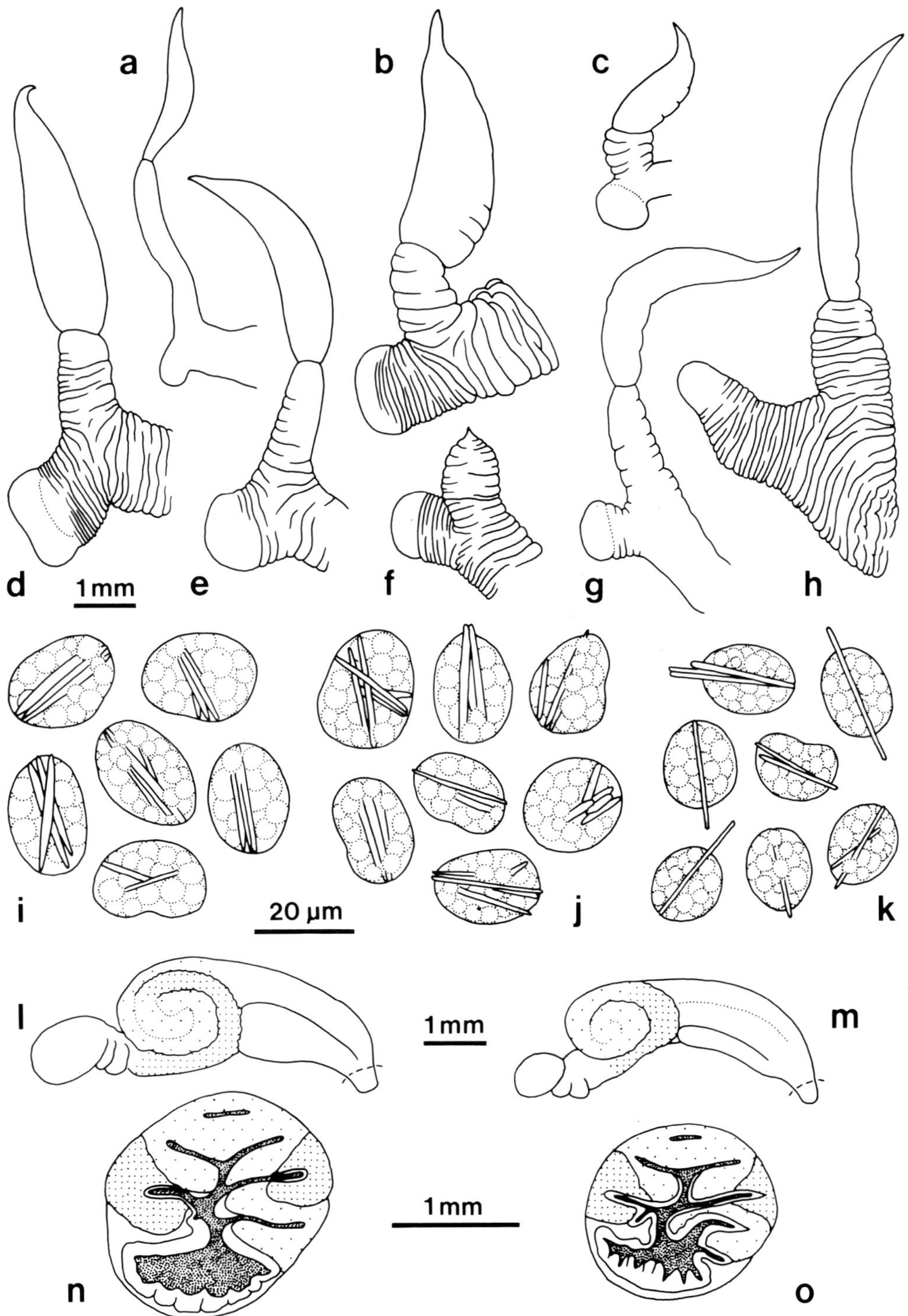


**Fig. 46** *Littoraria* (*Littorinopsis*) *intermedia:* **(a–e,g,h)** penes, relaxed; **(f)** penis, contracted; (a) Nxaxo R., S. Africa; **(b)** Penang, Malaysia; **(c)** Magnetic I., Qld.; **(d)** Coconut I., Oahu, Hawaiian Is; **(e)** Vaiore Bay, Tahiti; **(f,g)** Magnetic I., Qld.; **(h)** Gove, N. T.; **(i–k)** sperm nurse cells; **(i)** Ubin I., Singapore; **(j)** Magnetic I., Qld.; **(k)** Penang, Malaysia; **(l–o)** pallial oviducts, with transverse sections, Magnetic I., Qld.

absent; spiral indistinct externally, egg groove not usually darkly pigmented. Straight section to 2.5(3.4) mm, pale brown, terminating in a papilla. Bursa long, to 2.1(2.8) mm, anterior. Development ovoviviparous.

*Radula* (Fig. 45f). Length to 26 mm; relative length 0.76–1.07. Saw-toothed type; central rachidian cusp square, edge slightly pointed; cusps of paired teeth almost equilaterally triangular; lateral with gap anterior to main cusp.

DISTRIBUTION. *Habitat.* On roots and trunks throughout *Rhizophora* forests, 0.3–2.0 m (rarely to 3.5 m) above the ground. Occasional on trunks in *Bruguiera* forests and in *Avicennia* fringe. Not infrequent on sheltered rocks. An oceanic species.

*Range* (Fig. 47). Throughout tropical and subtropical Indo-Pacific, including South Africa, Red Sea and Persian Gulf, but excluding Bay of Bengal, north-western Australia and much of mainland South East Asia and Borneo. Extends as far eastwards as the Hawaiian Islands and Polynesia; a single shell is doubtfully recorded from the Galapagos Islands.

*Records.* **South Africa:** Umtata R. mouth (NM); Nxaxo R. mouth (ANSP); Isipingo (BMNH); Durban (BMNH, NM); St Lucia, Zululand (NMW, NM); **Mozambique:** Inhaca I. (NM); Beira (BMNH, NMW); **Tanzania:** North Reef, Dar-es-Salaam (AMS); Chukwani, Zanzibar (ANSP); **Madagascar:** Tuléar (AMS, MCZ); Nossi Bé (ANSP); Ile Ste Marie (NMV); **Mauritius** (BMNH, NMW); Poste de Flacq (ANSP); **Seychelles:** Mahé I. (BMNH); **Aldabra Atoll** (BMNH); **Chagos Arch.**: Diego Garcia (BMNH); **Ethiopia:** Melita Bay, Gulf of Zula (ANSP); Dahlak Kebir I. (USNM); **Egypt:** Halieb (USNM); Hurghada (RNHL); **Israel:** Eilat (USNM); **Jordan:** Aqaba (DGR); **Yemen:** Aden (BMNH); **Iran:** Bundar Abbas (BMNH); **India:** Bombay (USNM, MCZ); Vengurla, N. of Goa (USNM); Netravata R., Mangalore (USNM); Beypore Estuary (BMNH); between Wallarpat I. and Bolghatti, Cochin Harbour (ANSP); Manalī I., off Mandapam (MCZ); Tuticorin (USNM); Madras (BMNH, USNM); **Sri Lanka:** Galle (BMNH); Trincomalee (BMNH); **Andaman Is:** Port Blair (BMNH); **Nicobar Is** (BMNH); **Cocos-Keeling Is** (AMS, WAM); **Burma:** King I., Mergui Arch. (BMNH); **Thailand:** Ao Nam-Bor, Phuket I. (DGR); Tanga I., Butang Group (USNM); **Malaysia: Peninsula:** Penang (DGR); Sungei Merbok estuary, Kedah (DGR); Perhentian I., 50 km S. of Kota Bharu (ANSP); **Sabah:** Po Bui I., Sandakan (USNM); **Singapore:** Loyang Besar (WAM); Jelutong, Ubin I. (DGR); **Vietnam:** Chilins, Vung Tau district (ANSP); **Indonesia: Sumatra:** Belawan R. (RNHL); Weh I. (RNHL); Sinabang, Simeulue I. (RNHL); Bai I., Batu Is (USNM); Sanding I., Mentawai Is (USNM); **Java:** Njamoek Besar, Djakarta Bay (RNHL); **Lesser Sunda Is:** Sanur Beach, Bali (AMS); Maumere, Flores (USNM); Tanahdjampea I. (RNHL); Kambang I., 20 km SW. of Kupang, Timor (RNHL); **Sulawesi:** Manado (MCZ); Makassar (RNHL); **Moluccas:** Ternate, Halmahera (WAM); Kasirota I. (MCZ); Taliabu I., Sula Is (RNHL); Ambon (RNHL); Tamberfane, Aru Is (WAM); **Irian Jaya:** Sorendidori, Schouten Is (ANSP); Japen I. (RNHL, ANSP); Djajapura (RNHL); **Philippines:** Sanga Sanga I., Sulu Arch. (RNHL, ANSP); Jolo I., Sulu Arch. (ANSP); Makibojoc Bay, Bohol I. (WAM); Viejo Victorias, Negros I. (USNM); Batangas Bay, Luzon I. (WAM); Legaspi, Albay Prov., Luzon I. (ANSP); Panukalan, NW. Polillo I. (WAM); Linapacan I. (USNM); **Japan:** Toguchi, Kitanakagusuku, Okinawa I., Ryukyu Is (BMNH); Kii Peninsula, Honshu (Habe, 1964); **Papua New Guinea:** Madang (AMS); Collingwood Bay (AMS); Losuia, Kiriwina Is, Trobriand Is (AMS); **Bismarck Arch.:** New Britain (AMS); Koruniat I., Admiralty Is (ANSP); Bouin, Bougainville I. (ANSP); **Australia: N.T.:** Cape Don (AMS); Port Essington (AMS); Gove, Arnhemland (DGR); Groot Eylandt (WAM); **Qld.:** Thursday I., Torres Str. (DGR, QM); Moa I., Torres Str. (DGR); Quintell Beach, Iron Range (QM); Lizard I. (AMS, ANSP); Annan R. (AMS); Double I., Cairns (NMV); Dunk I. (QM); Magnetic I. (DGR); Hayman I. (AMS); N. Keppel I. (NMV); Heron I. (AMS); Caloundra (QM); Dunwich, Stradbroke I. (QM, NMV); Currumbin Creek, N. of Coolangatta (AMS); **New Caledonia:** Ile des Pins (AMS); Nouméa (AMS); Hienghène (ANSP); **New Hebrides:** Aneiteum and Tana (BMNH); Port Stanley, Malekula (BMNH); Pakea, Banks Group (AMS); **Solomon Is:** Santa Cruz I., Vanikoro Is (USNM); Auki, Malaita (BMNH); **Caroline Is:** Babelthuap I., Palau (USNM, ANSP); Ulali I., Truk (BPBM); Kapingamarangi (USNM); Kusaie (BPBM); **Mariana Is:** Tanapag Harbour, Saipan (ANSP); Guam (USNM); **Marshall Is:** Enybor I., Jaluit Atoll (USNM); **Gilbert Is:** Apaiang (MCZ); **Ellice Is:** Vaitupu (USNM); **Wallis Is:** Rotuma I. (BMNH); between Luanna and Fungalei Is (USNM); **Fiji:** Deuba, Viti Levu (AMS); Suva (RNHL, USNM); Lautoka, Viti Levu (MCZ); **Samoa Is:** Upolu I. (ANSP, BPBM); Tutuila I. (ANSP, USNM); **Society Is:** Tahaa (USNM); Raiatea (USNM, ANSP); Moorea

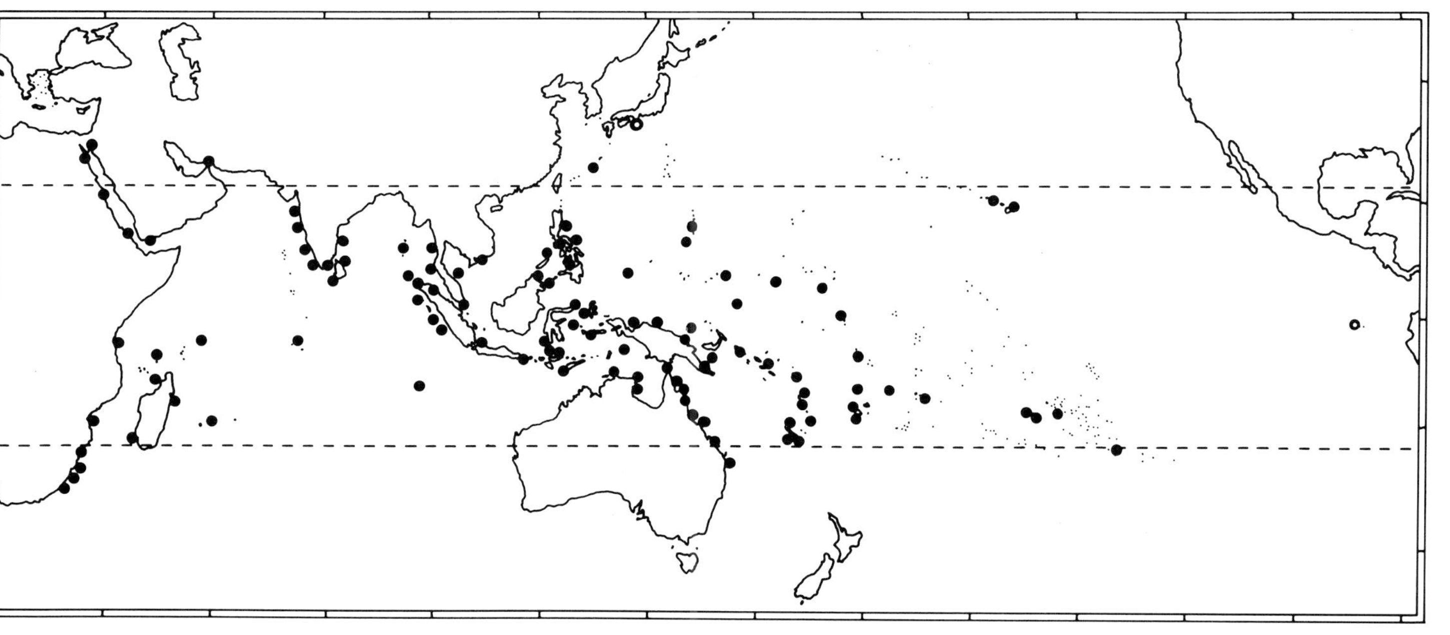

**Fig. 47** Distribution of *Littoraria* (*Littorinopsis*) *intermedia*.

(USNM); Vaiore Bay, Tahiti (USNM); **Tuamotu Is:** Anaa (USNM); **Gambier Is:** Mangareva I. (AMS, WAM); **Hawaiian Is:** Hilo, Hawaii (BPBM); Coconut I., Kaneohe Bay, Oahu (DGR, AMS, USNM); Pearl Harbour, Oahu (NMV, ANSP); **Galapagos Is:** Wreck Bay, Chatham I. (ANSP, doubtful record).

REMARKS. *L. intermedia* is the most widely distributed of the species in the genus which are associated with mangroves. Throughout its range penial shape remains relatively constant and is always distinctive. The penial sperm groove is closed as a duct, a feature found elsewhere in the subgenus only in *L. subvittata* and *L. philippiana*. Sperm nurse cells were found to be similar, and diagnostic of the species, in specimens from North Queensland; Gove, Northern Territory; Phuket Island, south-western Thailand; Singapore; Penang, Malaysia and Oahu, Hawaiian Islands, although some variation was shown in the degree to which the rods project from the cell.

Several rather characteristic geographical forms of the shell can be recognized, but do not deserve taxonomic recognition. Specimens from the Hawaiian Islands and Polynesia (as far west as Samoa) are often of larger size, the grooves narrow and shallow, and the coloration pallid or with continuous axial stripes (Fig. 43c,e,i); the lectotype is of this form. The most characteristic shell type is found in Australia, the western Pacific islands, South East Asia and the Indian Ocean, usually on roots of *Rhizophora* trees. In this type the coloration is dark, the pattern is of regular axial series of black dashes and the primary grooves are conspicuous (Fig. 44c,h,i). Occasional giant forms of this type occur, such as the lectotype of *Litorina scabra* var. *articulata* (Fig. 43g). Specimens from south-western Thailand, Penang and India have rather thin shells, the colour pattern is dark but irregular and the shells are sometimes of large size (Figs 43b, 44e,f). Such shells occur together with the common form found on *Rhizophora* at Phuket Island, south-western Thailand; intermediates can be found and there are no differences in the anatomy of the two forms. In the Persian Gulf, Red Sea and parts of the western Indian Ocean the shells are often small and the secondary sculpture well developed, the shell colour is whitish or pink, and the pigmented pattern pale and indistinct, or even absent (Fig. 43f,h). Similar forms are found in other regions on rocks, and this may be an ecotype rather than a geographical form.

In several parts of the range shells are sometimes found with an orange pink ground colour, usually with a normal pattern of darker pigment. This colour form is especially frequent in Arnhemland, Northern Territory, where it occurs on *Rhizophora* trees and is probably a local genetic form. Although variable in colour, the species is not polymorphic in the descriptive sense employed here, for variation is continuous and the range of colour forms not great. Reference has already been made to variation between habitats. Shells from rocks are often small, pigmentation is pale and pinkish shells are not unusual. Specimens from *Rhizophora* forests are almost invariably darkly pigmented, but on *Avicennia* trunks the shells may be paler in colour.

Kuroda & Habe (1952) recorded the species from a latitude of 37°N on the west coast of Korea, although in the absence of a figure correct identification cannot be certain. Amongst other *Littoraria* species considered in this monograph only *L. luteola* in Australia reaches such a high latitude. The distribution of the species demonstrates its preference for oceanic situations, since it is absent from such sheltered and turbid areas as Bengal, the Gulf of Siam and north-western Borneo. The distribution of *L. intermedia* in Australia deserves comment. The species is frequent on peninsulas and islands on the east coast, but the westernmost record is from the Coburg Peninsula, Northern Territory. This is curious, since the species is abundant elsewhere in the Indian Ocean, and both *L. scabra* and *L. pallescens*, the two other almost equally widely distributed species, are recorded further west in Australia. Conceivably, it may be significant that *L. cingulata* and *L. sulculosa* occur on the north-west coast, and that their ranges do not overlap that of *L. intermedia*. From the little information available it appears that *L. sulculosa* occupies a range of habitats and tidal levels similar to that of *L. intermedia*, but it is not known whether competitive exclusion could occur.

SIMILAR SPECIES. *L. intermedia* has been persistently confused with *L. articulata* in the literature, but can be immediately distinguished from the latter, as from *L. strigata*, by characters of the penis, sperm nurse cells and oviduct. Using shell characters, *L. intermedia* may be separated from these two species by the larger size, darker ground colour and interior, and by the greater number of axial series of dashes on the last whorl. *L. vespacea* is closer in shell pattern, but in *L. intermedia* the whorls are less inflated, the spire taller, and the columella less deeply excavated. Again anatomical characters are distinctive of each species. Small specimens of *L. scabra* are sometimes similar, but *L. intermedia* is distinguished by the purple columella, characteristically enlarged posterior rib adjacent to the suture, sculptured apical whorls of the postlarval shell, absence of spiral microsculpture in the grooves, and by the shape of the penis. In *L. subvittata* the penial form is similar, although the filament is smaller and lacks the prominent basal constriction of *L. intermedia*. However, shells of the species differ, that of *L. subvittata* being thinner, more elongate and of a distinctive pattern. Shells of *L. intermedia* are approached most closely by some uncommon forms of *L. pallescens;* while the pattern of *L. intermedia* is usually more regular, axial alignment of dashes more complete and grooves narrower, a very few specimens are separable only by means of penial characters. Dark forms of *L. lutea* should not be confused with *L. intermedia*, since in the former the arrangement of narrow, intercalated, secondary riblets is diagnostic.

## *Littoraria* (*Littorinopsis*) *subvittata* n. sp.

TYPES. Holotype: BMNH 198346, Aldabra Atoll, Indian Ocean. *Paratypes:* BMNH, AMS, USNM, NM.

ETYMOLOGY. Latin: *sub-*, almost, *vittatus*, longitudinally striped.

? *Litorina intermedia* var. *strigata* Philippi, 1847, vol. 2: 223, *Litorina* pl. 5, fig. 9 [in part; not conspecific with lectotype here designated, p. 209]
*Littorina ahenea*—Sowerby, 1892: 36; Bartsch, 1915: 120; Turton, 1932: 131 [all not Reeve, 1857]
*Littorina borbonica* Barnard, 1951: 100, pl. 13, fig. 13 [*nomen nudum*]
*Littorina* (*Littorinopsis*) *scabra scabra*—Rosewater, 1970: 456–461, pl. 352, figs 20, 21 [in part; not Linnaeus, 1758]; Kilburn, 1972: 403 [not Linnaeus, 1758]

NOMENCLATURE. One of the three shells illustrated by Philippi (1847) as *Litorina intermedia* var. *strigata* resembles this species, although no locality was given. (The designation of a lectotype establishes *L. strigata* (Philippi) as a different species, p. 209). It seems likely that records of '*Littorina ahenea*' from South Africa refer to *L. subvittata*, for Reeve's species was based upon the rather similar colour form of *L. angulifera* with prominent axial stripes on the shell. *L. angulifera* does not occur in South Africa and of the *Littoraria* species found there, only *L. subvittata* bears a resemblance to it. In a popular text book, Barnard (1951) figured *Littorina borbonica* and although only a drawing, there is little doubt that the shell is the species here described. Despite careful search, no other reference to *Littorina borbonica* has been traced. The accompanying text is so brief that the name must be considered a *nomen nudum* until an earlier description is discovered.

DIAGNOSIS. Shell: spire rather tall; peripheral keel usually weak or absent; columella of moderate width, excavated, pinkish brown; primary grooves 10–12; ribs remain low, numbering 40–60 on last whorl; grooves narrow; microsculpture indistinct; colour variable, usually cream with red brown dashes conspicuously aligned to produce axial stripes from suture to base, which number 8–12 on last whorl. Animal: penis bifurcate, filament large and uniformly tapering, sperm groove closed as a duct; ovoviviparous.

SHELL (Fig. 48). *Shape.* Height 13–34 mm. Teleoconch 7–9 whorls. Shell of moderate thickness, solid. Spire tall, outline only slightly convex; whorls lightly rounded, sutures distinct. Peripheral keel usually absent, but periphery accentuated by colour pattern; occasional large shells are keeled, but keel becomes obsolete on last whorl. Adult lip hardly thickened, only slightly or not at all flared. Varices 0–2, occasional. Columella of moderate width, excavated; pillar convex, pinched at base. Sexual dimorphism: males smaller, relatively lower spire and larger aperture, broader shell.

*Dimensions*: Table 15.

*Sculpture* (Fig. 49a–d). Protoconch normal. First teleoconch whorl smooth. Primary grooves 10–12, equally spaced; posterior 1–2 sometimes deeper. Secondary grooves form on whorl 6, at mid

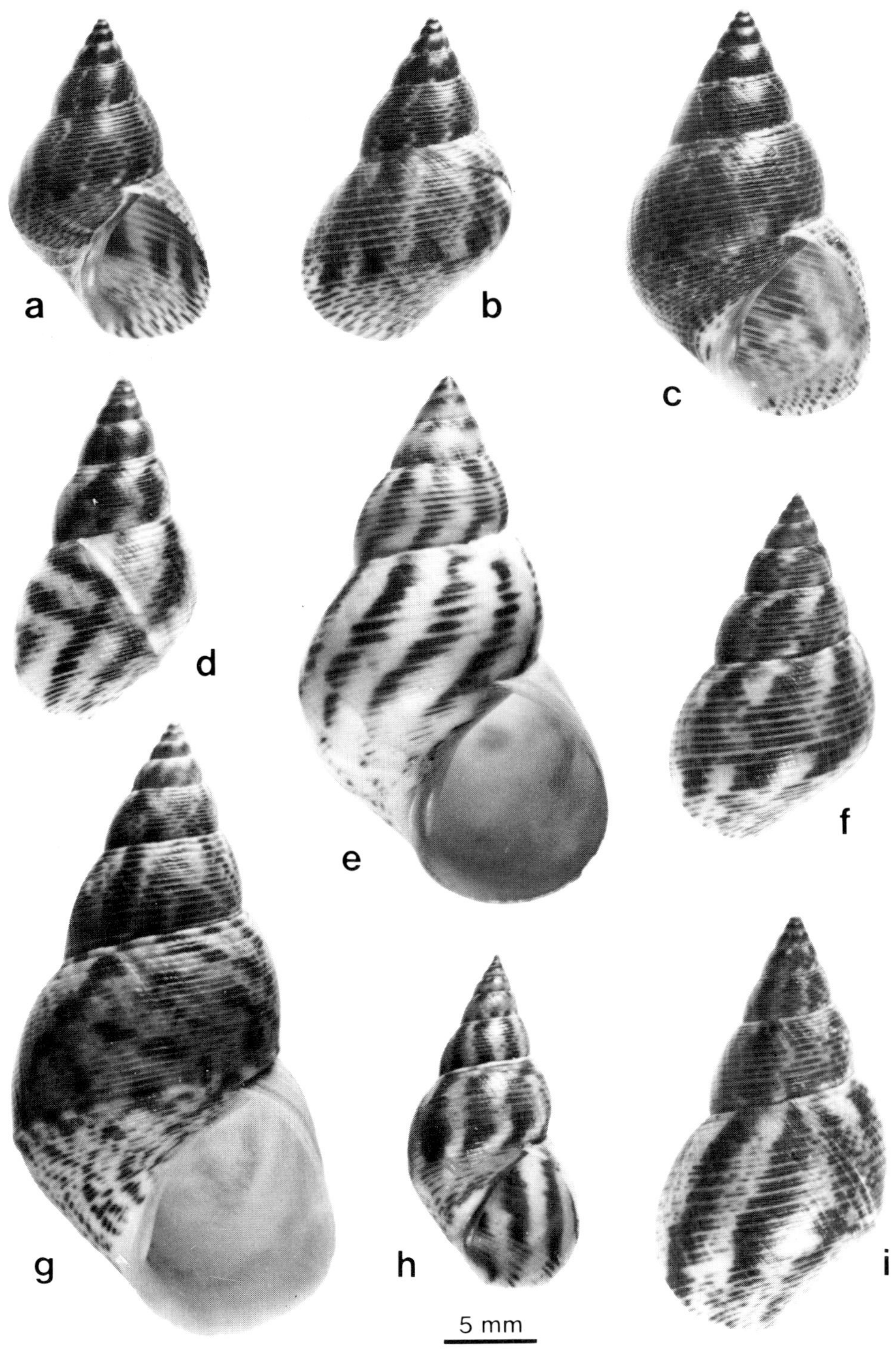


**Fig. 48** *Littoraria* (*Littorinopsis*) *subvittata:* **(a,b)** holotype, ♂, Aldabra (BMNH 198346); **(c,d)** paratypes, Aldabra (BMNH); **(e)** Port Shepstone, S. Africa (NM); **(f)** Lake St Lucia, S. Africa (NM); **(g)** Mauritius (AMS); **(h)** Tuléar, Madagascar (AMS); **(i)** paratype, Aldabra (BMNH).

**Table 15** Dimensions of *Littoraria* (*Littorinopsis*) *subvittata*.

| Specimen | Locality | Sex | Primary grooves | H (mm) | B (mm) | LA (mm) | WA (mm) | C (mm) | P | S | SH |
|---|---|---|---|---|---|---|---|---|---|---|---|
| *L. subvittata* holotype, BMNH 198346 | Aldabra | ♂ | 11 | 16.5 | 11.1 | 9.4 | 7.2 | 1.9 | 1.49 | 0.77 | 1.76 |
| BMNH | Aldabra | ♂ | 10 | 16.7 | 10.6 | 8.5 | 6.5 | 1.0 | 1.58 | 0.76 | 1.96 |
| BMNH | Aldabra | ♂ | 11 | 16.1 | 9.5 | 8.3 | 6.4 | 1.0 | 1.69 | 0.77 | 1.94 |
| BMNH | Aldabra | ♀ | 11 | 22.3 | 12.0 | 10.1 | 7.6 | 1.0 | 1.86 | 0.75 | 2.21 |
| BMNH | Aldabra | ♀ | 10 | 19.2 | 10.8 | 9.0 | 6.5 | 1.0 | 1.78 | 0.72 | 2.13 |
| AMS C.43106A | Mauritius | | 12 | 32.1 | 18.1 | 15.4 | 11.3 | 2.4 | 1.77 | 0.73 | 2.08 |
| AMS C.43106A | Mauritius | | 10 | 24.3 | 14.2 | 12.0 | 8.9 | 1.6 | 1.71 | 0.74 | 2.03 |
| AMS C.100070C | Tuléar, Madagascar | | 11 | 16.9 | 10.0 | 8.4 | 6.4 | 1.2 | 1.69 | 0.76 | 2.01 |
| AMS C.100070C | Tuléar, Madagascar | | 11 | 16.6 | 10.7 | 9.0 | 6.6 | 1.2 | 1.55 | 0.73 | 1.84 |
| AMS C.100070C | Tuléar, Madagascar | | 11 | 13.8 | 8.0 | 6.5 | 4.7 | 0.6 | 1.73 | 0.72 | 2.12 |
| BMNH, mean of 10 | Aldabra | ♂ | | 15.14 | | | | | 1.624 | 0.728 | 1.914 |
| standard error | | | | 0.41 | | | | | 0.021 | 0.011 | 0.018 |
| BMNH, mean of 10 | Aldabra | ♀ | | 18.49 | | | | | 1.762 | 0.737 | 2.119 |
| standard error | | | | 0.76 | | | | | 0.028 | 0.007 | 0.017 |
| statustic t or U | | | | 3.877 | | | | | 90 | 56 | 100 |
| probability | | | | 0.001 | | | | | 0.002 | 0.684 | <.001 |

point of each primary rib; ribs at suture and periphery divide later, and consequently may appear more prominent than others on last two whorls. Ribs flat, of equal width, totalling 40–60 on last whorl; all grooves remain as narrow impressed lines. Sculpture becomes almost obsolete after first major varix. Microsculpture of faint axial growth striae over whole surface.

*Colour*. Variable; ground colour shades of chrome yellow, cream, orange or pale brown, overlain by dark red brown or blackish brown pattern of dashes on ribs. Pattern usually conspicuously aligned to form oblique axial stripes from suture to base, numbering 8–12 on last whorl, although alignment may be less perfect on shoulder and base. Degree of pigmentation variable: dashes may join to form continuous spiral lines; axial stripes may be narrow, and faint or absent below periphery, but shells never lack pigment entirely. Pattern frequently reduced or absent after first varix. External pattern visible within aperture. Columella usually pinkish brown, seldom white.

ANIMAL. *Colour*. Pigmentation grey to black; sides of foot pale mottled to dark grey; head darkest, sometimes with pale central streak; tentacles banded, unpigmented stripe each side of base.

*Penis* (Fig. 50a–g). Length to 7.5 mm. Base bifurcate; limb bearing glandular disc short; disc rather irregular. Filament long, uniformly tapering. Sperm groove closed as a duct. Base and filament off white or cream, glandular disc cream to pale brown.

*Sperm*. Nurse cells (Fig. 50l,m) 8–10(15) $\mu$m; rounded to oval; no rods visible; yolk granules large.

*Pallial oviduct* (Fig. 50h–k). Length to 3.0 mm. Spiral section to 1.9 mm diam., $2\frac{1}{2}$ whorls; opaque albumen gland $\frac{1}{3}$ whorl, not visible externally, white; translucent albumen gland off white; capsule glands absent; spiral distinct externally, especially if egg groove is pigmented. Straight section to 1.3 mm, pale; terminating in a papilla. Bursa to 1.1 mm, anterior. Development ovoviviparous.

*Radula* (Fig. 49f). Length to 19 mm; relative length 0.82–0.89. Saw-toothed type; central rachidian cusp rather elongate and pointed; cusps of paired teeth almost equilaterally triangular; lateral with gap anterior to main cusp.

DISTRIBUTION. *Habitat*. Mangroves, marsh grass, wooden pilings and sheltered rocks. Probably an oceanic species.

*Range* (Fig. 51). Indian Ocean, from Aden to eastern Cape Province, South Africa, and east to Mauritius, Maldive Islands and Cocos-Keeling Islands.

*Records*. **South Africa:** Zwartkops R. mouth, Algoa Bay (NM); Port Alfred (BMNH, AMS, MCZ); Gonubie R. mouth, East London (ANSP); Umtata R. mouth (NM); Port St Johns (DGR); Nxaxo R.

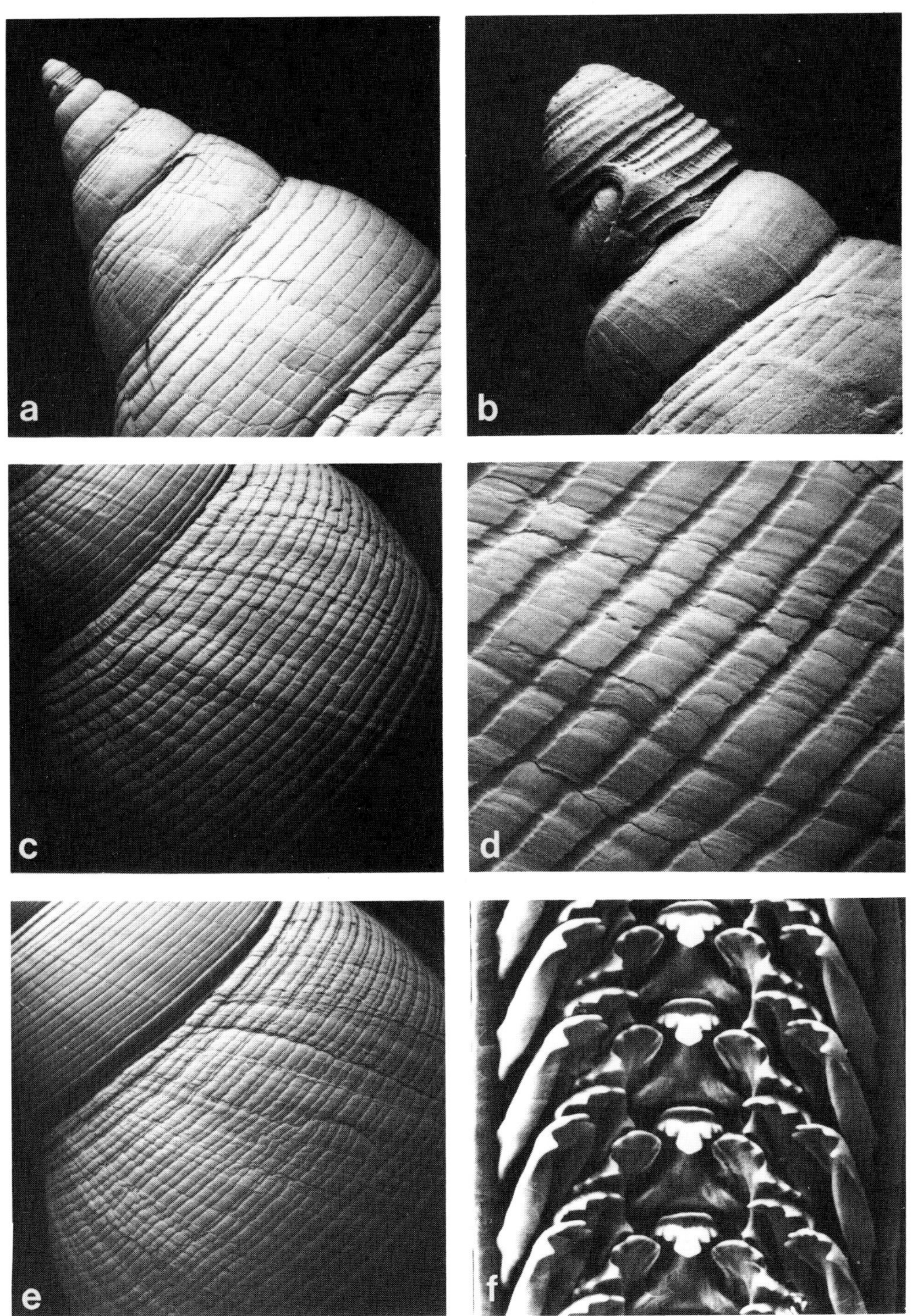


**Fig. 49** **(a–d)** *Littoraria* (*Littorinopsis*) *subvittata*, Aldabra: **(a)** spire (× 18); **(b)** protoconch (× 80); **(c)** last whorl (× 8); **(d)** detail (× 34). **(e)** *Littoraria* (*Littorinopsis*) *angulifera*, Sanibel I., Florida, last whorl (× 7). **(f)** *Littoraria* (*Littorinopsis*) *subvittata*, Aldabra, radula (× 250).

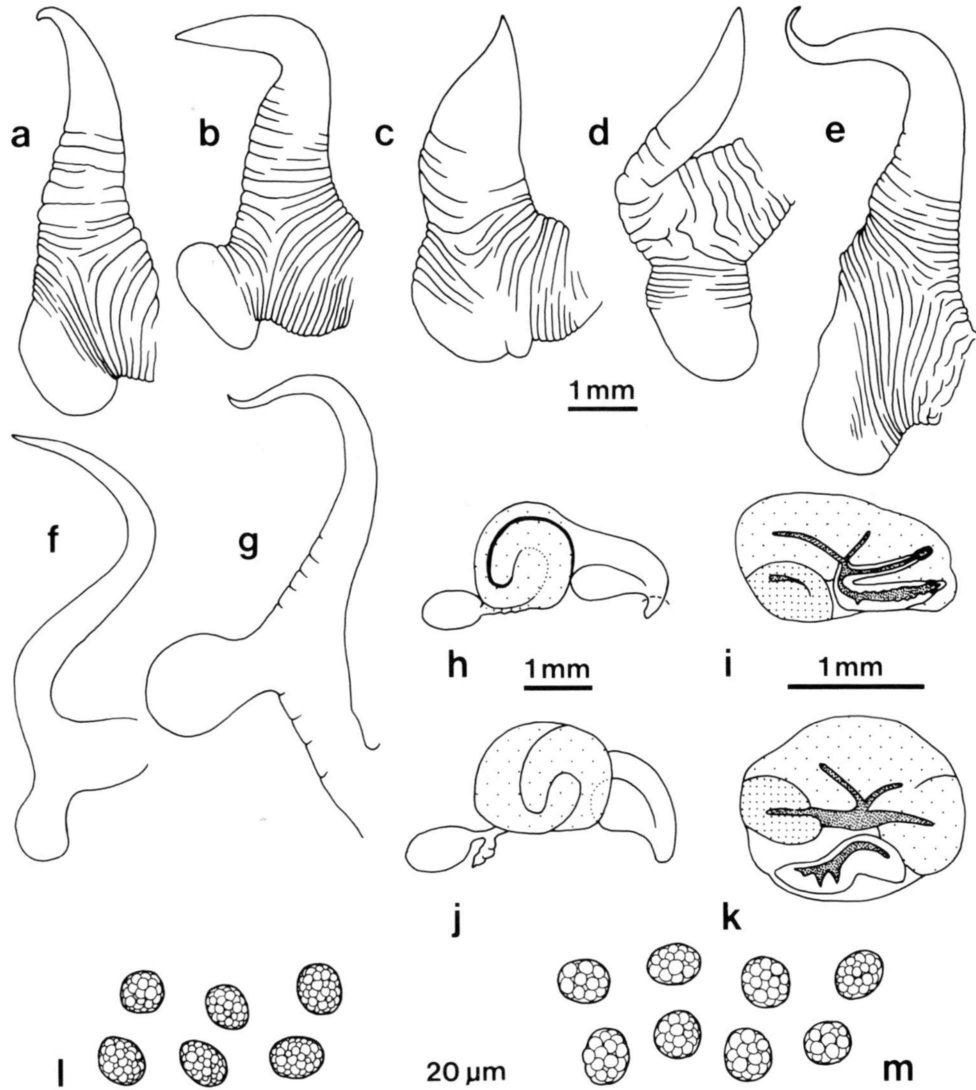


**Fig. 50** *Littoraria* (*Littorinopsis*) *subvittata*: **(a–g)** penes; **(a–d)** Aldabra; **(e)** Buffalo Harbour, S. Africa; **(f,g)** Nxaxo R., S. Africa; **(h-k)** pallial oviducts, with transverse sections, Aldabra; **(l,m)** sperm nurse cells, Aldabra.

mouth (ANSP); Port Shepstone (NM); Durban Bay (BMNH, WAM, NM, RNHL); St Lucia, Zululand (NMW, NM); **Mozambique:** Lourenço Marques (DGR); Inhaca I., Delagoa Bay (NM); Lunga Bay to Memba Bay (NM); Ilha do Ibo (AMS); **Tanzania:** Bagamoyo (MCZ); Chukwani, Zanzibar (ANSP); Tanga (BMNH, MCZ); **Kenya:** Gazi (NMW); Mombasa (BMNH); Mkunumbi (BMNH); **Somalia:** Mogadisho (MCZ); **Yemen:** Aden (BMNH); **Madagascar:** Tuléar (AMS, MCZ); Ile Ste Marie (MCZ); Nossi Bé (ANSP); **Aldabra Atoll** (BMNH); **Seychelles:** Menai I., Cosmoledo Atoll (USNM); St Joseph I., Amirante Is (BMNH); **Mauritius** (BMNH, AMS, MCZ); Camisard R. mouth (ANSP); **Maldive Is:** Male Harbour (ANSP); **Cocos-Keeling Is:** SE. side of North Lagoon, West I. (ANSP).

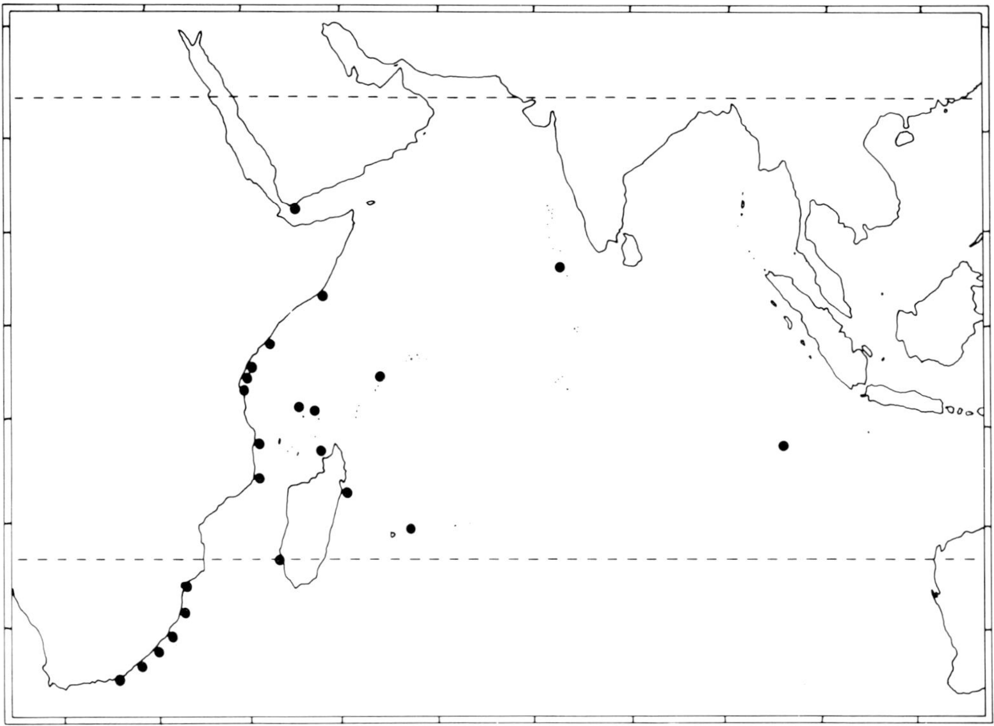

**Fig. 51** Distribution of *Littoraria* (*Littorinopsis*) *subvittata*.

REMARKS. It is surprising that such a common, widely distributed and fairly distinctive species should have remained unrecognized. It is sympatric with *L. scabra*, *L. intermedia* and *L. pallescens* over most of its range, and has perhaps been confused with *L. scabra* in the early faunistic lists for this region (e.g. von Martens, 1880; Smith, 1906; Dautzenberg, 1923, 1929; Viader, 1937). *L. subvittata* is in most respects rather constant over its range, although size varies considerably. The sperm nurse cells illustrated (Fig. 50l,m) are unusually small for the genus; these were drawn from material preserved in formalin and it is not known if they are typical.

SIMILAR SPECIES. Large specimens of *L. subvittata* can be confused with *L. scabra*, but are distinguished by their higher spire, more numerous ribs on the last whorl, narrower grooves which lack microsculpture, brown columella and by penial characters. Most specimens of *L. intermedia* have fewer ribs on the last whorl, but rarely east African shells may approach small specimens of *L. subvittata* and penial characters must then be used for separation. The excavated columella and closed penial groove distinguish *L. subvittata* from *L. ardouiniana*. *L. angulifera* from the tropical Atlantic (Figs 4o, 49e, 99d–f) shows more numerous ribs on the last whorl, a wide and usually lilac columella, lower spire and a penis with an open sperm groove and larger filament.

## *Littoraria* (*Littorinopsis*) *filosa* (Sowerby, 1832)

*Littorina filosa* Sowerby, 1832: *Littorina* fig. 5 ['S. America' in error, here corrected to E. coast of Queensland, Australia; lectotype figure here designated]; Mitchell, 1838: pl. 2, fig. 5; Reeve, 1857: *Littorina* pl. 5, figs 24 a–c; Nevill, 1885: 148 [in part]

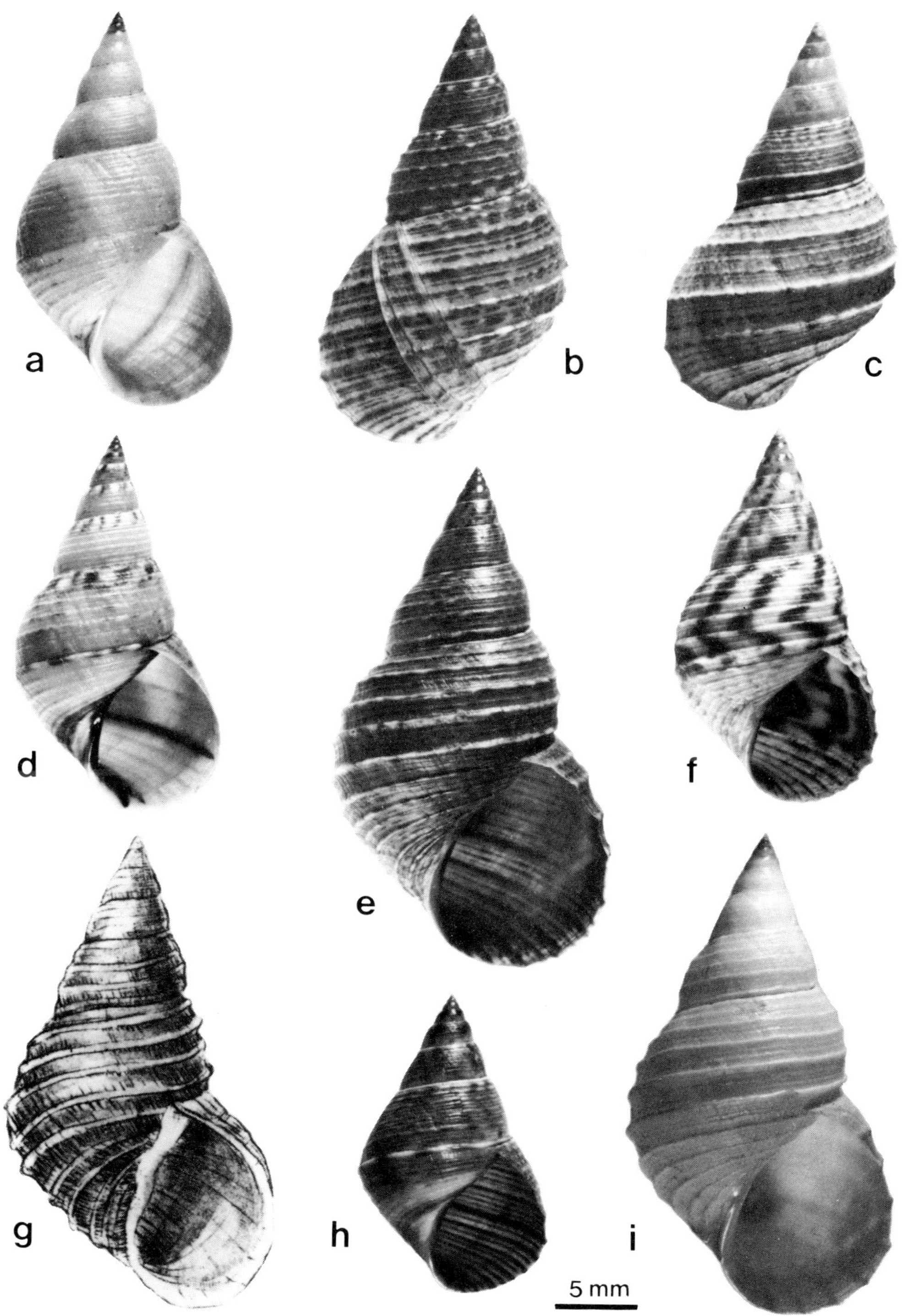


**Fig. 52** *Littoraria* (*Littorinopsis*) *filosa:* **(a)** ♀, Broome, W. A. (DGR); **(b)** ♂, Thursday I., Qld. (DGR); **(c)** Weipa, Qld. (BMNH); **(d)** ♂, Broome, W. A. (DGR); **(e)** ♀, Gove, N. T. (DGR); **(f)** ♀, Magnetic I., Qld. (DGR); **(g)** lectotype figure of *Littorina filosa* Sowerby, locality unknown (Sowerby, 1832, *Littorina* fig. 5); **(h)** ♀, Moa I., Qld. (DGR); **(i)** ♀, Gove, N. T. (DGR).

*Litorina filosa*—Philippi, 1848, vol. 3: 61, *Litorina* pl. 7, figs 1,2 [not pl. 6, fig. 4, corrected to *Litorina pallescens* in index]; Weinkauff, 1882: 57–58, pl. 7, figs 10,11
*Littorina (Melaraphe) scabra* var. *filosa*—Tryon, 1887: 244, pl. 43, figs 31, 31a [in part]
*Littorina (Littorinopsis) carinifera*—von Martens, 1897: 198 [in part; not Menke, 1830]
*Littorina (Littoraria) luteola*—Shikama & Horikoshi, 1963: pl. 16, fig. 7 [not Quoy & Gaimard, 1832]
*Littorina (Littorinopsis) scabra scabra*—Rosewater, 1970: 456–461, pl. 325, fig. 3, pl. 352, figs 22, 23, 33, 34 [in part; not Linnaeus, 1758]
*Littorina scabra*—Wilson & Gillett, 1979: 52, pl. 8, fig. 5c [in part; not Linnaeus, 1758]

NOMENCLATURE. The name '*filosa*' is not uncommon in the older literature, but does not seem to have been correctly applied since 1882. Since then it has sometimes been used for other brightly coloured or carinate species, especially *L. pallescens*. Sowerby's figure leaves no doubt as to the identity of the species, for the elongate outline and strong carinae are unique in the family. The locality was erroneously given as 'S. America'. However, Mitchell (1838: 15) reported that a collection of 'fossil shells from the basin of the Hunter, &c' made in 1831 was sent to Sowerby for description, and listed '*Turbo filosa* Sowerby (new species)', while figuring an apparently recent shell as '*Littorina filosa*'. The specimens were stated to have been deposited in the Australian Museum, Sydney. It seems likely to that this was the origin of Sowerby's shell, but no type specimen could be located in the AMS or BMNH. The locality given by Mitchell seems to apply to the small collection of fossils; it is unlikely that the shell could have originated near the Hunter River (Newcastle, N.S.W.), since only one specimen has been recorded from the state. The type locality has therefore been corrected to eastern Queensland.

DIAGNOSIS. Shell: thin; columella narrow, rounded; primary grooves 5–7(9), spacing markedly unequal, increasing anteriorly; last whorl with 9–11 strong, narrow carinae, the spaces between with numerous irregularly spaced spiral grooves; colour highly polymorphic, brown, yellow or orange pink, with or without pattern of brown dashes aligned at suture to form short stripes, numbering 10–12 on last whorl. Animal: penis bifurcate, filament large, narrowing only at tip; ovoviviparous.

SHELL (Frontispiece, Fig. 52). *Shape*. Height 16–35 mm. Teleoconch 7–8 whorls. Shell thin, translucent. Spire outline only slightly convex; whorls rounded; sutures distinct. Peripheral keel conspicuous until whorl 7, last whorl more uniformly rounded, but strongly carinate. Adult lip thin, slightly flared. Varices 0–2(4), not common. Columella narrow, rounded, not excavated; pillar straight to slightly convex. Sexual dimorphism: not significant; males tend to be a little smaller, adult lip is more strongly flared and varices correspondingly more frequent.

*Dimensions*: Table 16.

*Sculpture* (Fig. 53c,e,f). Protoconch normal. First 2 whorls of teleoconch smooth. Primary grooves 5–7; spacing unequal, showing 3- or 4-fold increase from suture to periphery; primary grooves remain as narrow, incised lines. Secondary sculpture develops on whorls 5–6, grooves appearing at anterior $\frac{1}{4}$–$\frac{1}{2}$ rib width. Tertiary and higher orders of grooves develop in same manner on primary and secondary ribs of whorls 6 and later. On whorls 7–8 each primary rib becomes a strong, narrow carina, numbering 9–11 in total; peripheral carina most prominent; spacing between carinae up to 5 times carina width, widest near periphery. Secondary ribs do not usually become carinate. Microsculpture: carinae smooth; spaces between with 2–4 flat secondary and tertiary ribs, between which run irregular fine spiral striae. Entire surface between carinae crossed by fine axial growth striae, which may give a microscopically decussate or pitted pattern where they cross the spiral microsculpture. Specimens from northern Western Australia, and some from the Lesser Sunda Is, differ from the foregoing description as follows: primary grooves 7–9, spacing subequal, only a 2-fold increase towards the periphery; total number of primary ribs on last whorl 12–15, of which usually only that at periphery is sufficiently developed to be termed carinate.

*Colour* (Frontispiece). Polymorphic; yellow and shades of brown most frequent, pink shells not uncommon. Ground colour chrome yellow, brown or orange pink. Apical 3 whorls with dark brown suture. Pattern dark brown, fading to lilac grey in large, old shells. Shells with sparse pattern: brown dashes confined to primary and secondary ribs, axially aligned near suture to form short stripes,

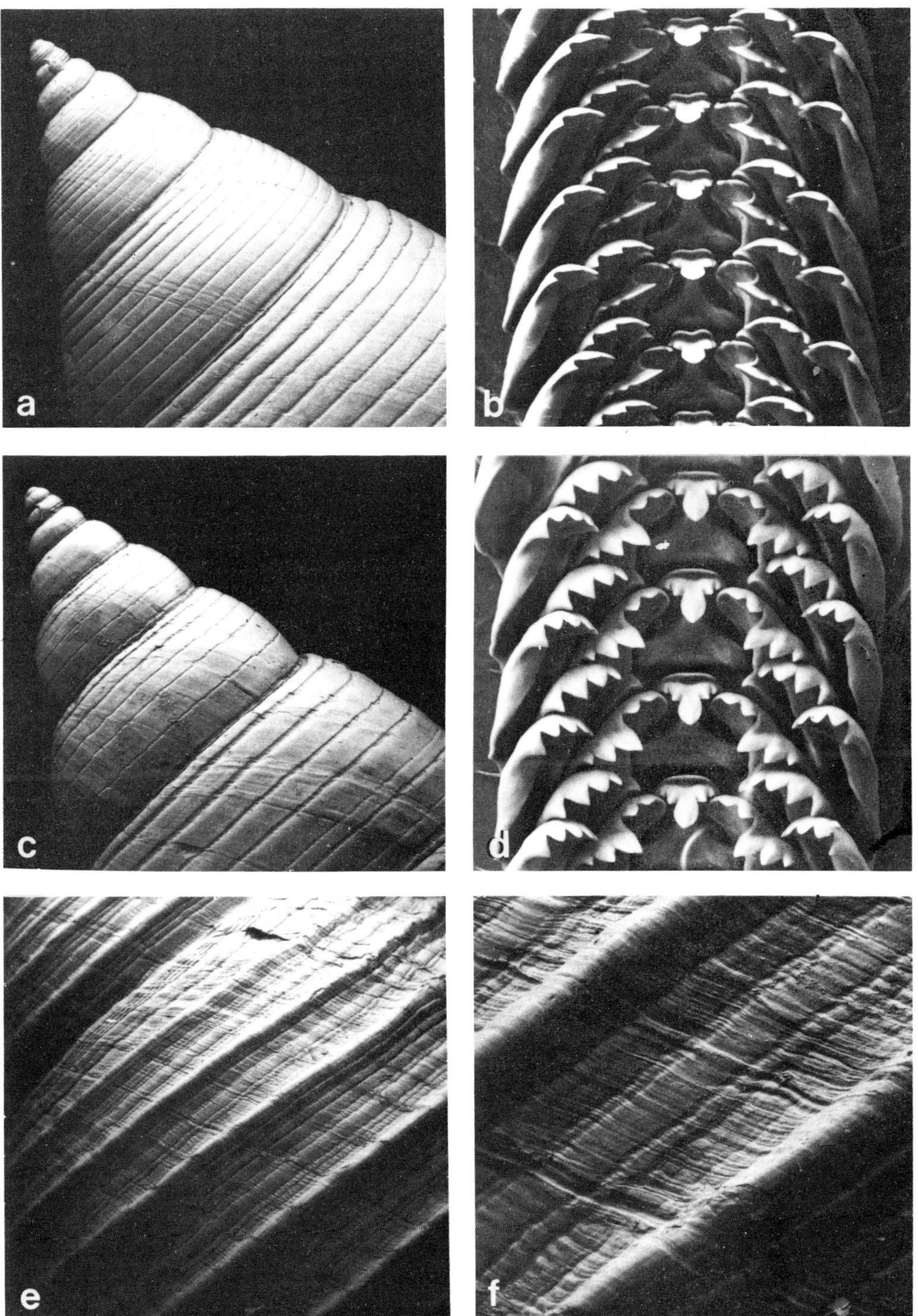


**Fig. 53** **(a,b)** *Littoraria* (*Littorinopsis*) *pallescens:* **(a)** spire ( × 17), Phuket I., Thailand; **(b)** radula ( × 200), Gove, N. T. **(c–f)** *Littoraria* (*Littorinopsis*) *filosa*, Magnetic I., Qld.: **(c)** spire ( × 20); **(d)** radula ( × 230); **(e)** last whorl ( × 8); **(f)** detail ( × 32).

**Table 16** Dimensions of *Littoraria (Littorinopsis) filosa*.

| Specimen | Locality | Sex | Primary grooves | H (mm) | B (mm) | LA (mm) | WA (mm) | C (mm) | P | S | SH |
|---|---|---|---|---|---|---|---|---|---|---|---|
| *Littorina filosa* lectotype figure (Sowerby, 1832) | [eastern Qld.] | | | 27.8 | 16.5 | 12.0 | 9.0 | | 1.68 | 0.75 | 2.32 |
| DGR | Thursday I., Qld. | ♂ | 7 | 29.0 | 18.0 | 15.9 | 11.7 | 1.6 | 1.61 | 0.74 | 1.82 |
| DGR | Thursday I., Qld. | ♀ | 5 | 34.6 | 21.9 | 17.5 | 13.0 | 1.9 | 1.58 | 0.74 | 1.98 |
| DGR | Magnetic I., Qld. | ♂ | 6 | 23.6 | 13.8 | 11.9 | 8.8 | 1.0 | 1.71 | 0.74 | 1.98 |
| DGR | Magnetic I., Qld. | ♀ | 6 | 26.5 | 16.0 | 13.1 | 10.3 | 1.3 | 1.66 | 0.79 | 2.02 |
| DGR | Darwin, N. T. | ♂ | 5 | 24.2 | 14.9 | 12.8 | 9.9 | 1.2 | 1.62 | 0.77 | 1.89 |
| DGR | Darwin, N. T. | ♀ | 5 | 24.1 | 16.0 | 12.8 | 9.2 | 1.2 | 1.51 | 0.72 | 1.88 |
| DGR | Broome, W. A. | ♂ | 8 | 23.3 | 13.8 | 12.6 | 9.3 | 1.4 | 1.69 | 0.74 | 1.85 |
| DGR | Broome, W. A. | ♀ | 8 | 27.5 | 15.7 | 13.3 | 9.5 | 1.4 | 1.75 | 0.71 | 2.07 |
| DGR, mean of 10 | Darwin, N. T. | ♂ | | 24.60 | | | | | 1.614 | 0.762 | 1.939 |
| standard error | | | | 0.37 | | | | | 0.016 | 0.007 | 0.013 |
| DGR, mean of 10 | Darwin, N T. | ♀ | | 25.20 | | | | | 1.580 | 0.759 | 1.978 |
| standard error | | | | 0.36 | | | | | 0.015 | 0.007 | 0.023 |
| statistic t or U | | | | 1.167 | | | | | 69 | 54 | 64.5 |
| probability | | | | 0.259 | | | | | 0.166 | 0.796 | 0.297 |

numbering 10–12 on last whorl; base pale but for dark band around columella. Dark shells: entire surface may be suffused or marbled with brown, occasionally alignment of dashes is complete, producing oblique stripes from suture to base. Pink and yellow shells commonly lack brown patterning and may fade almost to white. In all shells carinae are composed of more opaque material, appearing paler than rest of surface. Brown pattern visible within aperture. Columella white in shells lacking pattern, otherwise purple or pinkish brown. Shells from northern Western Australia frequently show dark pigment only near sutures, on peripheral keel, and in a band around the columella (a pattern rare elsewhere), columella and parietal callus almost always very dark purple brown, spiral white stripe at base of pillar, outlined by narrow purple band within aperture; only in those few yellow or pink shells lacking dark pigment are the columella and callus white.

ANIMAL. *Colour*. Pigmentation correlated with shell colour, from unpigmented to dark grey or black; sides of foot may be pale grey mottled; head darkest; tentacles may be banded, white stripe on each side of base. Unpigmented animals cream, red buccal mass.

*Penis* (Fig. 54a–f). Length to 12 mm. Base bifurcate; limb bearing glandular disc not long. Filament large, narrowing only near slightly mucronate tip. Sperm groove open. Base and filament off white; glandular disc opaque white or cream.

*Sperm*. Eupyrene sperm 100–120 $\mu$m. Nurse cells (Fig. 54j–l) 12–20 $\mu$m; oval; rods 1(2), central, not or scarcely projecting, parallel sided, ends rounded; yolk granules moderately large. (Rarely, abnormal individuals may have rounded nurse cells lacking rods, or with small, rounded rod pieces).

*Pallial oviduct* (Fig. 54g–i). Length to 5.3 mm. Spiral section to 2.4 mm diam., $3\frac{1}{2}$ whorls; opaque albumen gland $\frac{1}{2}$–$\frac{2}{3}$ whorl, white or cream; translucent albumen gland off white to pale brown; capsule glands absent; spiral indistinct externally; egg groove not usually darkly pigmented. Straight section relatively long, to 3.6 mm; off white to fawn; terminating in a papilla. Bursa long, to 2.6 mm, anterior. Development ovoviviparous.

*Radula* (Fig. 53d). Length to 17 mm; relative length 0.58–0.98. Saw-toothed type; central rachidian cusp elongate and pointed; cusps of paired teeth equilaterally triangular; lateral with slight gap aterior to main cusp.

DISTRIBUTION. *Habitat*. *Avicennia* fringe, especially on leaves, but also on trunks, 0.2–3.0 m above ground; usually scarce on leaves and trunks in *Rhizophora* forest; occasional in *Ceriops* zone; also on *Aegialitis*. A continental species.

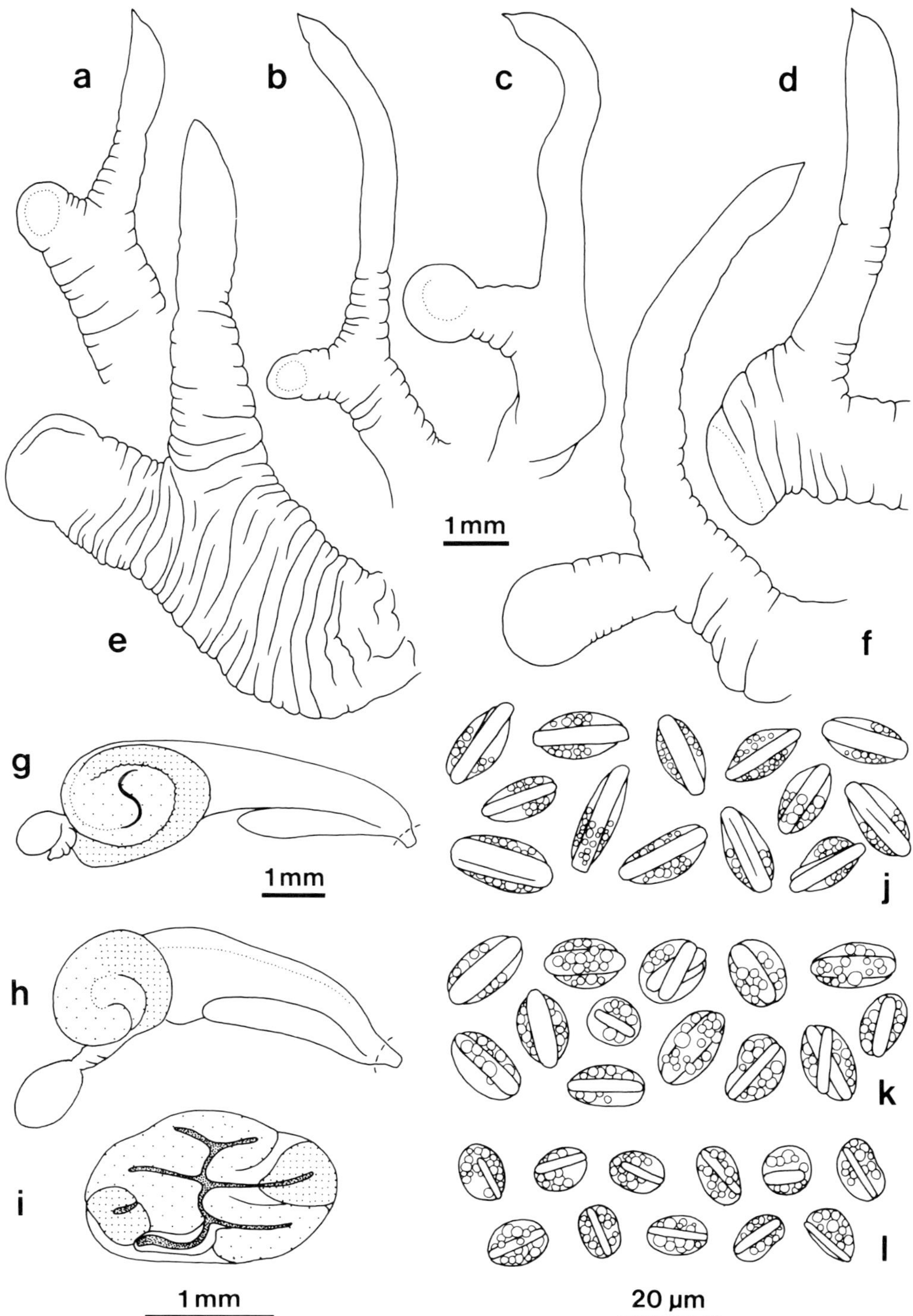


**Fig. 54** *Littoraria* (*Littorinopsis*) *filosa:* **(a–f)** penes; **(a)** Broome, W. A.; **(b–d)** Magnetic I., Qld.; **(e)** Broome, W, A,; **(f)** Magnetic I., Qld.; **(g–i)** pallial oviducts, with transverse section; **(g)** Darwin, N. T.; **(h,i)** Gove, N. T.; **(j–l)** sperm nurse cells; **(j)** Broome, W. A.; **(k,l)** Magnetic I., Qld.

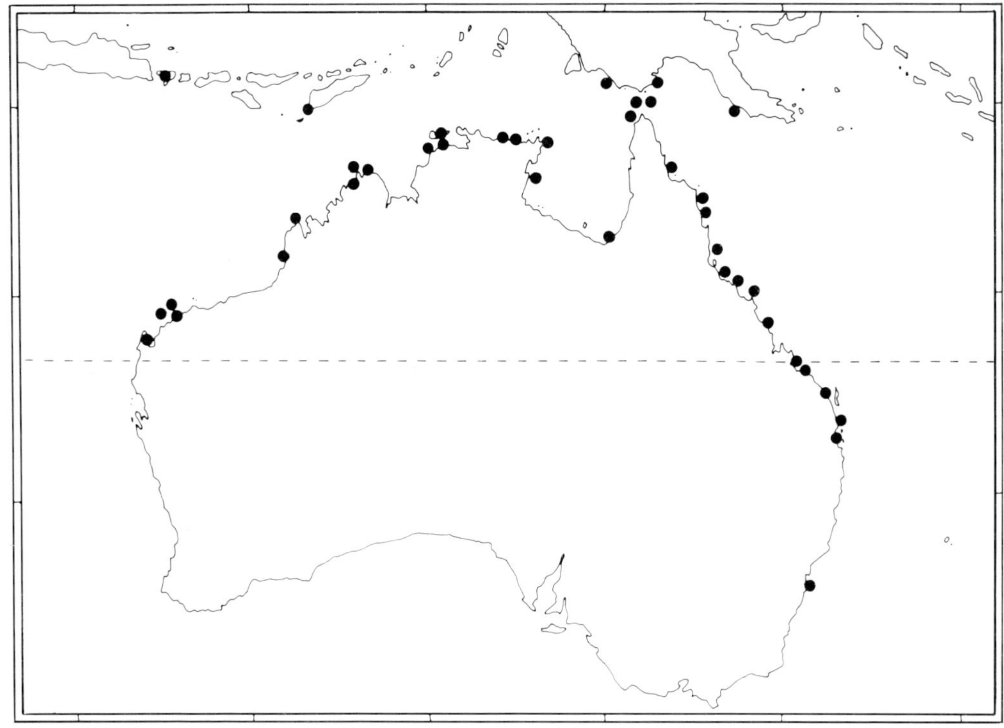

**Fig. 55** Distribution of *Littoraria* (*Littorinopsis*) *filosa*.

*Range* (Fig. 55). From Botany Bay, New South Wales, around east and north coasts of Australia, to Exmouth Gulf, Western Australia; south coast Papua New Guinea; Lesser Sunda Islands (Nineteenth Century material from Philippines and Solomon Islands almost certainly incorrectly localized.)

*Records*. **Indonesia: Lesser Sunda Is:** Bali (AMS); Bolang, SW. Timor (RNHL); **Irian Jaya:** Merauke (AMS, RNHL); **Papua New Guinea:** Oriomo R. (MCZ); Port Moresby (AMS); **Australia: W.A.:** Bay of Rest, Exmouth Gulf (WAM); Barrow I. (WAM); Monte Bello Is (WAM); 10 km W. of Cape Preston (WAM); Broome (DGR, AMS, WAM, NMV); Cape Leveque (WAM, NMV); Port Warrender, Admiralty Gulf (WAM); Troughton I. (BMNH); Vansittart Bay (AMS); Napier Broome Bay (WAM); **N.T.:** Grose I. (AMS); Ludmilla Creek, Darwin (DGR); Cape Gambier, Melville I. (AMS); Boucaut Bay (AMS); Glyde R., Arnhemland (AMS); Nhulunbuy, Gove Pen. (DGR); Groote Eylandt (AMS); **Qld:** Karumba (AMS); Thursday I. (DGR, QM); Moa I. (DGR); Murray Is (AMS); Claremont Is (QM); Cooktown (AMS); Port Douglas (AMS); Dunk I. (AMS, QM); Magnetic I. (DGR); Bowen (AMS, NMV); Hayman I. (AMS); Sarina (AMS); Yeppoon (AMS, NMV); Port Curtis (AMS); Pialba (QM); Maroochydore (AMS); Wellington Point, Moreton Bay (QM); **N.S.W.:** Kurnell Pen. (DGR).

Remarks. Of all *Littoraria* species, *L. filosa* exhibits the most striking colour polymorphism. The proportions of colour morphs show some variation both with habitat and geographical area, but all populations are highly polymorphic. There is little variation in shell shape on the east coast of Australia, but in the Torres Strait, Northern Territory and Papua New Guinea the shells are of large size, rather broad and solid. In this region many of the yellow and pink shells are speckled with brown pigment, whereas entirely unpigmented yellow and pink shells are common elsewhere. The most characteristic geographical form (Fig. 52a,d) occurs from Cape Leveque to Exmouth Gulf, Western Australia, and also in Timor, and is described above. Collections from the Admiralty Gulf, Western Australia, and further north and east, are normal. The differences in sculpture and coloration are not considered sufficient to warrant taxonomic recognition, since anatomical features remain constant, and occasional carinate shells occur in this region of north-western Australia, while rather smooth shells are occasionally found elsewhere.

From Moreton Bay southwards *L. fillosa* appears to be replaced as the dominant species on *Avicennia* leaves by *L. luteola*, while to the north of Australia *L. pallescens* takes over the habitat.

SIMILAR SPECIES. The typical carinate form may be confused with *L. philippiana* and *L. pallescens*, but is distinguished by its narrow, rounded columella and by the smaller number and unequal spacing of the primary ribs. *L. filosa* and *L. cingulata* are separated by the form of the primary, secondary and microsculpture, by colour and by penial characters. *L. luteola* is superficially similar, but again sculptural details are diagnostic

## *Littoraria (Littorinopsis) cingulata cingulata* (Philippi, 1846)

*Littorina cingulata* Philippi, 1846: 142 [north coast of Australia; lectotype (Rosewater, 1970) BMNH 1968352]; Reeve, 1857: *Littorina* pl. 6, fig. 25
*Litorina cingulata*—Philippi, 1847, vol. 3: 11–12, *Litorina* pl. 6, fig. 5; Weinkauff, 1882: 58, pl. 7, fig. 13
*Littorina filosa*—Nevill, 1885: 148 [in part; not Sowerby, 1832]
*Littorina (Melaraphe) scabra* var. *filosa*—Tryon, 1887: 244, pl. 42, fig. 29 [in part; not Sowerby, 1832]
*Littorina (Littorinopsis) scabra scabra*—Rosewater, 1970: 456–461, pl. 352, figs 7, 26, 27, 30 [in part; not Linnaeus, 1758]

NOMENCLATURE. The species was described and later figured by Philippi from specimens in the Cuming Collection; the lectotype designated by Rosewater (1970) is the figured specimen.

DIAGNOSIS. Shell: columella rounded, usually purplish; primary grooves 5–6, primary ribs rounded and prominent, 11–13 on last whorl, separated by grooves $\frac{1}{2}$–1 times rib width; secondary sculpture of minor riblets on last whorl only; microsculpture of regular spiral striae in grooves; colour cream with marbled pattern of orange brown on ribs, apical whorls often blue grey. Animal: penis bifurcate, base thick, filament fairly short and tapering; ovoviviparous.

SHELL (Fig. 56). *Shape*. Height 13–25 mm. Teleoconch 7.5–8.5 whorls. Shell thick, solid. Spire outline straight; whorls rounded; sutures impressed. Peripheral keel marked by largest rib in young shells, becoming less conspicuous on last whorl. Adult lip slightly thickened and flared. Varices unusual, 0–2. Columella of moderate width, rounded; sometimes very slightly excavated in largest shells; pillar gently concave. Sexual dimorphism: males smaller, with relatively lower spire and larger aperture, adult lip a little more flared.

*Dimensions*: Table 17.

*Sculpture* (Fig. 57a,c,d). Protoconch normal. First 2 whorls of teleoconch smooth. Primary grooves 5 6(7); spacing almost equal, posterior 1 or 2 ribs may be a little narrower. Secondary sculpture confined to whorl 8, where 1–3 riblets develop in each primary groove, and primary ribs may become divided by single narrow grooves. Primary ribs rounded and prominent, largest at periphery, 11–13 in total on last whorl, spacing between is $\frac{1}{2}$–1 times rib width. Secondary ribs and grooves remain narrow and inconspicuous. Microsculpture: primary ribs smooth but for faint axial growth striae; regular spiral striae confined to grooves.

*Colour*. Rather constant. Ground colour whitish, cream or blue grey. First 6 whorls often conspicuously blue grey, with ribs whitish, especially at sutures. Irregular marbled pattern of orange brown or dark brown confined to ribs; pattern sometimes faint and ribs unicolorous orange brown; sometimes aligned to form axial stripes, especially on spire. Aperture with purple brown lines or dashes corresponding to ribs of body whorl, darkest at margin, glazed with whitish, pale pink or cream

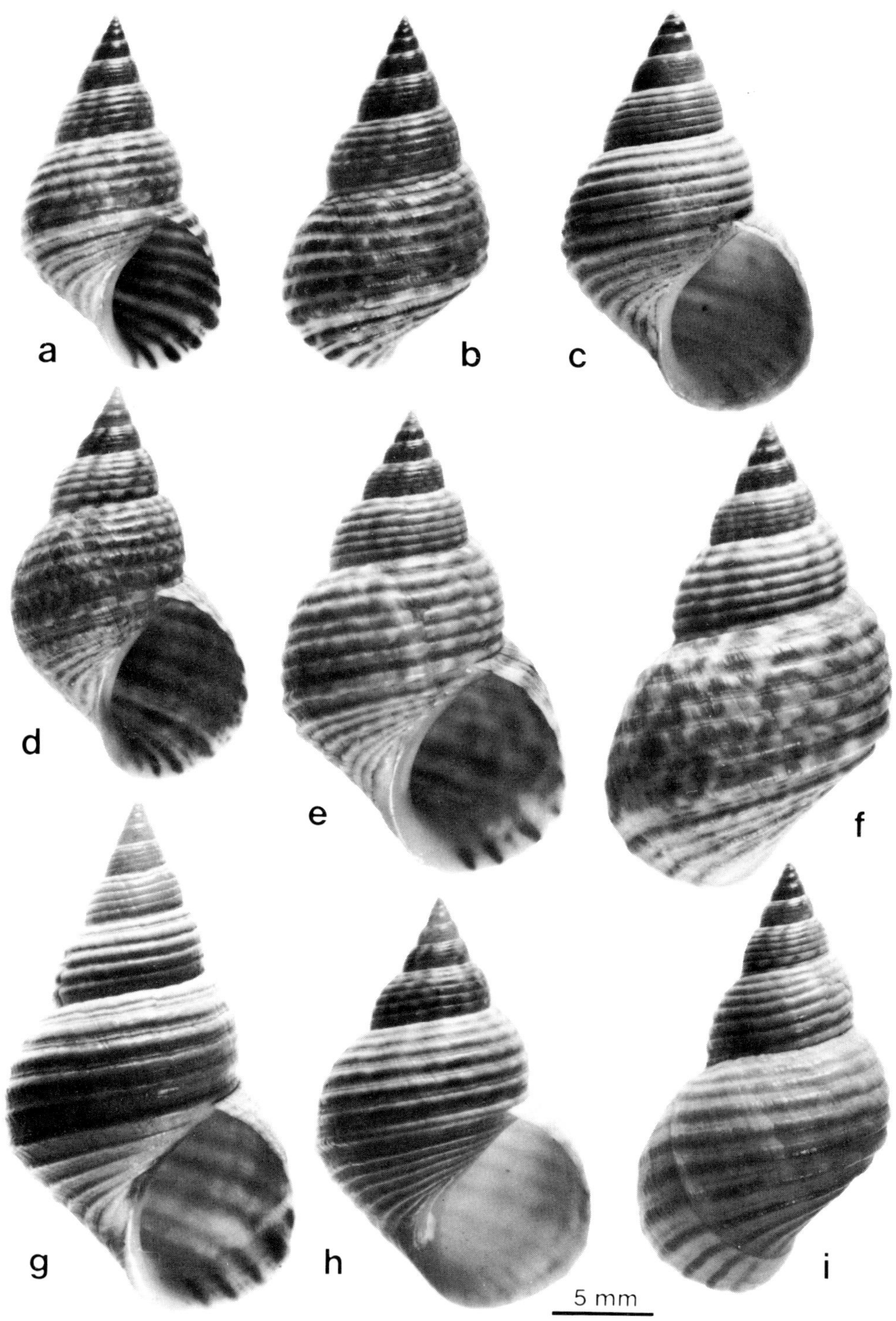


**Fig. 56** *Littoraria* (*Littorinopsis*) *cingulata cingulata:* **(a,b)** ♂, Broome, W. A. (DGR); **(c)** lectotype of *Littorina cingulata* Philippi, N. Australia (BMNH 1968352); **(d–f)** Broome, W. A. (DGR); **(d)** ♂; **(e,f)** ♀; **(g)** Cape Leveque, W. A. (NMV); **(h)** Broome, W. A. (NMV); **(i)** Sunday I., W.A. (WAM).

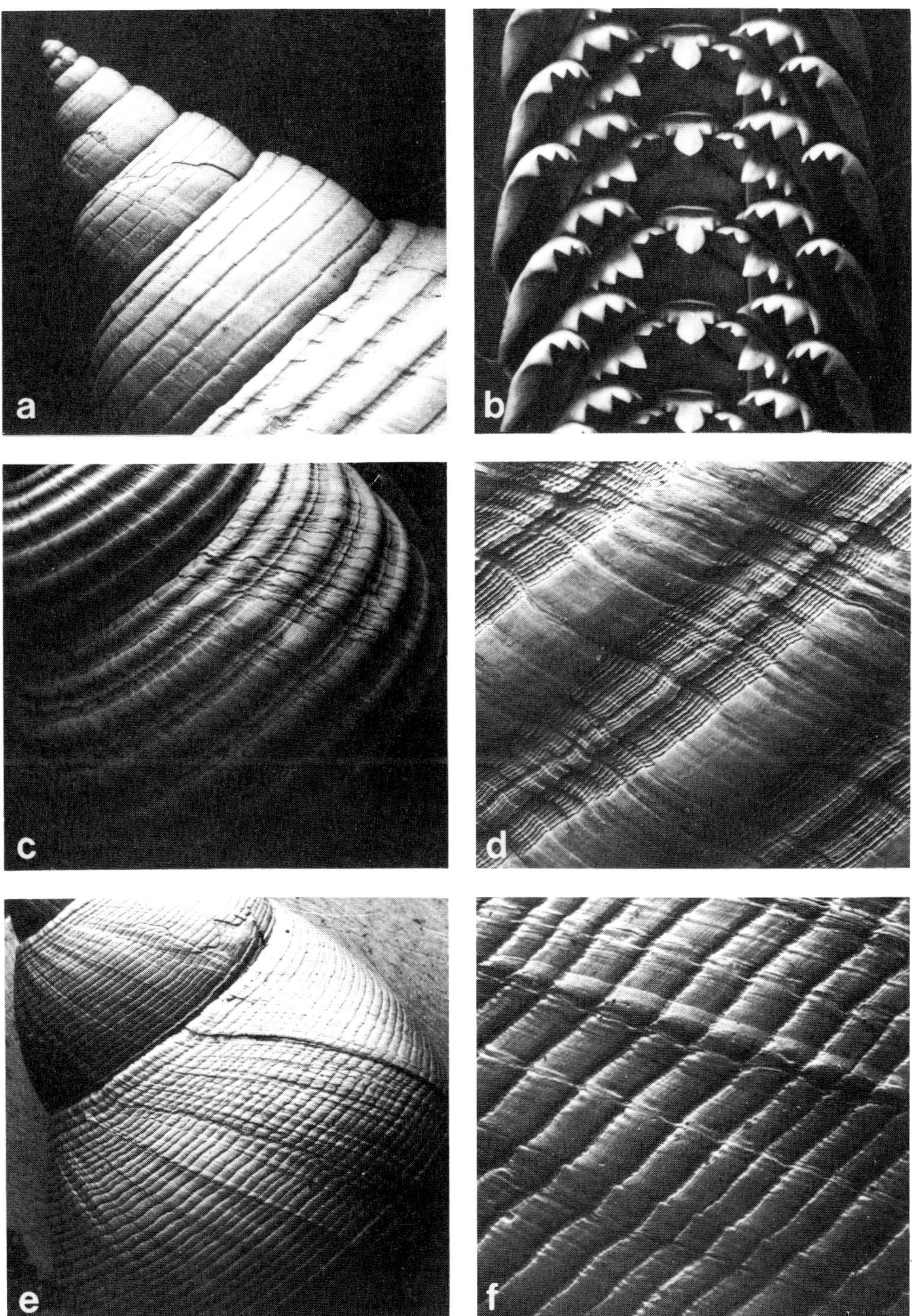


**Fig. 57** **(a–d)** *Littoraria* (*Littorinopsis*) *cingulata cingulata*, Broome, W. A.: **(a)** spire ( × 17); **(b)** radula ( × 220); **(c)** last whorl ( × 8); **(d)** detail ( × 32). **(e–f)** *Littoraria* (*Littorinopsis*) *delicatula*, Port Canning, Bengal, India: **(e)** last whorl ( × 9); **(f)** detail ( × 32).

**Table 17** Dimensions of *Littoraria* (*Littorinopsis*) *cingulata cingulata*.

| Specimen | Locality | Sex | Primary grooves | H (mm) | B (mm) | LA (mm) | WA (mm) | C (mm) | P | S | SH |
|---|---|---|---|---|---|---|---|---|---|---|---|
| *Littorina cingulata* lectotype, BMNH 1968354 | N. coast of Australia | | 6 | 19.2 | 12.2 | 10.0 | 7.5 | 1.5 | 1.57 | 0.75 | 1.92 |
| DGR | Broome, W. A. | ♂ | 6 | 18.8 | 11.6 | 9.4 | 7.0 | 1.1 | 1.62 | 0.74 | 2.00 |
| DGR | Broome, W. A. | ♂ | 6 | 17.6 | 11.5 | 9.4 | 6.5 | 1.1 | 1.53 | 0.69 | 1.87 |
| DGR | Broome, W. A. | ♀ | 5 | 22.4 | 14.1 | 11.1 | 8.0 | 1.4 | 1.59 | 0.72 | 2.02 |
| DGR | Broome, W. A. | ♀ | 6 | 19.7 | 12.2 | 9.6 | 7.1 | 1.3 | 1.61 | 0.74 | 2.05 |
| DGR, mean of 10 | Broome, W. A. | ♂ | | 17.14 | | | | | 1.569 | 0.742 | 1.923 |
| standard error | | | | 0.27 | | | | | 0.027 | 0.007 | 0.029 |
| DGR, mean of 10 | Broome, W. A. | ♀ | | 17.92 | | | | | 1.618 | 0.742 | 2.035 |
| standard error | | | | 0.18 | | | | | 0.017 | 0.007 | 0.019 |
| statistic t or U | | | | 2.366 | | | | | 74 | 53 | 83 |
| probability | | | | 0.029 | | | | | 0.076 | 0.854 | 0.012 |

callus within. Columella dark purple, pale lilac pink, or sometimes white. Parietal callus often conspicuously purple, especially at posterior limit of aperture.

ANIMAL. *Colour*. Pigmentation pale grey to black; sides of foot pale mottled; head darkest, sometimes with pale central streak; tentacles banded, unpigmented stripe each side of base.

*Penis* (Fig. 58a–e). Length to 6.0 mm. Base thick, bifurcate; limb bearing glandular disc usually short. Filament fairly short, tapering. Sperm groove open. Base fawn, filament cream, glandular disc pale brown.

*Sperm*. Eupyrene sperm 147–164 $\mu$m. Nurse cells (Fig. 58j–l) 12–15 $\mu$m; rounded to oval; rods single, central or asymmetrically placed, not projecting, parallel sided with rounded ends, or oval; yolk granules small.

*Pallial oviduct* (Fig. 58f–i). Length to 5.0 mm. Spiral section to 2.0 mm diam., $3\frac{1}{2}$ whorls; opaque albumen gland $\frac{2}{3}$–1 whorl, white; translucent albumen gland off white; capsule glands absent; spiral indistinct externally, egg groove not darkly pigmented. Straight section to 3.4 mm, white, terminating in a papilla. Bursa long, to 3.2 mm, anterior. Development assumed ovoviviparous.

*Radula* (Fig. 57b). Length to 15.5 mm; relative length 0.70–0.94. Saw-toothed type; central rachidian cusp rather elongate, pointed; cusps of paired teeth equilaterally triangular; lateral with no gap anterior to main cusp.

DISTRIBUTION. *Habitat*. At Broome, Western Australia, occurs near back of forest entirely of *Avicennia*, 0–2.0 m above ground, on trunks and sometimes leaves; occasionally on *Rhizophora* roots; rarely on sheltered rocks.

*Range* (Fig. 59). North-western Australia, from mouth of King Sound to Exmouth Gulf.

*Records*. **Australia: W.A.:** Bay of Rest, Exmouth Gulf (WAM); Barrow I. (WAM, AMS, USNM); Cossack (BMNH); Port Hedland (AMS); Broome (DGR, WAM, NMV); West I., Lacepede Group (AMS); Cape Leveque (NMV); Sunday I. (WAM); Buccaneer Arch. (AMS).

REMARKS. As a result of the remote and restricted distribution of this species, specimens are rare in museum collections outside Australia. For this reason and perhaps also because of confusion with the sympatric *L. filosa* and *L. sulculosa*, it has not been listed as a distinct species for a century. *L. cingulata cingulata* shows constant shell characters throughout its range. The relationship of the typical form with the geographical subspecies *L. cingulata pristissini* is discussed in the remarks on the latter.

SIMILAR SPECIES. Although characters of penis, sperm nurse cells and pallial oviduct place *L. cingulata cingulata* and *L. sulculosa* in different subgenera, their shells are closely similar. They can, however,

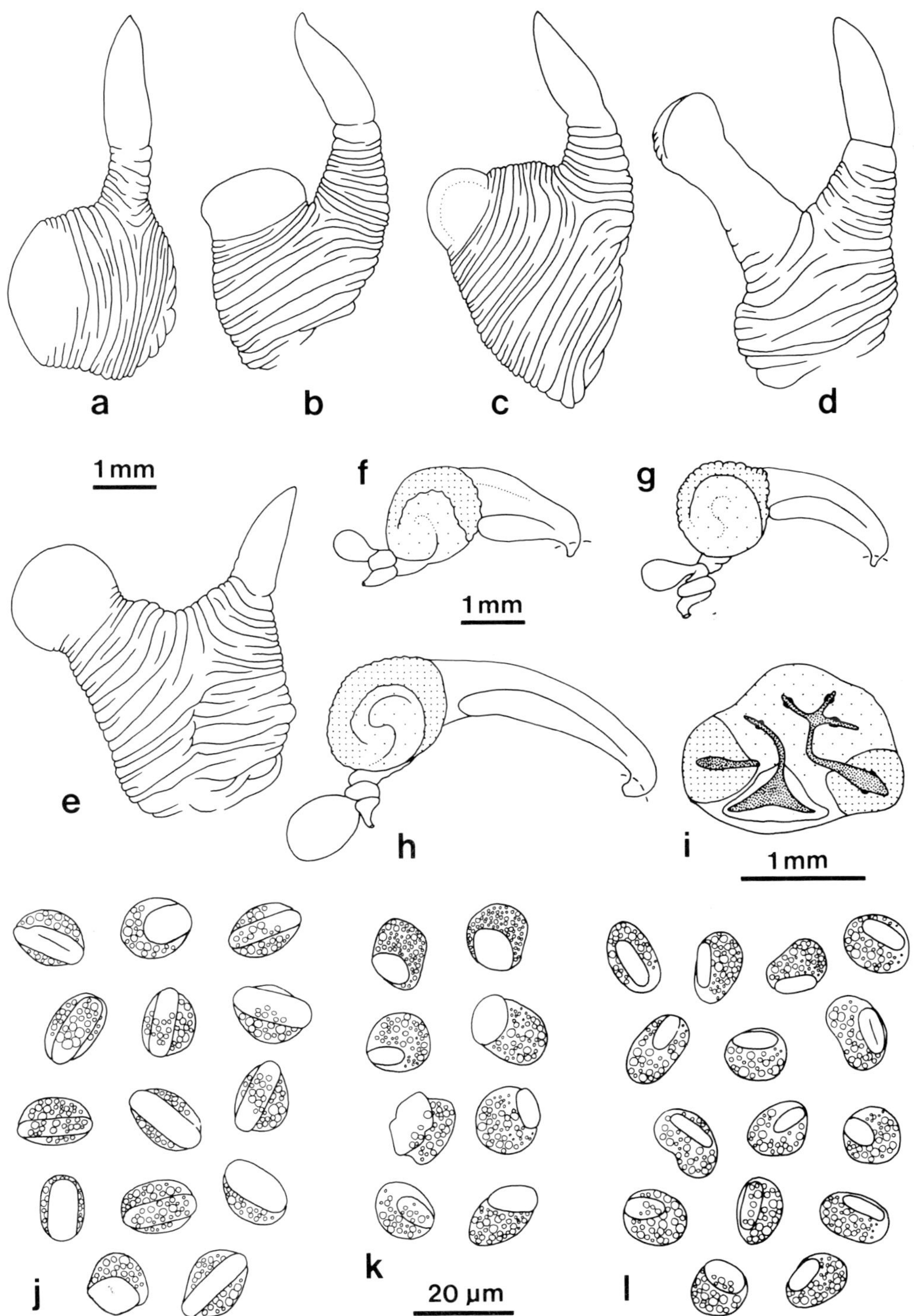


**Fig. 58** *Littoraria* (*Littorinopsis*) *cingulata cingulata*, Broome, W. A.: **(a–e)** penes; **(f–i)** pallial oviducts, with transverse section; **(j–l)** sperm nurse cells.

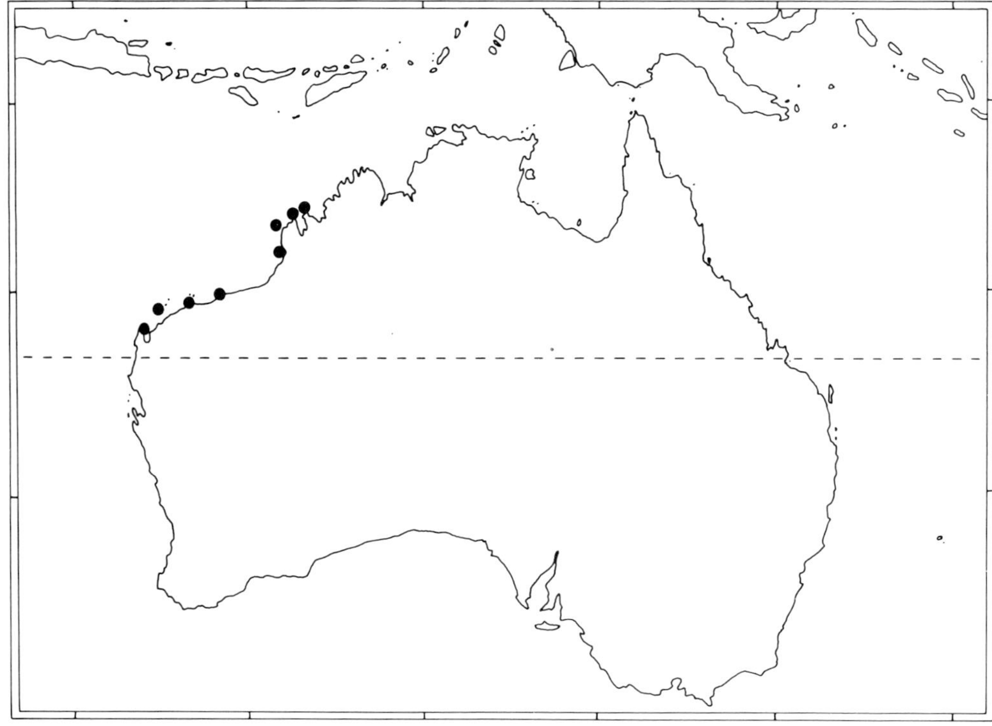

**Fig. 59** Distribution of *Littoraria (Littorinopsis) cingulata cingulata.*

always be immediately separated by the microsculpture between the conspicuous ribs, of axial striae in *L. sulculosa* and spiral striae in *L. c. cingulata*. In the field, *L. c. cingulata* is distinguished by its diffusely marbled pattern, columella often flushed with purple, 11–13 primary ribs on last whorl and thinner shell; in comparison, *L. sulculosa* has a pattern of long dashes or continuous spiral lines, usually a white columella, 9–11 primary ribs on last whorl and a thicker shell. *L. filosa* is only superficially similar, the shell is much thinner, the ribs narrow and often carinate, the microsculpture irregular and the species is colour polymorphic. Separation of the two subspecies of *L. cinguata* is usually straightforward, based on the more numerous ribs and inflated whorls of *L. c. pristissini*.

## *Littoraria (Littorinopsis) cingulata pristissini* n. subsp.

Types. Holotype: AMS C.138323, Little Lagoon, Denham, Shark Bay, Western Australia. Paratypes: AMS, WAM, BMNH, USNM.

Etymology. Latin: *pristis*, shark, and *sinus*, bay, after type locality.

Diagnosis. Shell: outline slightly convex, whorls inflated, periphery rounded; columella rounded, lilac or white; primary grooves 7–9; development of secondary and tertiary sculpture variable, usually 40–60 low ribs on last whorl; grooves narrow; microsculpture usually of strong spiral striae in grooves; colour polymorphic, brown predominating, often 10–17 axial stripes on last whorl. Animal: penis bifurcate, base thick, filament fairly short and tapering; ovoviviparous.

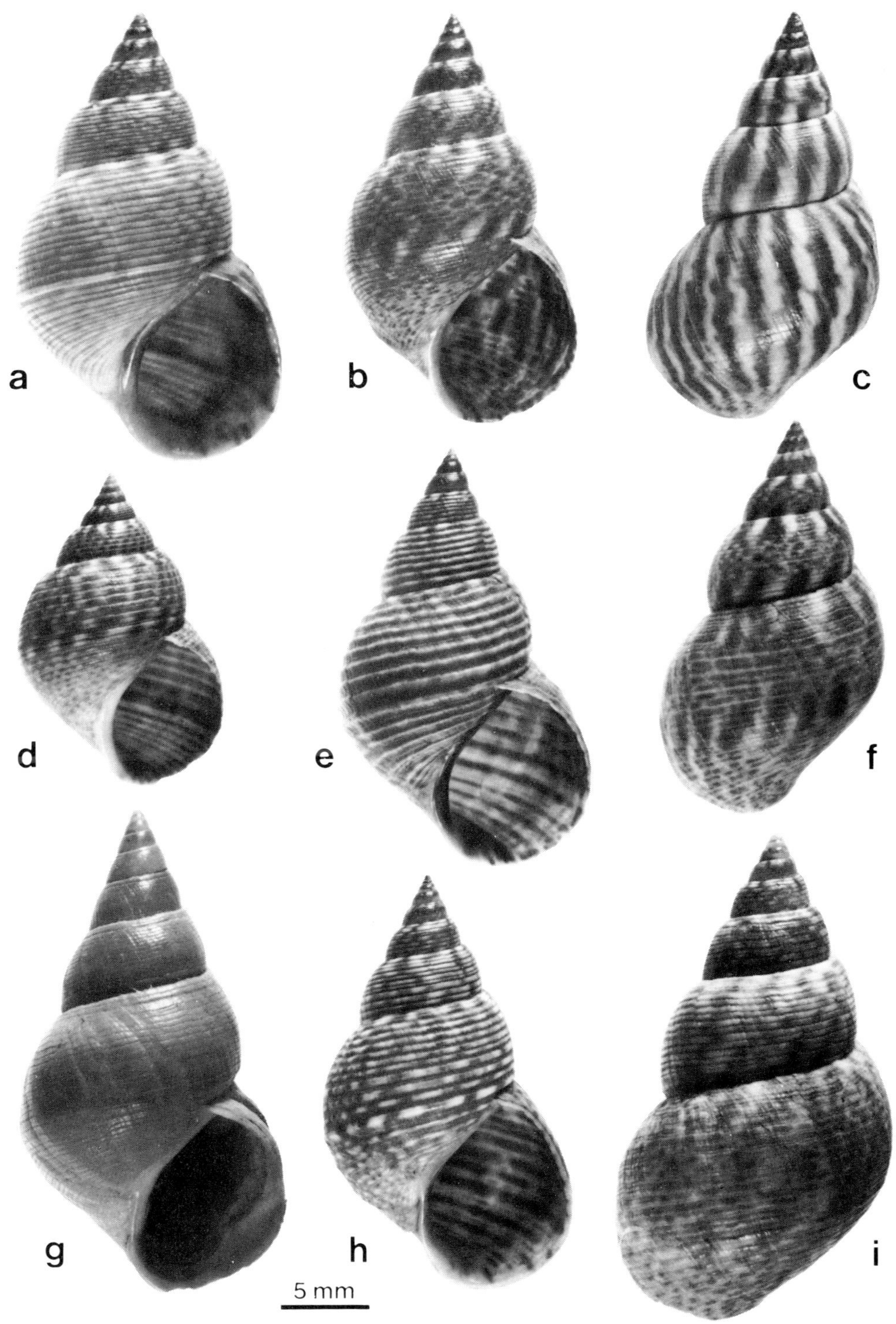


**Fig. 60** *Littoraria* (*Littorinopsis*) *cingulata pristissini*: **(a–c)** Denham, W. A.; **(a)** paratype, ♀ (BMNH); **(b)** holotype, ♂ (AMS C.138323); **(c)** paratype, ♀ (BMNH); **(d)** Herald Bight, W.A. (WAM); **(e)** paratype, ♂, Denham, W. A. (BMNH); **(f)** Herald Bight, W.A. (WAM); **(g)** Dirk Hartog I., W.A.; **(h)** paratype, ♂, Denham, W.A. (BMNH); **(i)** Dirk Hartog I., W.A. (WAM).

SHELL (Fig. 60). *Shape*. Height 17–30 mm. Teleoconch 8–8.5 whorls. Shell of moderate thickness, solid. Spire outline slightly convex; whorls inflated; sutures impressed. Peripheral keel weak until whorl 7, usually absent on last whorl, but sometimes marked by an enlarged rib. Adult lip slightly thickened, not flared. Varices absent, or present only as strong growth lines. Columella of moderate width, rounded, slightly flattened or hollowed in largest shells; pillar concave. Sexual dimorphism: males smaller, relatively lower spire and larger aperture.

*Dimensions*: Table 18.

*Sculpture* (Fig. 61a,c–f). Protoconch normal. First 2 whorls of teleoconch smooth. Primary grooves (6)7–9(10), equally spaced. Secondary grooves form at $\frac{1}{2}$ width of primary ribs on whorls 5 and 6; posterior rib may remain undivided until whorl 7, therefore appearing the most prominent on whorls 4–6. Tertiary grooves form in the same way on whorl 7; ribs may also be intercalated in primary grooves. Depending upon development of tertiary sculpture, ribs on last whorl total 40–60; usually all remain low and flat, peripheral rib sometimes a little enlarged; grooves narrow, less than $\frac{1}{2}$ rib width. Rarely secondary sculpture is delayed or absent, last whorl then sculptured by as few as 14–16 raised, rounded ridges, separated by primary grooves of up to twice rib width, sometimes with intercalated secondary riblets. At opposite extreme, shells may be almost smooth, with only posterior primary groove and basal grooves developed. Microsculpture of fine, irregular axial growth striae over whole surface; spiral striae sometimes absent, but usually strong and regular in grooves of shells in which groove width is greatest, rarely developed over whole shell surface.

*Colour*. Polymorphic; shades of brown predominate, pink occasional, yellow rare. Ground colour whitish, orange pink or cream yellow. In dark shells pattern is of red brown or dark brown flecks densely scattered over surface, sometimes aligned only at sutures, especially on spire; first 6 whorls blue grey with conspicuous alternation of brown and white spots at sutures; not infrequently entire shell marked by 10–17 vertical or oblique continuous axial stripes; occasionally shell uniform reddish black throughout. Pink, white or yellow shells may lack pattern entirely, or show a uniform speckling of pale orange brown. Aperture marked by purple brown oblique stripes or irregular spots, corresponding to external pattern; rarely spiral lines; glazed by a lilac or cream callus within. Columella lilac, or white in palest shells.

ANIMAL. *Colour*. Pigmentation pale grey to black, correlated with shell colour; sides of foot pale, mottled; head darkest, sometimes pale in centre; tentacles banded, unpigmented stripe each side of base.

**Table 18** Dimensions of *Littoraria* (*Littorinopsis*) *cingulata pristissini*.

| Specimen | Locality | Sex | Primary grooves | H (mm) | B (mm) | LA (mm) | WA (mm) | C (mm) | P | S | SH |
|---|---|---|---|---|---|---|---|---|---|---|---|
| *L. c. pristissini* holotype, AMS C.138323 | Denham, W. A. | ♂ | 8 | 22.8 | 14.2 | 11.6 | 8.6 | 1.2 | 1.61 | 0.74 | 1.97 |
| DGR | Denham, W. A. | ♂ | 7 | 22.1 | 13.4 | 11.0 | 8.5 | 1.4 | 1.65 | 0.77 | 2.01 |
| DGR | Denham, W. A. | ♂ | 8 | 19.3 | 11.4 | 9.4 | 6.8 | 1.3 | 1.69 | 0.72 | 2.05 |
| DGR | Denham, W. A. | ♂ | 7 | 21.4 | 14.8 | 11.5 | 8.6 | 1.8 | 1.45 | 0.75 | 1.86 |
| DGR | Denham, W. A. | ♀ | 7 | 21.0 | 14.4 | 11.3 | 8.8 | 1.5 | 1.46 | 0.78 | 1.86 |
| DGR | Denham, W. A. | ♀ | 7 | 25.2 | 16.0 | 12.1 | 9.3 | 1.7 | 1.58 | 0.77 | 2.08 |
| DGR | Denham, W. A. | ♀ | 7 | 24.9 | 14.2 | 11.6 | 8.2 | 1.5 | 1.75 | 0.71 | 2.15 |
| DGR, mean of 10 | Denham, W. A. | ♂ | | 21.15 | | | | | 1.533 | 0.759 | 1.933 |
| standard error | | | | 0.59 | | | | | 0.043 | 0.006 | 0.021 |
| DGR, mean of 10 | Denham, W. A. | ♀ | | 23.02 | | | | | 1.622 | 0.758 | 2.069 |
| standard error | | | | 0.51 | | | | | 0.023 | 0.006 | 0.021 |
| statistic t or U | | | | 2.405 | | | | | 72 | 53.5 | 93.5 |
| probability | | | | 0.027 | | | | | 0.106 | 0.825 | <.001 |

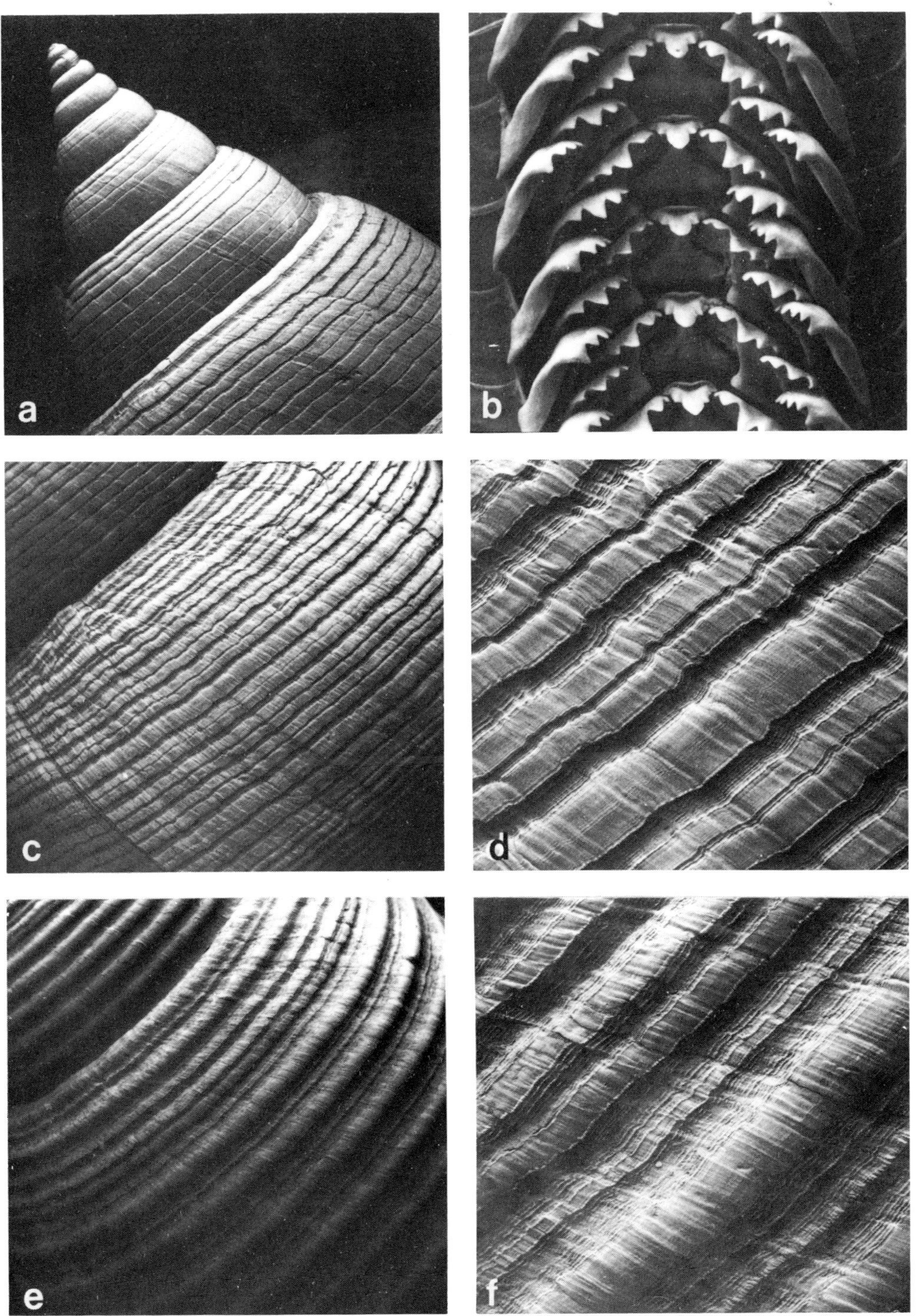


**Fig. 61** *Littoraria* (*Littorinopsis*) *cingulata pristissini*, Denham, W.A.: **(a)** spire ( × 17); **(b)** radula ( × 220); **(c)** last whorl ( × 7); **(d)** detail ( × 30); **(e)** last whorl ( × 7); **(f)** detail ( × 30).

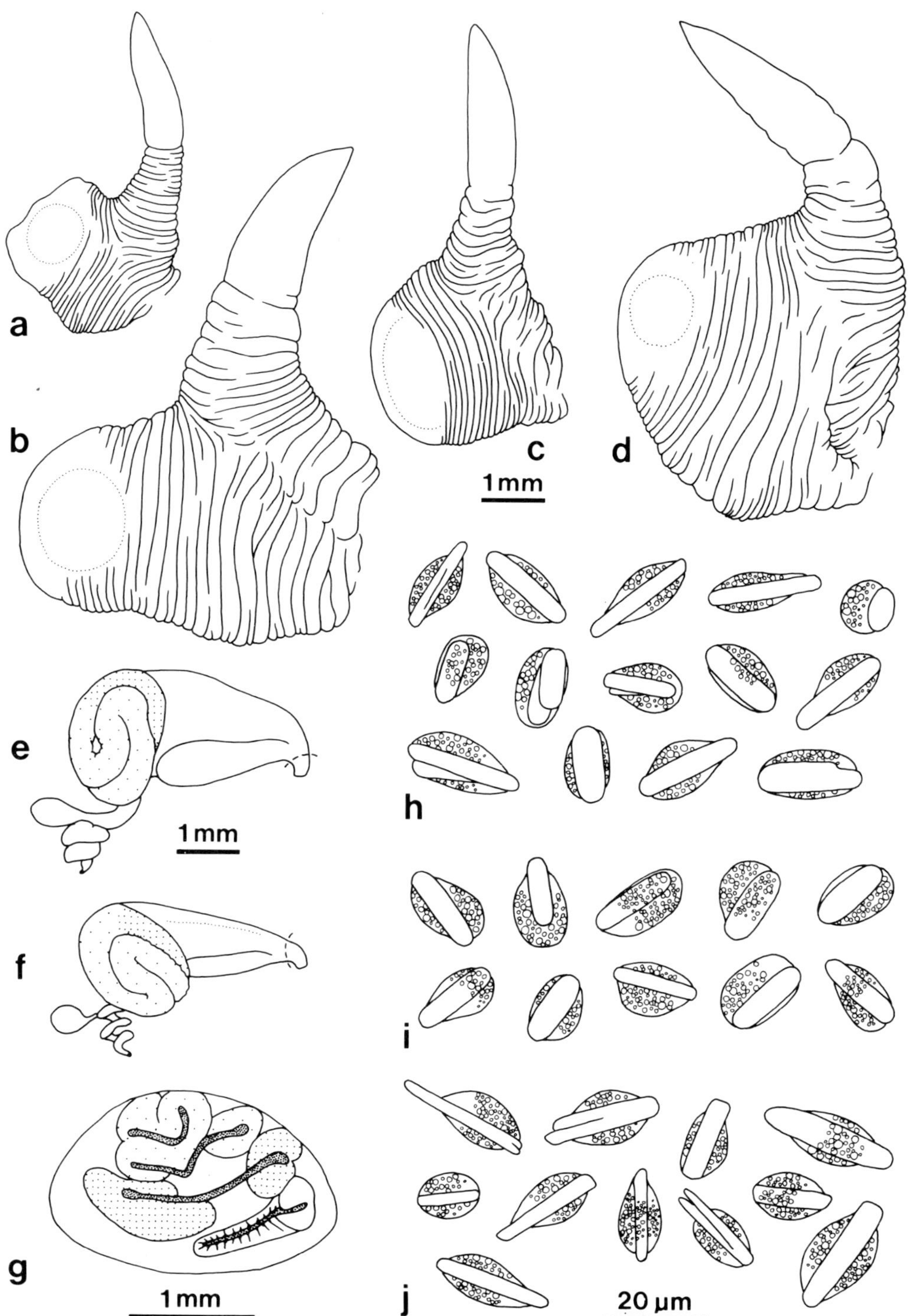


**Fig. 62** *Littoraria* (*Littorinopsis*) *cingulata pristissini*, Denham, W.A.: **(a–d)** penes; **(e–g)** pallial oviducts, with transverse section; **(h–j)** sperm nurse cells.

*Penis* (Fig. 62a–d). Length to 10.5 mm. Base thick, bifurcate, limb bearing glandular disc rather short, except when extremely relaxed. Filament tapering. Sperm groove open. Base cream to fawn; filament cream; glandular disc pale brown.

*Sperm* (Fig. 5c). Eupyrene sperm 154–171 $\mu$m. Nurse cells (Fig. 62h–j) 14–24 $\mu$m; oval to fusiform; rods 1–2, central, often projecting, parallel sided, ends bluntly rounded; yolk granules small.

*Pallial oviduct* (Fig. 62e–g). Length to 3.7 mm. Spiral section to 1.9 mm diam., $3\frac{1}{2}$ whorls; opaque albumen gland $\frac{1}{2}$–1 whorl, white; translucent albumen gland off white; capsule glands absent; spiral distinct externally; egg groove not pigmented. Straight section to 2.3 mm, off white, terminating in a papilla. Bursa long, to 2.2 mm, anterior. Reproduction assumed ovoviviparous.

*Radula* (Fig. 61b). Length to 21 mm; relative length 0.68–0.88. Saw-toothed type; central rachidian cusp square, edge rounded or slightly pointed; cusps of paired teeth equilaterally triangular; lateral with no gap anterior to main cusp.

DISTRIBUTION. *Habitat*. At type locality, on pneumatophores of *Avicennia* fringing creek between lagoon and sea, 0–0.2 m above ground; in broader *Avicennia* fringe found up to 0.6 m on trunks of outermost trees, occasionally on leaves, while found under logs at ground level further back; abundant amongst samphires at edges of sandy lagoons; occasional under stones at high tide level on sheltered rocky shores.

*Range* (Fig. 63). Restricted to Shark Bay, Western Australia.

*Records*. **Australia: W.A.:** Useless Inlet (AMS); Denham (DGR, AMS, WAM); Tetradon Loop, off Dirk Hartog I. (WAM); Herald Bight, Peron Peninsula (WAM); Monkey Mia (DGR, WAM); Faure I. (WAM); Carnarvon (DGR, AMS, WAM); Quobba Point (WAM); 32 km N. of Cardabia (subfossil, WAM).

REMARKS. Typical specimens of *L. cingulata pristissini* and *L. cingulata cingulata* are so distinctive that a close relationship between them could be doubted. The shells are usually easily separated. In *L. c. pristissini* the whorls are rounded, the aperture relatively small, the sculpture usually of 40–60 narrow spiral ribs on the last whorl, and the colour highly polymorphic, often with a pattern of axial stripes. In *L. c. cingulata* the shell is usually keeled, the last whorl sculptured by 11–13 rounded ribs and the colour is a uniform marbled orange brown. However, the sculpture of *L. c. pristissini* is extremely variable, even within local populations, ranging from shells almost smooth to others with only 14–16 rounded ribs (Fig. 60e). The strongly ribbed shells comprise only a small fraction of the population (3.8%, $n = 367$, sampled from 4 localities) and are usually males (13 of 177 males, 1 of 190 females, $\chi^2 = 9.8$, $0.001 < P < 0.005$). The ribbed shells approach *L. c. cingulata* closely, not only in sculpture, but also in colour, which is never pink, yellow or axially striped as in typical *L. c. pristissini*. However, in shape and in number of primary grooves (usually 7 or more) the shells resemble the Shark Bay subspecies, with which they are linked by an entire range of intermediates. Despite the conchological differences, anatomically the two subspecies are closely similar. Small differences shown by *L. c. pristissini* include the more obvious spiral form of the oviduct, the less elongate central cusp of the rachidian tooth and the more oval sperm cells with longer rods. There is no difference in the form of the penis.

*L. c. pristissini* is the only species of *Littoraria* to be found in Shark Bay and on the basis of available museum specimens seems to be geographically isolated from *L. c. cingulata*, which does not occur further south than Exmouth Gulf. This discontinuity is probably real, for according to G.W. Kendrick (WAM, pers. comm.) the only area of mangroves between Exmouth Gulf and the Gascoyne River is located 30 km south from North West Cape on the western coast, and he has been unable to find any living molluscs there. Little is known of coastal currents in the area, but they are said to be variable and weak (Phillips *et al.*, 1979) or southwards across the mouth of Shark Bay in autumn and winter (Legeckis & Cresswell, 1981). It is thus conceivable that the 300 km stretch of inhospitable coast between Shark Bay and Exmouth Gulf effectively isolates the two subspecies at the present time, although both probably have a lengthy planktonic larval phase (p. 66).

Since these two closely related forms are allopatric, their taxonomic status can only be judged on the basis of degree of morphological difference as compared with other examples of inter- and

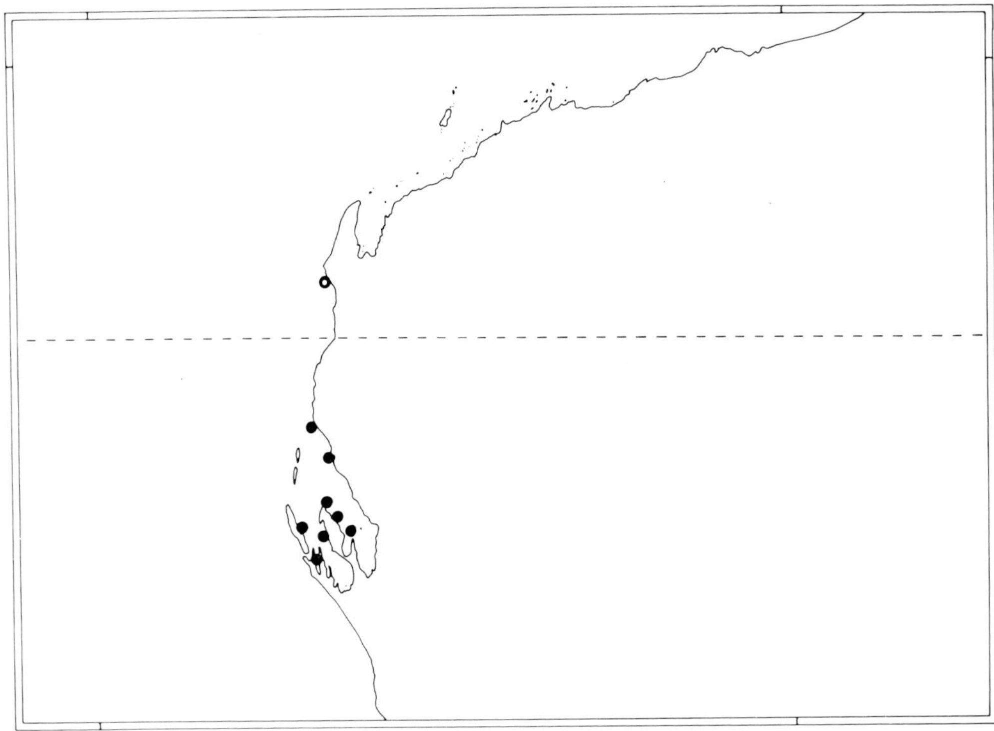

**Fig. 63** Distribution of *Littoraria* (*Littorinopsis*) *cingulata pristissini*. Open circle, subfossil occurrence.

intraspecific variation in the genus (Mayr, 1969, pp. 196–197). Several species of *Littoraria*, including *L. pallescens*, *L. carinifera* and *L. filosa*, show regional variation in shell sculpture, but not to such a degree as *L. cingulata*. Shell colour is known to be influenced by substrate in some *Littoraria* species (p. 18), but in *L. cingulata* the habitats of the two forms, chiefly on *Avicennia* bushes, are similar. It would seem that the polymorphism of the Shark Bay population may be determined genetically. Amongst other species, only *L. intermedia* is polymorphic over a part of its range and then the polymorphism is less striking than that of *L. c. pristissini*. The anatomical differences enumerated above are no greater than the normal range of intraspecific variation in other *Littoraria* species. Clearly the combination of shell sculpture, colour polymorphism and anatomical details distinguishing *L. c. pristissini* are more significant than other cases of intraspecific variation and are worthy of taxonomic recognition. Throughout the present work penis shape has been used as a diagnostic character, constant within species, which suggests that it might be used in species recognition by the animals themselves during copulation (p. 25). If this is so, the identity of the penes in the two forms of *L. cingulata* is circumstantial evidence that, were their ranges to meet, there would be no barrier to interbreeding. Since the taxa are closely related, taxonomically distinct, but allopatric, the category of subspecies is appropriate.

The known distributions of these two subspecies suggest that differentiation may have taken place following the isolation of the Shark Bay population from a once continuously distributed ancestral species. In the middle Holocene there were areas of mangroves in lagoons and embayments along the coast from Shark Bay to Exmouth Gulf, which have since almost completely disappeared (G.W. Kendrick, pers. comm.). Two subfossil shells of *L. c. pristissini* have been seen from a locality 32 km north of Cardabia (WAM), showing that this subspecies did in the past extend further north than Shark Bay. This time period might seem too short for such a level of differentiation to be achieved between isolated populations. However, the ancestral population might well have shown geographical variation before isolation was complete, as in the case of *L. filosa* in north-western Australia.

Alternatively, more ancient episodes of habitat restriction may have caused the initial isolation. It is noteworthy that local populations of *L. c. pristissini* are highly variable in shell sculpture and shape. This might suggest that differentiation from the nominate subspecies is relatively recent, or even that isolation is incomplete, with some gene flow from the northern populations. Further fossil material may provide evidence of the history of these subspecies.

SIMILAR SPECIES. The relationship between *L. c. pristissini* and the nominate subspecies is discussed above. *L. angulifera* and *L. subvittata* are similar in shell characters, but in each the columella is excavated. Specimens might be confused with *L. ardouiniana*, but in *L. c. pristissini* the primary grooves are fewer, varices are absent and the microsculpture is often more prominent.

## *Littoraria* (*Littorinopsis*) *luteola* (Quoy & Gaimard, 1833)

*Littorina luteola* Quoy & Gaimard, 1833: 477–478, pl. 33, figs 4–7 [Port Jackson, near Sydney [N.S.W., Australia]; lectotype (Rosewater, 1970) MNHNP]; Deshayes & Milne Edwards, 1843: 210–211
*Melarhaphe luteola*—Hedley, 1918a: M51
*Melarapha luteola*—Iredale & McMichael, 1962: 38
*Littorina scabra*—Angas, 1871: 95; Whitelegge, 1889: 264 [both not Linnaeus, 1758]; Muggeridge, 1979 [not Linnaeus, 1758; reproductive biology]
*Littorina* (*Littorinopsis*) *scabra scabra*—Rosewater, 1970: 456–461, pl. 352, fig. 9 [in part; not Linnaeus, 1758]
*Littorina filosa* var. *subcingulata* Nevill, 1885: 149 [Port Jackson [N.S.W., Australia]; lectotype here designated ZSI, 17.7 mm]
*Littorina* (*Melaraphe*) *scabra* var. *filosa*—Tryon, 1887: 244 [in part; not Sowerby, 1832]
*Littorina* (*Melaraphe*) *flammea*—Tryon, 1887: 245, pl. 43, fig. 36 [in part; not Philippi, 1847]

NOMENCLATURE. Rosewater (1970) designated a lectotype from the syntypic series in MNHNP. This specimen was unavailable for loan, but Rosewater's figure leaves no doubt as to its identity. Figure 64a herein is a specimen collected by Quoy & Gaimard, which may perhaps be a paralectotype (P. Bouchet, pers. comm.), although labelled 'Port Western'.

DIAGNOSIS. Shell: thin; spire elongate, outline convex; lip flared, varices 0–2; columella narrow, rounded, purple; primary grooves 8–10; 18–23 rounded ribs on body whorl, of which 2 at periphery are most prominent; secondary sculpture may be absent; colour polymorphic, dashes of dark pigment may be aligned to form short stripes at suture and periphery numbering 9–11 on last whorl. Animal: penis bifurcate, glandular disc large, filament small and tapering; ovoviviparous.

SHELL (Frontispiece, Fig. 64). *Shape*. Height 8–23(28) mm. Teleoconch 7–9 whorls. Shell thin and delicate. Spire elongate, outline distinctly convex; whorls rounded or slightly turreted; sutures impressed. Whorls weakly keeled, although periphery is made prominent by 2 strong ribs, and often also by colour pattern. Adult lip thin and flared. Varices 0–2(4). Columella narrow, rounded, not excavated; pillar concave. Sexual dimorphism: males smaller, relatively lower spire and larger aperture, broader shell, adult lip more strongly flared and varices more frequent.

*Dimensions*: Table 19.

*Sculpture* (Fig. 65a,c–f). Protoconch normal. First whorl of teleoconch smooth. Primary grooves 8–10(11); spacing almost equidistant, but for posterior 3 grooves which are somewhat further apart; posterior 1 or 2 grooves deeper and wider than the rest, consequently posterior 2 or 3 ribs of spire whorls are the most prominent, this prominence often accentuated by colour pattern; posterior rib slightly narrower than the 2 ribs below. Secondary sculpture may develop on whorl 7, where larger ribs near suture and periphery divide at $\frac{1}{2}$ width; small shells often lack secondary sculpture; all ribs may divide in large shells. Occasionally single, narrow secondary riblets are intercalated in

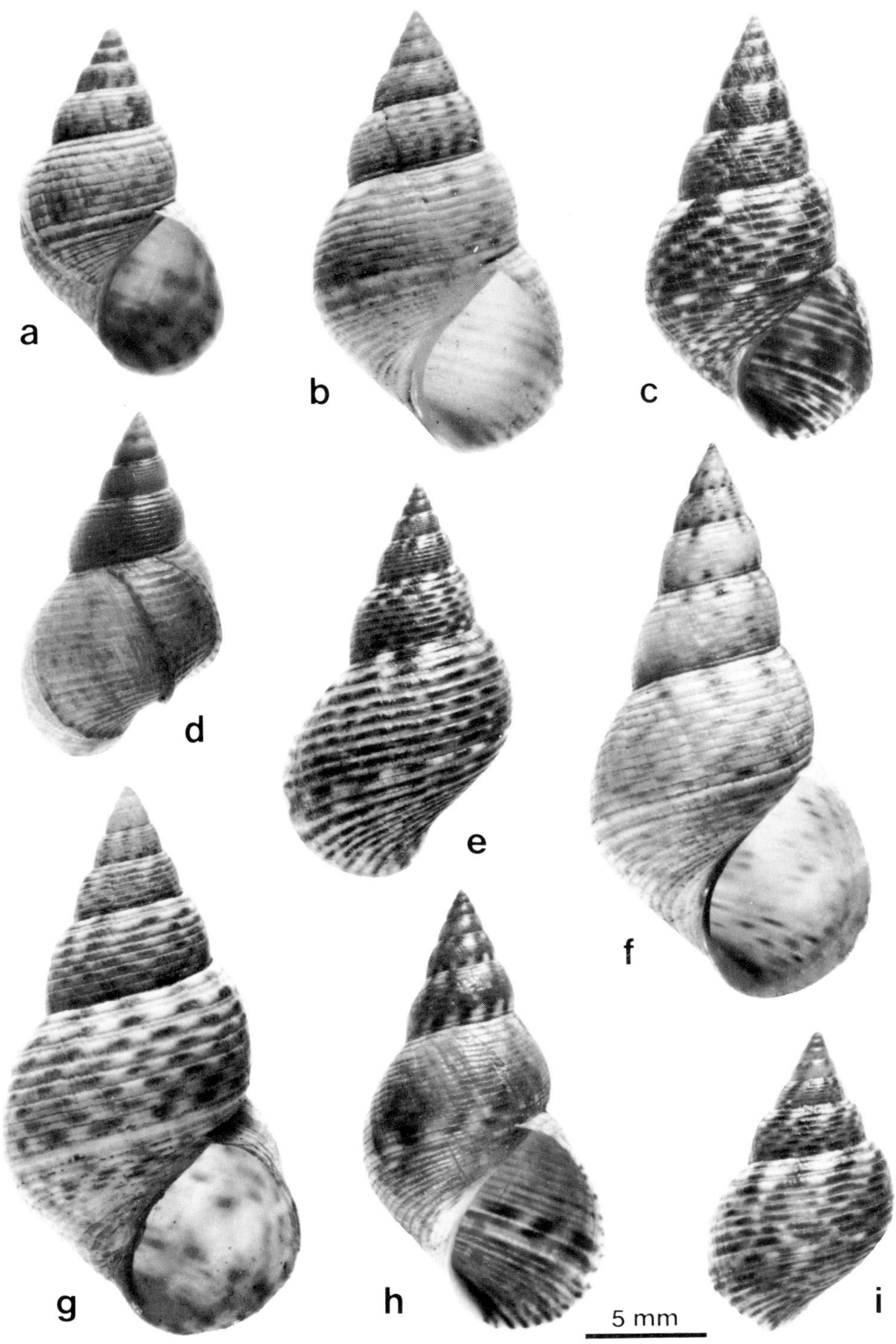


**Fig. 64** *Littoraria* (*Littorinopsis*) *luteola:* **(a)** specimen collected by Quoy & Gaimard, Port Western (MNHNP); **(b)** lectotype of *Littorina filosa* var. *subcingulata* Nevill, Port Jackson, N.S.W. (ZSI); **(c)** ♀, Kurnell, N.S.W. (DGR); **(d)** Thursday I., Qld. (QM); **(e,f)** Kurnell, N.S.W. (DGR); **(e)** ♂; **(f)** ♀; **(g)** Uralla Bay, N.S.W. (NMV); **(h)** ♂, Kurnell, N.S.W. (DGR); **(i)** Tallebudgera, SE. Qld. (QM).

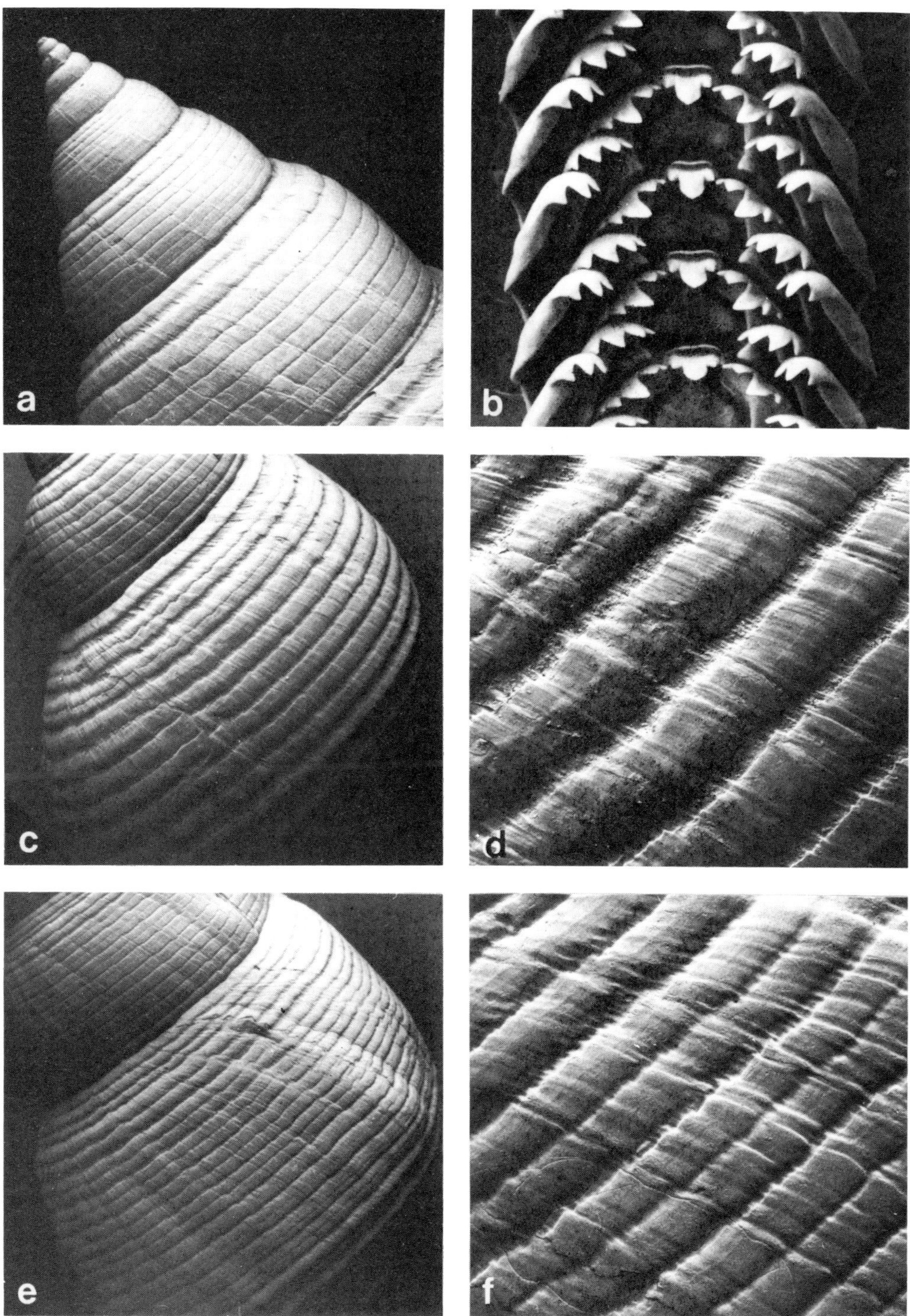


**Fig. 65** *Littoraria* (*Littorinopsis*) *luteola*, Kurnell, N.S.W.: **(a)** spire ( × 18); **(b)** radula ( × 230); **(c)** last whorl ( × 8); **(d)** detail ( × 33); **(e)** last whorl ( × 9); **(f)** detail ( × 39).

**Table 19** Dimensions of *Littoraria* (*Littorinopsis*) *luteola*.

| Specimen | Locality | Sex | Primary grooves | H (mm) | B (mm) | LA (mm) | WA (mm) | C (mm) | P | S | SH |
|---|---|---|---|---|---|---|---|---|---|---|---|
| *Littorina luteola* lectotype[1], MNHNP | Port Jackson, N.S.W. | | | 16.3 | | | | | | | |
| MNHNP[2] | Port Western | | 8 | 14.3 | 9.1 | 7.3 | 5.8 | | 1.57 | 0.79 | 1.96 |
| *Littorina filosa* var. *subcingulata* lectotype, ZSI | Port Jackson, N.S.W. | | 9 | 17.7 | 10.8 | 8.8 | 6.2 | 0.9 | 1.64 | 0.70 | 2.01 |
| DGR | Kurnell, N. S. W. | ♀ | 8 | 22.0 | 11.9 | 9.9 | 7.1 | 0.9 | 1.85 | 0.72 | 2.22 |
| DGR | Kurnell, N. S. W. | ♀ | 9 | 17.4 | 9.5 | 8.0 | 5.9 | 0.8 | 1.83 | 0.74 | 2.18 |
| DGR | Kurnell, N. S. W. | ♂ | 9 | 14.8 | 8.7 | 8.2 | 6.0 | 0.7 | 1.70 | 0.73 | 1.80 |
| DGR | Kurnell, N. S. W. | ♂ | 9 | 16.9 | 10.3 | 9.3 | 6.9 | 0.8 | 1.64 | 0.74 | 1.82 |
| DGR | Magnetic I., Qld. | ♀ | 9 | 19.4 | 10.2 | 8.8 | 6.1 | 1.0 | 1.90 | 0.69 | 2.20 |
| DGR | Magnetic I., Qld. | ♂ | 11 | 9.5 | 5.6 | 4.9 | 3.7 | 0.5 | 1.70 | 0.76 | 1.94 |
| DGR, mean of 10 | Kurnell, N. S. W. | ♂ | | 12.98 | | | | | 1.644 | 0.744 | 1.827 |
| standard error | | | | 0.65 | | | | | 0.026 | 0.011 | 0.023 |
| DGR, mean of 10 | Kurnell, N. S. W. | ♀ | | 16.06 | | | | | 1.751 | 0.726 | 2.102 |
| standard error | | | | 1.14 | | | | | 0.031 | 0.008 | 0.038 |
| statistic t or U | | | | 2.335 | | | | | 81 | 67 | 98 |
| probability | | | | 0.031 | | | | | 0.018 | 0.218 | <.001 |

[1] – shell height from Rosewater (1970, p. 459); [2] – specimen collected by Quoy & Gaimard.

widest primary grooves at periphery and near suture. Tertiary sculpture usually absent. Body whorl sculptured by rounded primary and secondary ribs numbering 18–23(28); most prominent are 2 ribs at periphery (peripheral rib and first basal rib), which remain undivided and are separated by widest of primary grooves, equal to rib width. Posterior groove up to $\frac{1}{2}$ average rib width; other grooves narrow, less than $\frac{1}{3}$ average rib width. Microsculpture of axial growth striae covering surface, strongest in widest grooves; faint spiral striae often visible, on ribs only; occasional large shells show single rows of beads in widest grooves.

*Colour* (Frontispiece). Polymorphic; brown most frequent, yellow or red not uncommon. Ground colour pale brown, cream yellow or dark orange pink. Pattern of dark brown dashes on ribs; dashes axially aligned on 3 large subsutural ribs and 2 large peripheral ribs, making these parts of the whorls conspicuous by alternation of pale and dark colour stripes, which number 9–11 around suture of last whorl; elsewhere dashes may be so dense as to almost cover surface, may be sparsely scattered in a marbled pattern, or all pigmentation may be faint or even absent. External pattern clearly visible within aperture. Columella purple or red brown; spiral white stripe at base of pillar within inner apertural lip, outlined by narrow purple band within aperture.

ANIMAL. *Colour*. Pigmentation pale to dark grey or black, correlated with shell colour; sides of foot mottled; head darkest, sometimes with pale central streak; tentacles banded, unpigmented stripe each side of base.

*Penis* (Fig. 66a–e). Length to 3.9 mm. Base bifurcate; glandular disc large, somewhat irregular. Filament small, tapering, tip slightly mucronate. Sperm groove open. Base and filament off white, opaque white zone between filament and base; glandular disc white to pale brown.

*Sperm*. Eupyrene sperm 114–137 μm. Nurse cells (Fig. 66f–h) 17–26 μm; oval to rounded; rods 1–2(4), often asymmetrically placed, somewhat projecting, broad, rectangular and usually with rounded ends, or even oval; yolk granules large.

*Pallial oviduct* (Fig. 66i–n). Length to 3.2 mm. Spiral section to 1.9 mm diam., $2\frac{1}{2}$ whorls; opaque albumen gland $\frac{1}{2}$ whorl, white; translucent albumen gland cream to fawn; capsule glands absent; spiral indistinct externally; internal structure rather characteristic, with thick, red epithelial lining and last revolution of egg groove meeting lowermost chamber rather far forward; egg groove not usually darkly pigmented. Straight section relatively short, to 2.0 mm, fawn, reddish epithelial lining

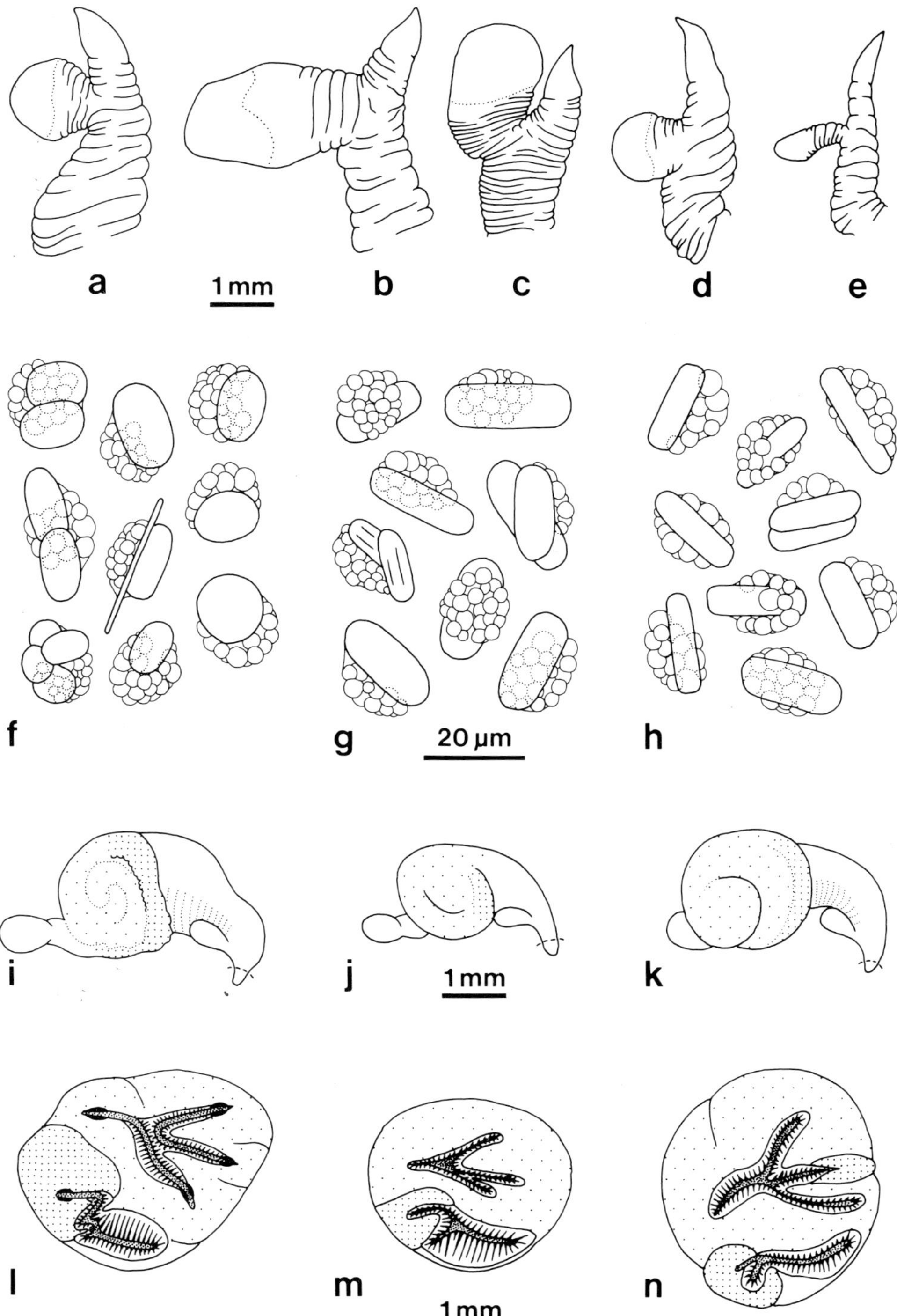


**Fig. 66** *Littoraria* (*Littorinopsis*) *luteola:* **(a–e)** penes; **(a,b)** Kurnell, N.S.W.; **(c)** Pittwater, N.S.W.; **(d)** Kurnell, N.S.W.; **(e)** Magnetic I., Qld.; **(f–h)** sperm nurse cells; **(f,g)** Pittwater, N.S.W.; **(h)** Kurnell, N.S.W.; **(i–n)** pallial oviducts, with transverse sections, Kurnell, N.S.W.

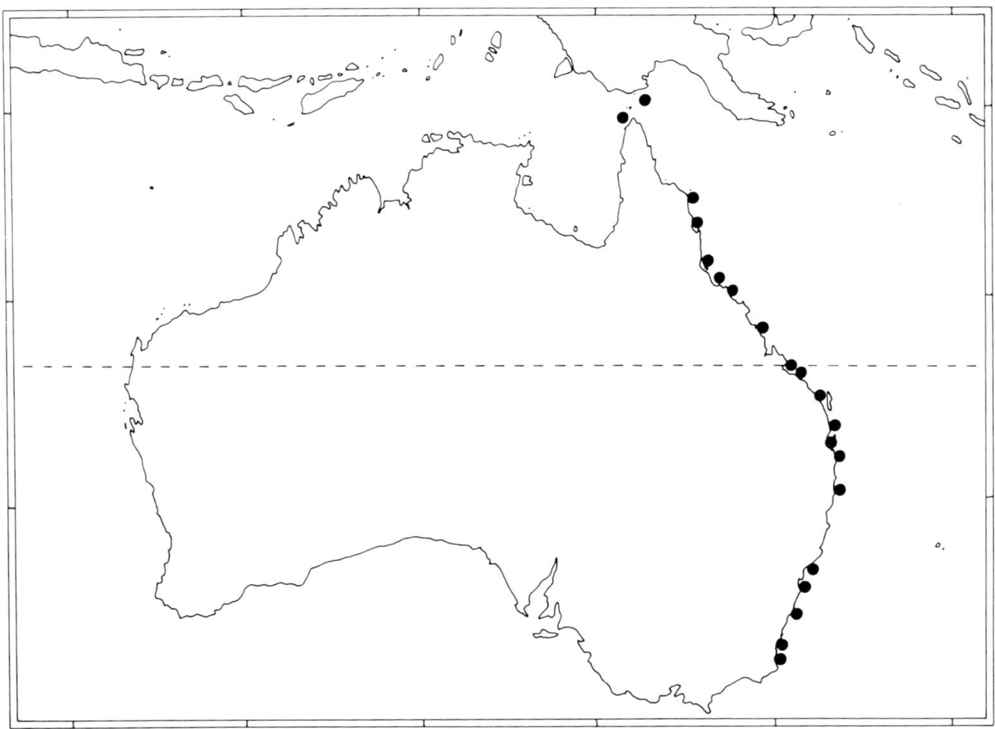

**Fig. 67** Distribution of *Littoraria* (*Littorinopsis*) *luteola*.

visible posteriorly; terminating in a papilla. Bursa small, to 0.6 mm, anterior. Development ovoviviparous.

*Radula* (Fig. 65b). Length to 11 mm; relative length 0.52–0.69. Saw-toothed type; central rachidian cusp square, edge slightly pointed; cusps of paired teeth equilaterally triangular; lateral with only slight gap anterior to main cusp.

DISTRIBUTION. *Habitat.* Leaves and trunks of *Avicennia*; in forests consisting entirely of *Avicennia* (southern N.S.W.), occurs from seaward edge to back of forest, 0.4–2.2 m above ground; in more complex forests (Qld.) occurs in *Avicennia* fringe and *Ceriops* zone; also found at back of swamps in saltmarsh vegetation and on and under driftwood at ground level. A continental species.

*Range* (Fig. 67). Endemic to eastern Australia, from southern New South Wales to Torres Strait Islands. (Old material from Solomon Islands, New Britain and other Pacific islands almost certainly incorrectly localized.)

*Records.* **Australia: Qld.:** Thursday I. (QM); Murray I. (QM); Cooktown (AMS); Yule Point (WAM); Dunk I. (QM); Missionary Bay, Hinchinbrook I. (DGR); Magnetic I. (DGR); Bowen (AMS); Sarina (AMS); Yeppoon (AMS); Eurimbula Creek, S. of Bustard Head (AMS); Pialba (QM); Caloundra (QM); Dunwich, Stradbroke I. (WAM, NMV); Tallebudgera, near Coolangatta (QM); **N.S.W.:** Brunswick Heads (AMS); Yamba (AMS); Port Stephens (AMS); Palm Beach, Pittwater (AMS); Kurnell Pen. (DGR); Huskisson, Jervis Bay (AMS); Narooma (AMS, NMV); Merimbula (AMS).

REMARKS. This species is little known and has not been recognized as distinct by authors outside Australia for about a century. *L. luteola* shows little geographical variation over its range, but sexual dimorphism is very marked. Shells from the leaves of *Avicennia* trees tend to be highly colour polymorphic, while those from bare trunks within thick forests and from driftwood in salt marshes are usually dark brown in colour. *L. luteola* extends as far into the temperate zone as any of the mangrove associated species of *Littoraria*, reaching 37°S (*L. intermedia* is recorded from 37°N by Kuroda & Habe, 1952); mangroves are, however, found at still higher latitudes. This is by far the most abundant and often the only species of *Littoraria* in the mangroves of New South Wales, but becomes scarce north of about Rockhampton in Queensland, where it is gradually replaced as the dominant species on *Avicennia* leaves by *L. filosa*. The reproductive biology of the species has been studied in detail by Muggeridge (1979).

SIMILAR SPECIES. The narrow and rounded columella immediately distinguishes *L. luteola* from the larger *L. philippiana* and from the thicker shelled *L. intermedia*. In the same geographical range only *L. filosa* has a similar columella, but is larger, broader, and is usually strongly carinate. Confusion is most likely with small specimens of *L. ardouiniana*, although the distributions do not overlap; *L. ardouiniana* is distinguished by its finer and more numerous spiral ribs, and by the lack of the 2 large peripheral ribs which are characteristic of *L. luteola*. The form of the oviduct is unique to *L. luteola*.

## *Littoraria* (*Littorinopsis*) *ardouiniana* (Heude, 1885)

*Leptopoma* (?) *ardouinianum* Heude, 1885: 95–96, pl. 25, figs 8,8a [A-long, Tonkin [Vietnam]; type in Academia Sinica, Peking? (P. Bouchet, pers. comm.)]
*Littorina* (*Lamellilitorina*) *ardouiniana*—Tryon, 1887: 253, pl. 46, fig. 28
*Littorina ardouiniana*—Fischer, 1891: 170
*Melarhaphe scabra*—Yen, 1942: 195 [not Linnaeus, 1758]
*Littorina* (*Littorinopsis*) *scabra scabra*—Rosewater, 1970: 456–461 [in part; not Linnaeus, 1758]

NOMENCLATURE. The type of this species could not be obtained, but Heude's figure (Fig. 68d herein) leaves little doubt as to its identity. The fine spiral sculpture (Heude's fig. 8a), prominent varices, indistinct axial stripes and narrow columella are all clearly shown and are characteristic of the species.

DIAGNOSIS. Shell: moderately thin; spire tall, whorls rounded; lip flared; varices 0–3; columella fairly narrow, rounded; primary grooves 10–13; body whorl with 37–78 rounded primary and secondary ribs; colour polymorphic, brown dashes may be aligned to form oblique axial stripes numbering 8–11 on last whorl. Animal: penis bifurcate, filament large and broad with sharply tapering tip, base ochraceous; ovoviviparous.

SHELL (Fig. 68). *Shape*. Height 11–30(36) mm. Teleoconch 7–9 whorls. Shell moderately thin, becoming solid at large size. Spire tall, outline convex; whorls lightly rounded, sutures impressed. Peripheral keel absent on last whorl; in smaller shells a slight angulation may be accentuated by a rib more prominent than the others. Adult lip thin, flared. Varices 0–3(7). Columella fairly narrow, usually rounded, sometimes slightly flattened, not excavated; pillar straight to gently convex. Sexual dimorphism: males smaller, relatively lower spire and larger aperture, shell relatively wider.

*Dimensions*: Table 20.

*Sculpture* (Fig. 96e,f). Protoconch of normal shape, sculpture not seen. Probably all whorls of teleoconch sculptured by spiral grooves (apices of available specimens not well preserved). Primary grooves (9)10–13, equally spaced. Secondary sculpture may appear as early as whorls 5–6, more usually on whorl 7, where 1–4 ribs, generally on shoulder, become divided each by a secondary groove at $\frac{1}{3}$–$\frac{1}{2}$ width; rarely all ribs are thus divided. On whorls 7–8 1–3 small secondary riblets form by intercalation in each primary groove, usually appearing just before and continuing strongly after the first varix. Primary ribs of equal width; until last whorl all remain low, but for slightly more prominent peripheral rib; secondary riblets remain small, usually less than $\frac{1}{2}$ width of primary ribs; all ribs become more prominently raised and rounded on last whorl; total number of primary and secondary ribs beyond first varix 37–78. Primary grooves $\frac{1}{2}$–1 rib width on penultimate whorl,

**Fig. 68** *Littoraria* (*Littorinopsis*) *ardouiniana:* **(a,b)** Three Fathoms Cove, Hong Kong (BMNH); **(a)** ♂; **(b)** ♂; **(c)** ♀, Ubin I., Singapore (DGR); **(d)** figure of *Leptopoma* (?) *ardouinianum* Heude, Along, Tonkin (Heude, 1885, pl. 25, fig. 8); **(e–g)** Three Fathoms Cove, Hong Kong (BMNH); **(e,f)** ♀; **(g)** ♀; **(h)** Tai Po, S. China (BMNH); **(i)** China (BMNH).

**Table 20** Dimensions of *Littoraria* (*Littorinopsis*) *ardouiniana*.

| Specimen | Locality | Sex | Primary grooves | H (mm) | B (mm) | LA (mm) | WA (mm) | C (mm) | P | S | SH |
|---|---|---|---|---|---|---|---|---|---|---|---|
| *Leptopoma ardouinianum* orginal figure (Heude, 1885) | Along, Tonkin | | | 17.6 | 11.4 | 8.7 | 6.8 | | 1.54 | 0.78 | 2.02 |
| BMNH | China | | 13 | 17.7 | 10.3 | 8.9 | 6.5 | 0.8 | 1.72 | 0.73 | 1.99 |
| BMNH | China | | 13 | 14.9 | 8.5 | 7.4 | 5.6 | 0.7 | 1.75 | 0.76 | 2.01 |
| BMNH 1969391 | unknown | | 12 | 23.4 | 12.7 | 10.5 | 7.4 | 1.1 | 1.84 | 0.70 | 2.23 |
| BMNH | Okinawa, Japan | ♀ | 11 | 15.5 | 8.7 | 7.5 | 4.8 | 0.5 | 1.78 | 0.64 | 2.07 |
| BMNH 1947.8.15.112 | Tai Po, China | | 10 | 16.5 | 9.6 | 8.5 | 6.1 | 0.6 | 1.72 | 0.72 | 1.94 |
| AMS C.131826 | Tai Tam, Hong Kong | ♂ | 11 | 14.2 | 7.8 | 6.7 | 4.8 | 0.6 | 1.82 | 0.72 | 2.12 |
| BMNH | Three Fathoms Cove, Hong Kong | ♂ | 10 | 26.9 | 16.1 | 14.1 | 10.9 | 1.4 | 1.67 | 0.77 | 1.91 |
| BMNH | Three Fathoms Cove, Hong Kong | ♂ | 10 | 27.1 | 15.7 | 14.1 | 10.6 | 1.7 | 1.73 | 0.75 | 1.92 |
| BMNH | Three Fathoms Cove, Hong Kong | ♀ | 10 | 35.8 | 18.5 | 15.9 | 12.3 | 1.7 | 1.94 | 0.77 | 2.25 |
| BMNH | Three Fathoms Cove, Hong Kong | ♀ | 13 | 30.2 | 15.9 | 12.8 | 9.0 | 1.5 | 1.90 | 0.70 | 2.36 |
| BMNH, mean of 10 | Three Fathoms | ♂ | | 24.85 | | | | | 1.730 | 0.757 | 1.924 |
| standard error | Cove, Hong Kong | | | 0.44 | | | | | 0.017 | 0.006 | 0.009 |
| BMNH, mean of 10 | Three Fathoms | ♀ | | 28.29 | | | | | 1.845 | 0.740 | 2.236 |
| standard error | Cove, Hong Kong | | | 1.29 | | | | | 0.025 | 0.011 | 0.036 |
| statistic t or U | | | | 2.518 | | | | | 87 | 67 | 100 |
| probability | | | | 0.021 | | | | | 0.004 | 0.218 | <.001 |

widest at and just below periphery; up to 4 times rib width on last whorl, but crowded with secondary riblets. Microsculpture of faint, irregular axial growth striae covering surface, stronger in grooves; a few irregular spiral striae confined to grooves.

*Colour*. Polymorphic; brown patterned shells predominate. Ground colour pale yellow, pale brown or orange pink. Pattern of indistinct brown dashes and flecks on ribs, often aligned at suture and periphery, or over whole surface, to form irregular, oblique axial stripes, numbering 8–11 on last whorl, or sometimes producing a coarsely marbled pattern. Pink or yellow shells may lack dark pigment entirely, or may have short axial stripes only at suture and periphery; some shells show paler spiral zones at periphery and middle of base. Aperture cream, external pattern visible within, obscured by whitish or pink callus in larger shells. Columella usually pink or purple, may be entirely white in pale shells.

ANIMAL. *Colour*. Pigmentation grey to black; sides of foot darkly mottled; head darkest; tentacles banded, unpigmented stripe each side of base.

*Penis* (Fig. 69a–d). Length to 6.0 mm. Base bifurcate, limb bearing glandular disc not long. Filament large, broad, tip sharply tapering. Sperm groove open. Base pale ochraceous brown, filament cream, glandular disc opaque cream.

*Sperm*. Nurse cells (Fig. 69e–g) 16–29 $\mu$m, oval; rods 1–3, central, usually projecting, parallel sided, blunt; yolk granules small.

*Pallial oviduct* (Fig. 69h–k). Length to 5.1 mm. Spiral section to 2.3 mm diam., $2\frac{1}{2}$ whorls; opaque albumen gland $\frac{3}{4}$ whorl, cream; translucent albumen gland pale brown; capsule glands absent; epithelial lining reddish; spiral distinct externally; egg groove with a little black pigment. Straight section to 3.0 mm, pale brown, terminating in a papilla. Bursa long, to 2.2 mm, anterior. Development assumed ovoviviparous.

*Radula* (Fig. 96d). Length to 21 mm; relative length 0.59–0.90. Saw-toothed type; central rachidian cusp square, edge sometimes slightly pointed; cusps of paired teeth equilaterally triangular; lateral with no gap anterior to main cusp.

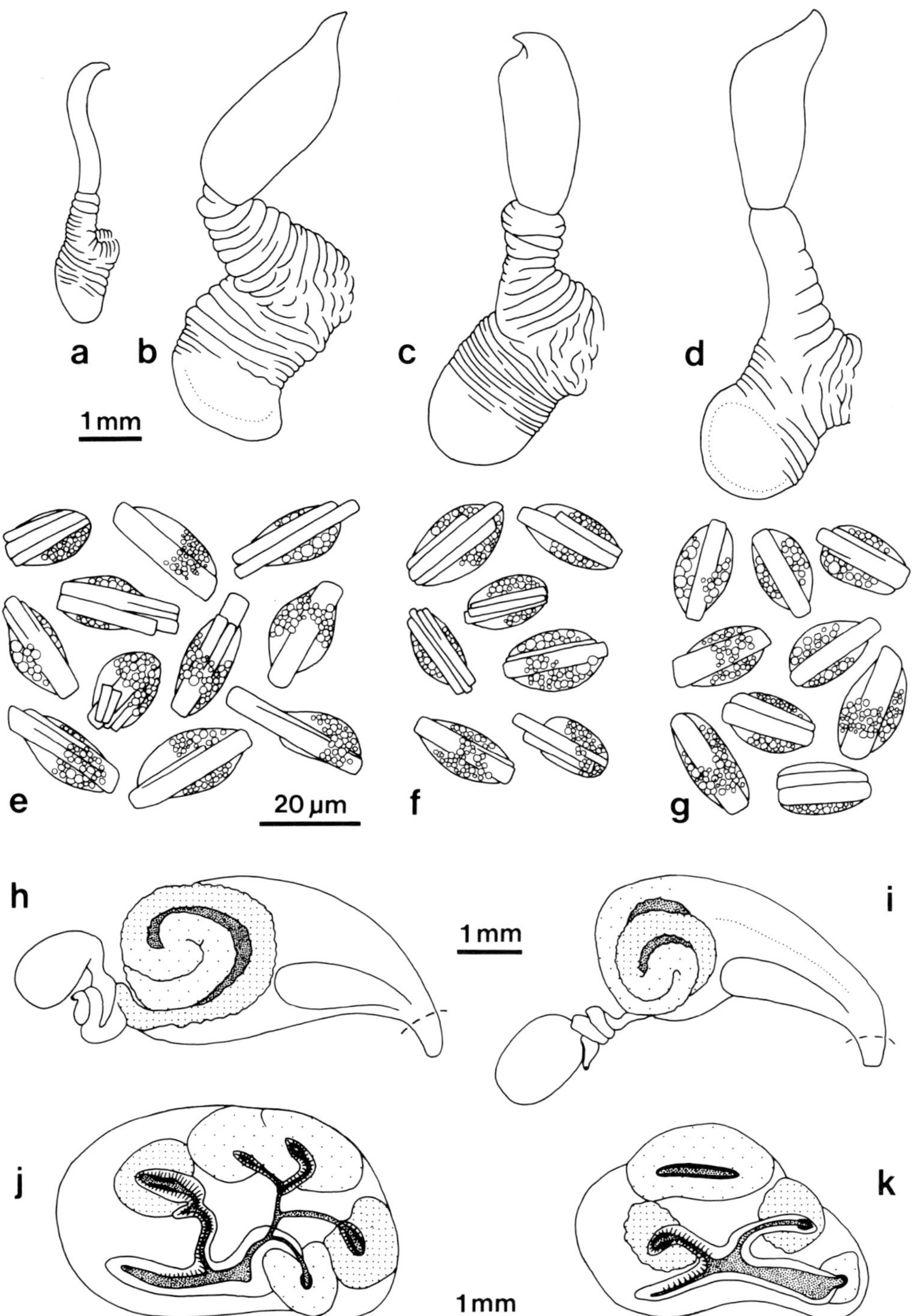


**Fig. 69** *Littoraria* (*Littorinopsis*) *ardouiniana:* **(a–d)** penes; **(a)** Tai Tam Harbour, Hong Kong; **(b–k)** Three Fathoms Cove, Hong Kong; **(e–g)** sperm nurse cells; **(h–k)** pallial oviducts, with transverse sections.

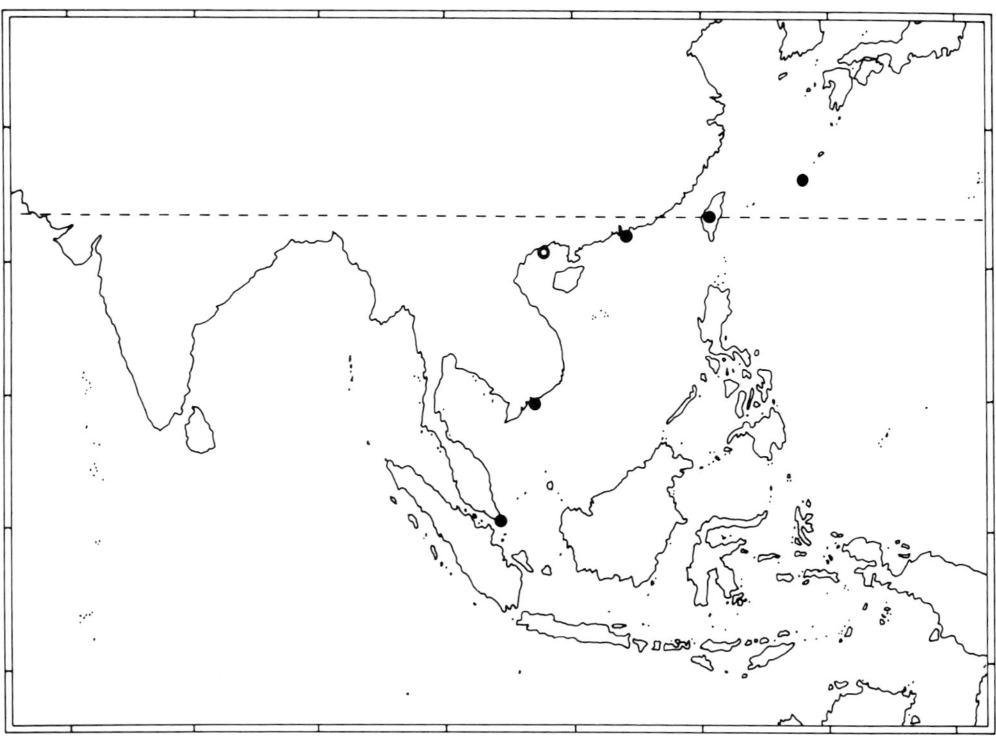

**Fig. 70** Distribution of *Littoraria* (*Littorinopsis*) *ardouiniana*.

DISTRIBUTION. *Habitat*. Leaves and trunks in *Avicennia* fringe, to 1.5 m above ground and probably higher. Also reported on *Aegiceras* and *Kandelia* (B.S. Morton and M. Nishihira, pers. comm.), on rocks (Heude, 1885; Fischer, 1891) and in marsh grass (collection in AMS, coll. W.F. Ponder & B.S. Morton). A continental species.

*Range* (Fig. 70). Okinawa Island to Singapore.

*Records*. **Singapore:** Jelutong, Ubin I. (DGR); **Vietnam:** Chilins, Vung Tau district (ANSP); Bay of Along (Heude, 1885); **Hong Kong:** Three Fathoms Cove, New Territories (BMNH); West Point (BMNH); Tai Tam Harbour (AMS); **Taiwan** (MCZ); **Japan;** Okkubi R., Okinawa I. (BMNH).

REMARKS. Specimens of *L. ardouiniana* are rare in museum collections and the species has apparently not been mentioned as distinct in the literature since 1891. The shell is highly colour polymorphic and this, together with the thin texture and prominent varices, suggests that the typical habitat may be at rather high levels on leaves of mangroves. There is considerable variation in adult shell size and in the number of ribs on the body whorl. The rather numerous varices led Tryon (1887) to place this species in the subgenus *Lamellilitorina;* however, details of shell sculpture, protoconch and oviduct preclude any close relationship with the only known member of that subgenus, *L. albicans*.

SIMILAR SPECIES. *L. ardouiniana* may be confused with a number of other species on account of its variation in size, sculpture and colour. Shells may superficially resemble those of *L. subvittata*, *L. lutea*, *L. philippiana* and *L. angulifera* in the rather fine spiral ribs and diffusely striped colour pattern, but specimens of *L. ardouiniana* are separated by the narrow and rounded columella pillar. Amongst species which share this character, *L. cingulata pristissini* is one of the closest in appearance, but the primary spiral grooves on the early whorls are more numerous in *L. ardouiniana*, varices are more prominent and microsculpture scarcely present. Small shells could be confused with *L. luteola* or *L. delicatula*. From the former, *L. ardouiniana* is distinguished by the more numerous and finer ribs on the last whorl and absence of the two enlarged ribs at the periphery. Some small specimens do approach rather closely to *L. delicatula*, but in *L. ardouiniana* the sculpture is of more prominent and unequal ribs, separated by wider grooves, the periphery is more rounded, the shell thicker and the columella not excavated. Penial characters and the form of the sperm nurse cells separate *L. ardouiniana* from all the similar species mentioned above, with the exception of *L. delicatula* in which the anatomy is not known.

## *Littoraria (Littorinopsis) delicatula* (Nevill, 1885)

*Littorina conica* var. *delicatula* Nevill, 1885: 149–150 [Port Canning and False Point, Bengal [India]; holotype ZSI]
*Littorina delicatula*—Annandale & Prashad, 1919: 246, fig. 2c [radula], pl. 20, fig. 4
*Littorina conica* var. *subintermedia* Nevill, 1885: 150 [Port Canning and False Point [Bengal, India]; lectotype ZSI, 16.5 mm, here designated]
*Littorina subintermedia*—Annandale & Prashad, 1919: 245–246, fig. 2b [radula], pl. 20, fig. 3
*Littorina (Melaraphe) undulata*—Tryon, 1887: 244 [in part; not Gray, 1839]
*Littorina (Littoraria) undulata*—Rosewater, 1970: 436–439 [in part; not Gray, 1839]
*Littorina (Littorinopsis) scabra scabra*—Rosewater, 1970: 456–461 [in part; not Linnaeus, 1758]

NOMENCLATURE. Nevill (1885) described both *delicatula* and *subintermedia* as varieties of *L. conica*, which indeed they do in some respects resemble and which, according to Nevill, occurs at the same locality. The name *delicatula* was proposed for the very thin shelled, brightly coloured form. The lectotype of *subintermedia* is a little more solid, the sculpture more coarse (38 grooves on last whorl) and the colour pattern darker, but the two seem, from shell characters alone, to be conspecific. Nevertheless, Annandale & Prashad (1919) separated the two as distinct species on the basis of shell shape, thickness, colour and radular characters. The radular cusps of *L. subintermedia* were said to be transversely striate, but to judge by an accompanying drawing of the radula of *L. melanostoma*, the accuracy of observation could be doubted. Until anatomical evidence is available, *L. subintermedia* is tentatively regarded as a synonym of *L. delicatula*.

DIAGNOSIS. Shell: thin, often extremely delicate; sutures only slightly impressed; peripheral keel becoming obsolete on last whorl; lip flared; primary grooves 11–14; 35–50 flat ribs on last whorl, separated by impressed lines; colour polymorphic, yellow, brown or pink, often with brown pattern in short axial stripes only at suture and periphery, numbering 9–11 on last whorl. Animal: unknown.

SHELL (Fig. 71). *Shape*. Height 13–19 mm. Teleoconch 6.5–7.5 whorls. Shell thin, translucent, often extremely delicate. Spire only slightly convex; whorls but lightly rounded; sutures only slightly impressed. Peripheral keel prominent in young shells, becoming obsolete on last whorl. Adult lip thin, flared; varices 0–1(3). Columella narrow, excavated; pillar straight to slightly concave. Sexual dimorphism: males smaller, relatively lower spire and larger aperture, adult lip more strongly flared, varices more frequent (after Nevill, 1885; and pers. obs. of dimorphic empty shells).

*Dimensions:* Table 21.

*Sculpture* (Fig. 57e–f). Protoconch normal. First whorl of teleoconch smooth. Primary grooves (10)11–14; spacing almost equal, posterior 3 or 4 a little closer together. Secondary grooves form on whorls 5–6, at $\frac{1}{2}$ width of primary ribs. Rib width remains almost equal throughout, only peripheral rib and sometimes sutural rib are a little wider and more prominent; ribs number 35–50 on last whorl, all are low and flattened. Primary and secondary grooves alike are narrow, impressed lines. Microsculpture of faint axial growth striae over whole surface.

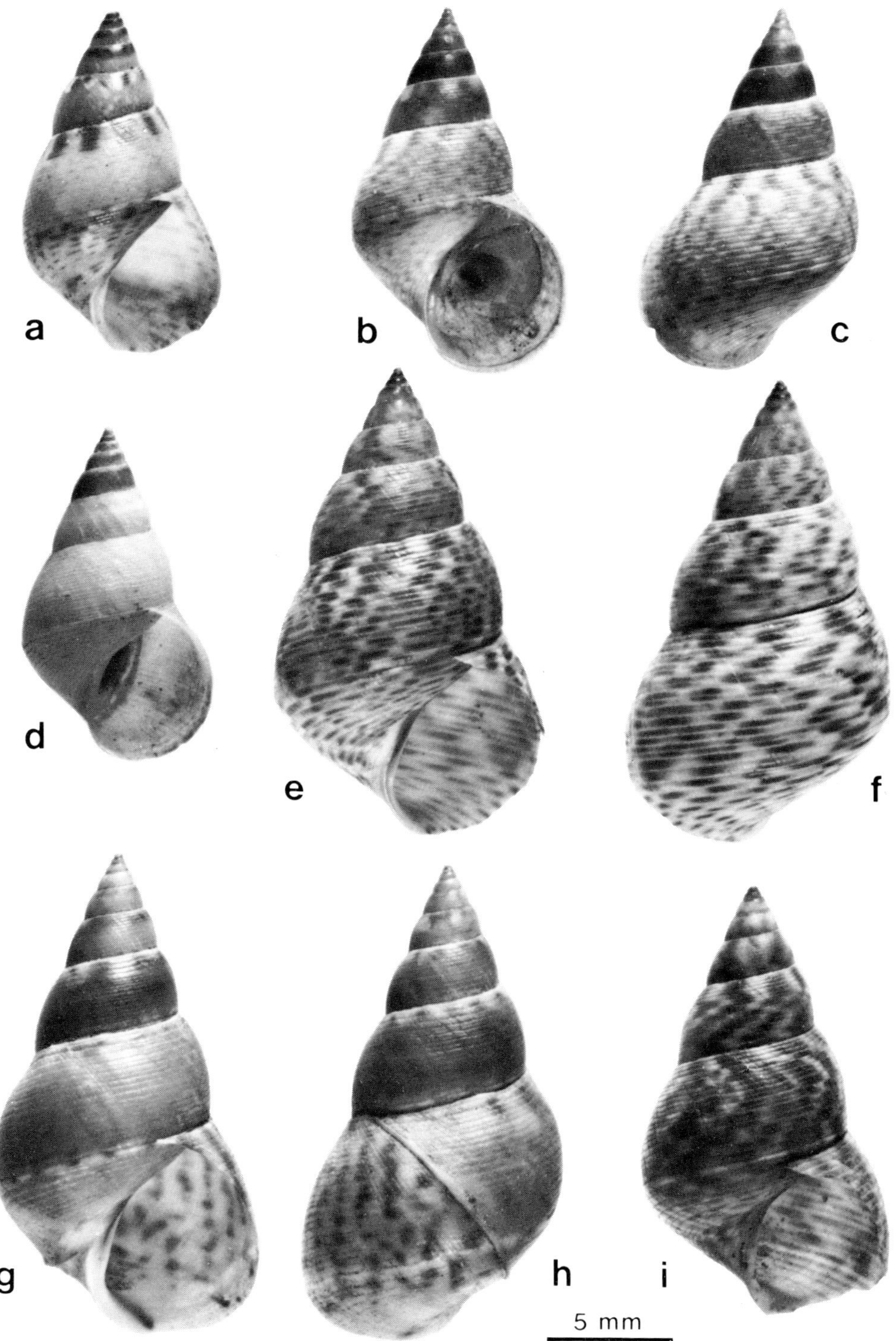


**Fig. 71** *Littoraria* (*Littorinopsis*) *delicatula*: **(a)** locality unknown (BMNH); **(b,c)** holotype of *Littorina conica* var. *delicatula* Nevill, Port Canning and False Point, Bengal, India (ZSI); **(d–f)** Port Canning, Bengal, India (BMNH); **(g,h)** locality unknown (BMNH); **(i)** lectotype of *Littorina conica* var. *subintermedia* Nevill, Port Canning and False Point, Bengal, India (ZSI).

**Table 21** Dimensions of *Littoraria* (*Littorinopsis*) *delicatula*.

| Specimen | Locality | Sex | Primary grooves | H (mm) | B (mm) | LA (mm) | WA (mm) | C (mm) | P | S | SH |
|---|---|---|---|---|---|---|---|---|---|---|---|
| *Littorina conica* var. *delicatula* holotype, ZSI | Port Canning and False Point, Bengal, India | ?♂ | 12 | 14.2 | 8.5 | 7.9 | 5.9 | 0.8 | 1.67 | 0.75 | 1.80 |
| *Littorina conica* var. *subintermedia* lectotype, ZSI | Port Canning and False Point, Bengal, India | ?♀ | 11 | 16.5 | 10.2 | 7.9 | 6.0 | 0.9 | 1.62 | 0.76 | 2.09 |
| BMNH acc. 1944 | Port Canning, Bengal | ?♂ | 14 | 14.2 | 8.9 | 7.8 | 5.7 | 0.8 | 1.60 | 0.73 | 1.82 |
| BMNH acc. 1944 | Port Canning, Bengal | ?♂ | 14 | 13.1 | 8.1 | 7.5 | 5.4 | 0.7 | 1.62 | 0.72 | 1.75 |
| BMNH acc. 1944 | Port Canning, Bengal | ?♀ | 13 | 14.7 | 8.9 | 7.0 | 5.7 | 0.8 | 1.65 | 0.81 | 2.10 |
| BMNH acc. 1944 | Port Canning, Bengal | ?♀ | 13 | 17.8 | 9.7 | 8.2 | 6.3 | 0.8 | 1.84 | 0.77 | 2.17 |
| BMNH acc. 1944 | unknown | ?♀ | 12 | 18.5 | 10.7 | 9.1 | 7.0 | 1.2 | 1.73 | 0.77 | 2.03 |
| BMNH | Port Canning, Bengal | ?♀ | 12 | 17.9 | 11.0 | 8.3 | 6.1 | 0.5 | 1.63 | 0.73 | 2.16 |
| BMNH 1969407 | Andaman Is | ?♀ | 10 | 18.5 | 11.3 | 9.4 | 7.0 | 1.4 | 1.64 | 0.74 | 1.97 |

Sex of shells based on description of dimorphism by Nevill (1885).

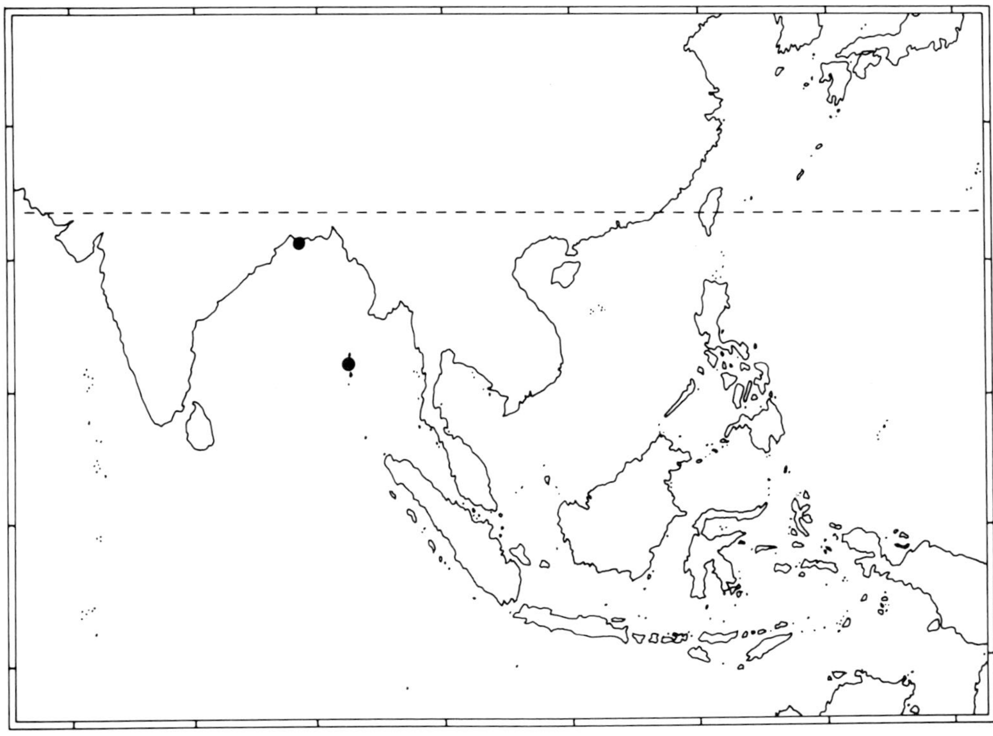

**Fig. 72** Distribution of *Littoraria* (*Littorinopsis*) *delicatula*.

*Colour*. Polymorphic; yellow, brown and pink shells occur. Ground colour pale yellow, cream or orange pink, overlain by brown dashes on ribs. Pattern variable; dark pigment often absent; pale shells may show brown dashes at suture and periphery only; darker shells heavily speckled, dashes aligned at suture and near periphery, forming short axial stripes, numbering 9–11 at suture of last whorl, a dark band just below periphery and around columella. Pattern clearly visible within aperture. Columella white or tinged with purple.

ANIMAL. Unknown.

DISTRIBUTION. *Habitat*. 'Trees and bushes far above high-tide mark' (Annandale & Prashad, 1919, p. 246). A habitat on leaves of mangroves would be predicted from the extreme fragility and bright coloration of the shell.

*Range* (Fig. 72). Bengal and Andaman Islands.

*Records*. **India:** Port Canning, Bengal (BMNH, USNM, ZSI); **Andaman Is** (BMNH).

REMARKS. This species is very rare in museum collections. Shell thickness is variable; specimens from Port Canning, Bengal, are mostly extremely fragile, while a shell from the Andaman Islands is solid. Thin shells and colour polymorphism are typical of species living at high levels on leaves (e.g. *L. filosa*, *L. luteola*, *L. albicans*), so it is likely that *L. delicatula* occurs in a similar habitat. Characters of the animal are unknown, but the thin, colour polymorphic and sometimes varicose shell supports classification in the subgenus *Littorinopsis*.

SIMILAR SPECIES. *L. delicatula* is most likely to be confused with small, thin shells of *L. conica*. However, *L. delicatula* has a taller spire, varices are sometimes present, fresh shells are brilliantly coloured (Nevill, 1885) and the protoconch is not papillose. If the subgeneric placement of *L. delicatula* is correct, anatomical characters should immediately separate it from *L. conica*. Small specimens of *L. ardouiniana* may be similar, but *L. delicatula* can be distinguished by its finer sculpture of almost equal ribs, separated by narrow grooves, by the more angular periphery, the thinner shell and the narrowly excavated columella. This last mentioned character is not always useful, since in specimens of *L. delicatula* with the thinnest shells the columella may be so narrow that the excavation is scarcely apparent, while in *L. ardouiniana* the columella is sometimes a little flattened. The ranges of *L. ardouiniana* and *L. delicatula* are not known to overlap. It is conceivable that additional specimens and anatomical evidence might indicate a closer relationship between these two species, but from present evidence they seem always to be distinct. *L. filosa*, *L. luteola* and *L. albicans* are only superficially similar to *L. delicatula*. Confusion with *L. undulata* should never arise.

## Subgenus *PALUSTORINA* n. subgen.

TYPE SPECIES: *Littoraria melanostoma* (Gray, 1839)

ETYMOLOGY. Condensation of Latin: *palustris*, swamp, and *Littorina*.

DIAGNOSIS. Shell usually solid; microsculpture, if present, of spiral striae on ribs and axial striae in grooves; shell coloration variable but not polymorphic; outer lip of aperture rarely flared; varices usually absent. Penial base simple, not bifurcate, glandular disc incorporated into base; penial sperm groove open. Sperm nurse cells flagellate. Bursa copulatrix opens near posterior end of straight section of pallial oviduct; capsule glands present. Development oviparous. Radula with 'hooded' rachidian tooth.

## *Littoraria* (*Palustorina*) *melanostoma* (Gray, 1839)

*Littorina melanostoma* Gray, 1839: 140 [Penang, Malaysia; lectotype (Rosewater, 1970) BMNH 1968364]; Reeve, 1857: *Littorina* pl. 9, figs 45a,b; Nevill, 1885: 151; von Martens, 1887: 170; Annandale & Prashad, 1919: 245, fig. 2a [radula]; Yen, 1933: 94; Berry & Chew, 1973 [reproductive biology, ecology]

*Litorina melanostoma*—Philippi, 1847, vol. 2: 224–225, *Litorina* pl. 5, fig. 16; Weinkauff, 1878: 41, pl. 4, fig. 19

*Litorina* (*Melaraphe*) *melanostoma*—Tryon, 1887: 245, pl. 43, figs 42, 43; Dautzenberg & Fischer, 1905: 148

*Littorina* (*Malaraphe*) *melanostoma*—Casto de Elera, 1896: 311

*Littorina* (*Littorinopsis*) *melanostoma*—von Martens, 1897: 199; Oostingh, 1923: 49–50, fig. 2; Rosewater, 1970: 462–464, pl. 325, figs 28, 29, pl. 355, figs 1–4

*Littoraria melanostoma*—Habe & Kosuge, 1966: pl. 6, fig. 8

*Littorinopsis melanostoma*—Brandt, 1974: 55, pl. 4, fig. 63

*Littorina melanostoma* var. *articulata* Nevill, 1885: 151 [Hong Kong; lectotype here designated ZSI, 12.1 mm; not Philippi, 1846]

Nomenclature. *Littorina melanostoma* var. *articulata* Nevill is simply a small form of the species; the lectotype is evidently adult, showing a somewhat thickened apertural lip, and resembles the dwarf forms found in salt marshes in Hong Kong.

Diagnosis. Shell: solid; spire tall, sides almost flat, sutures not impressed; periphery rounded; aperture elongate, somewhat quadrangular; columella narrow, rounded; primary grooves 6–8; secondary sculpture absent; ribs flat, numbering 15–17 on last whorl; grooves narrow; microsculpture of spiral striae on ribs, axial lines or pits in grooves; colour cream or yellow, usually with small brown spots aligned into axial series or stripes, numbering 27–40 on last whorl, aperture yellow, parietal callus and upper part of columella dark purple brown. Animal: penis not bifurcate, coarse annular wrinkles at posterior edge, filament fairly long; oviparous.

Shell (Fig. 73). *Shape*. Height (12)20–31 mm. Teleoconch whorls 6.5–7.5. Shell of moderate thickness, solid. Spire tall, outline straight or slightly convex; whorls almost flat; sutures not impressed. Periphery roundly angled in young shells, last whorl more evenly rounded. Aperture rather elongate, somewhat quadrangular; adult lip scarcely thickened, not flared. Varices absent. Columella narrow, rounded, not excavated; pillar concave. Sexual dimorphism: not significant.

*Dimensions*: Table 22.

*Sculpture* (Fig. 74a,c,d). Protoconch normal. All whorls of teleoconch sculptured with spiral grooves. Primary grooves 6–8, equally spaced or 2 posterior ribs narrower. Secondary sculpture absent. Ribs remain flat, of approximately equal width, often a little narrower at suture and periphery, numbering (13)15–17(19) on last whorl. Grooves narrow, less than $\frac{1}{8}$ width of ribs on last whorl. Microsculpture of strong, regular, spiral striae covering surface; stria at mid point of each rib may be a little more prominent, but never develops into a secondary groove. Faint axial growth lines are stronger in grooves, which may appear pitted on last whorl.

*Colour*. Rather constant. Ground colour pale yellow or cream, fading to white; very rarely pale pink. Pattern of numerous red brown vertical bars on ribs, aligned to form narrow axial stripes, 27–40 on last whorl; pattern sometimes reduced to lines of dots or to a scarcely visible mottling; in darkest shells stripes almost cover surface, but pattern becomes abruptly paler below periphery. Apical 5 whorls may be tinged lilac grey or pink. Aperture pale yellow, exterior pattern only visible in darkest shells. Parietal callus and upper part of columellar pillar dark purple brown, lower part of pillar white.

Animal. *Colour*. Pigmentation pale grey to black; sides of foot pale, mottled, and with conspicuous opaque white flecks; snout grey to black, rest of head pale; tentacles banded, broad unpigmented patch at base, dark stripe behind eye.

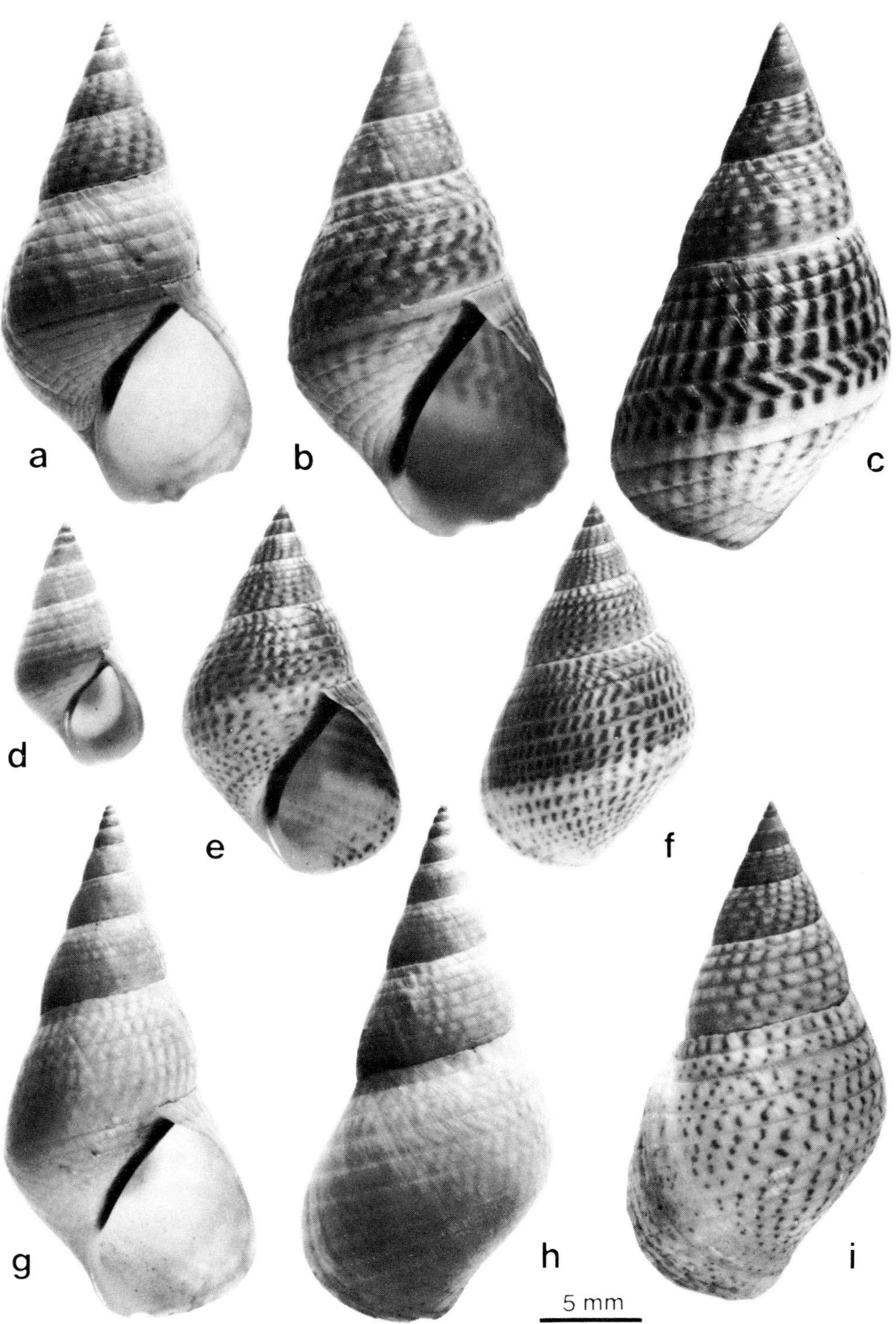


**Fig. 73** *Littoraria* (*Palustorina*) *melanostoma:* **(a)** lectotype of *Littorina melanostoma* Gray, Penang, Malaysia (BMNH 1968364); **(b,c)** Port Canning, Bengal, India (BMNH); **(d)** lectotype of *Littorina melanostoma* var. *articulata* Nevill, Hong Kong (ZSI) ; **(e,f)** ♀, Ubin I., Singapore (DGR); **(g,h)** Mergui Arch., Burma (BMNH); **(i)** ♂, Kanchanadit, Thailand (DGR).

**Table 22** Dimensions of *Littoraria* (*Palustorina*) *melanostoma*.

| Specimen | Locality | Sex | Primary grooves | H (mm) | B (mm) | LA (mm) | WA (mm) | C (mm) | P | S | SH |
|---|---|---|---|---|---|---|---|---|---|---|---|
| *Littorina melanostoma* lectotype, BMNH 1968364 | Penang, Malaysia | | 7 | 23.7 | 13.7 | 12.3 | 8.1 | | 1.73 | 0.66 | 1.93 |
| *Littorina melanostoma* var. *articulata* lectotype, ZSI | Hong Kong | | 6 | 12.1 | 7.1 | 6.5 | 4.5 | 0.7 | 1.70 | 0.69 | 1.86 |
| BMNH acc. 2176 | Port Canning, Bengal, India | | 7 | 29.9 | 16.5 | 14.6 | 10.0 | 1.8 | 1.79 | 0.68 | 2.05 |
| BMNH | Port Canning, Bengal, India | | 7 | 28.7 | 16.9 | 15.5 | 10.1 | 1.5 | 1.70 | 0.65 | 1.85 |
| DGR | Kanchanadit, Thailand | ♂ | 6 | 29.0 | 16.2 | 14.2 | 9.0 | 1.5 | 1.79 | 0.63 | 2.04 |
| DGR | Kanchanadit, Thailand | ♂ | 6 | 19.7 | 10.7 | 9.9 | 6.1 | 0.9 | 1.84 | 0.62 | 1.99 |
| DGR | Kanchanadit, Thailand | ♀ | 6 | 24.0 | 13.4 | 12.0 | 8.0 | 1.3 | 1.79 | 0.67 | 2.00 |
| DGR | Kanchanadit, Thailand | ♀ | 6 | 20.6 | 11.1 | 10.4 | 6.4 | 1.1 | 1.86 | 0.62 | 1.98 |
| DGR, mean of 10 | Kanchanadit, | ♂ | | 24.80 | | | | | 1.801 | 0.639 | 1.980 |
| standard error | Thailand | | | 0.58 | | | | | 0.019 | 0.007 | 0.019 |
| DGR, mean of 10 | Kanchanadit, | ♀ | | 23.83 | | | | | 1.753 | 0.647 | 1.952 |
| standard error | Thailand | | | 0.58 | | | | | 0.018 | 0.005 | 0.026 |
| statistic t or U | | | | 1.183 | | | | | 72 | 65 | 62 |
| probability | | | | 0.252 | | | | | 0.106 | 0.280 | 0.394 |

*Penis* (Fig. 75a–g). Length to 5.8 mm. Base simple, incorporating glandular disc; prominent annular wrinkles at posterior edge. Filament fairly long, tapering near tip. Sperm groove open. Base and filament white to cream, glandular disc opaque white.

*Sperm*. Eupyrene sperm 256–287 μm. Nurse cells (Fig. 75j–l) 33–50 μm, elongate oval, tapering at base; basal flagellum to 160 μm; rods 1–4, usually basal, narrowly elongate; yolk granules small, becoming larger towards base.

*Pallial oviduct* (Figs 7, 8). Length to 8.2 mm. Spiral section to 7.3 mm diam., $6\frac{1}{2}$ whorls; opaque albumen gland $\frac{1}{3}$ whorl, white; translucent albumen gland white to pale grey; opaque capsule gland $1\frac{1}{2}$ whorls, pale pink; translucent capsule gland red brown; spiral distinct externally; egg groove darkly pigmented. Straight section to 3.5 mm, pale grey to brown; no terminal papilla. Bursa posterior, extending into spiral section. Development oviparous (Berry & Chew, 1973).

*Egg capsules* (Fig. 75h,i, after Berry & Chew, 1973). Capsule 290–310 μm diam., symmetrically biconvex disc, with narrow circumferential flange; containing single ovum 120–140 μm diam.

*Radula* (Fig. 74b). Length to 26 mm; relative length 0.61–0.94. Chisel-toothed type; central rachidian cusp wide, same length as two small flanking cusps, giving almost continuous cutting edge; cusps of paired teeth extremely obliquely triangular, minor cusps reduced; lateral with gap anterior to main cusp.

DISTRIBUTION. *Habitat*. On trunks and sometimes leaves in *Avicennia* fringe, 0.3–1.8 m above ground, only on the outermost trees in very sheltered situations; most common at back of *Avicennia* zone and in outer parts of *Rhizophora* forest; roots and occasionally leaves in *Rhizophora* forest, 0.2–0.6 m above ground; scarce in *Bruguiera* zone. Juveniles occur mainly on leaves. Berry & Chew (1973) report juveniles amongst marsh grass. A continental species.

*Range* (Fig. 76). South East Asia, from southern India and Bay of Bengal to north coast of Java and southern China.

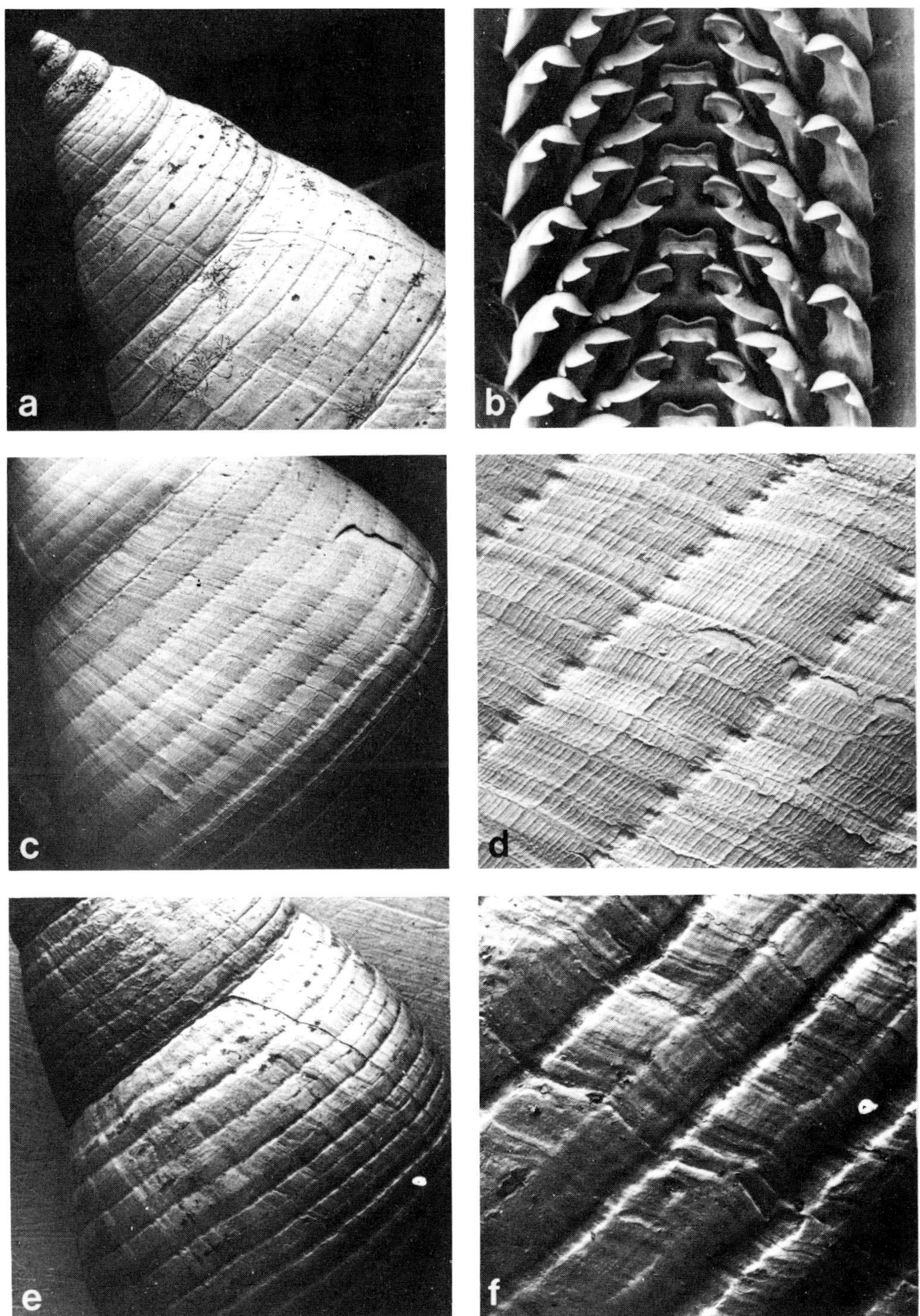


**Fig. 74** **(a–d)** *Littoraria* (*Palustorina*) *melanostoma:* **(a)** spire, Ubin I., Singapore ( × 20); **(b)** radula, Kanchanadit, Thailand ( × 170); **(c,d)** Ubin I., Singapore; **(c)** last whorl ( × 8); **(d)** detail ( × 31). **(e,f)** *Littoraria* (*Palustorina*) *flammea*, China: **(e)** last whorl ( × 8); **(f)** detail ( × 31).

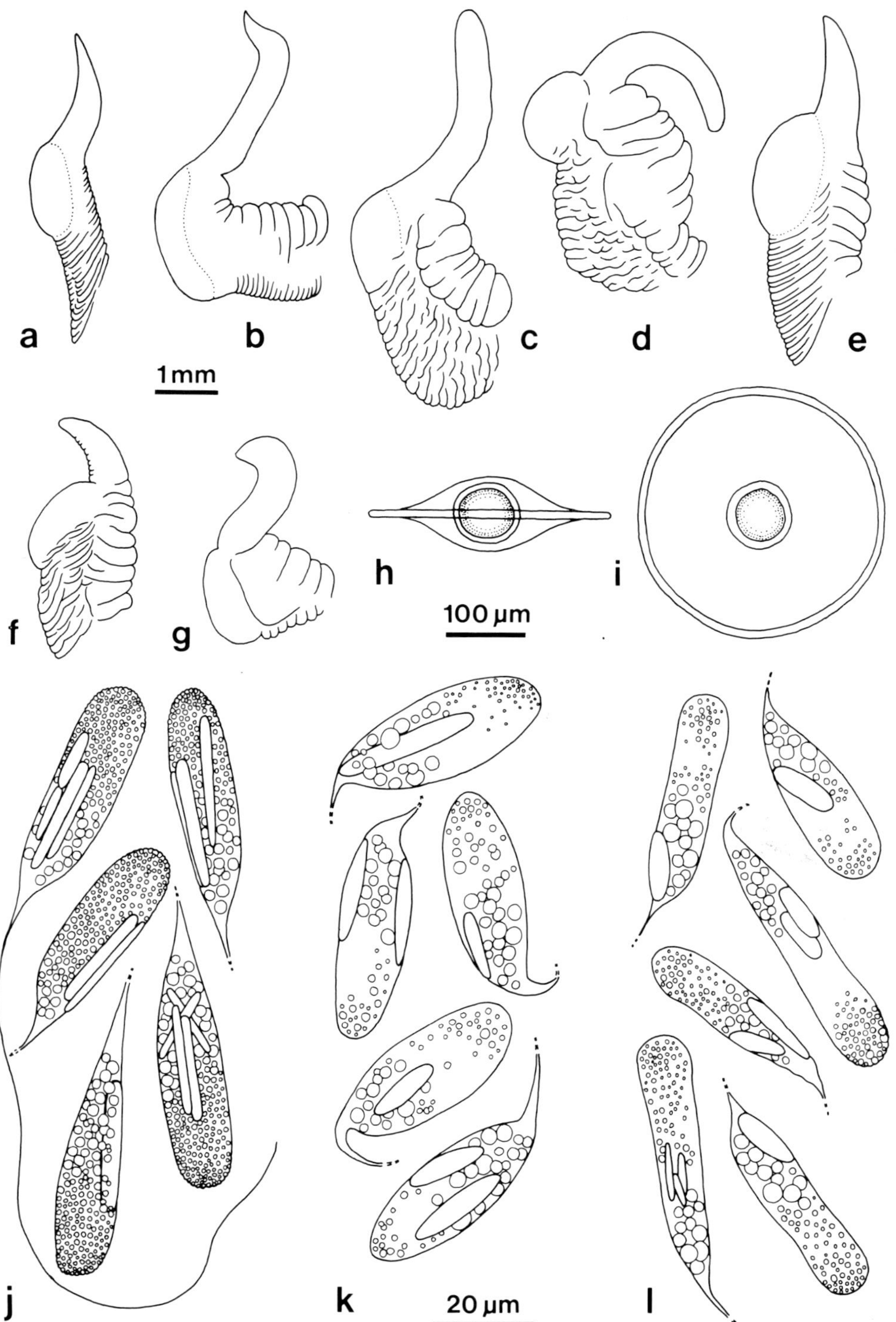


**Fig. 75** *Littoraria* (*Palustorina*) *melanostoma:* **(a–g)** penes; **(a,b)** Santubong, Sarawak; **(c)** Kanchanadit, Thailand; **(d)** Three Fathoms Cove, Hong Kong; **(e,f)** Ubin I., Singapore; **(g)** Three Fathoms Cove, Hong Kong; **(h,i)** egg capsule, W. Malaysia (after Berry & Chew, 1973); **(j–l)** sperm nurse cells, flagellum shown on one only; **(j)** Kanchanadit, Thailand; **(k,l)** Ubin I., Singapore.

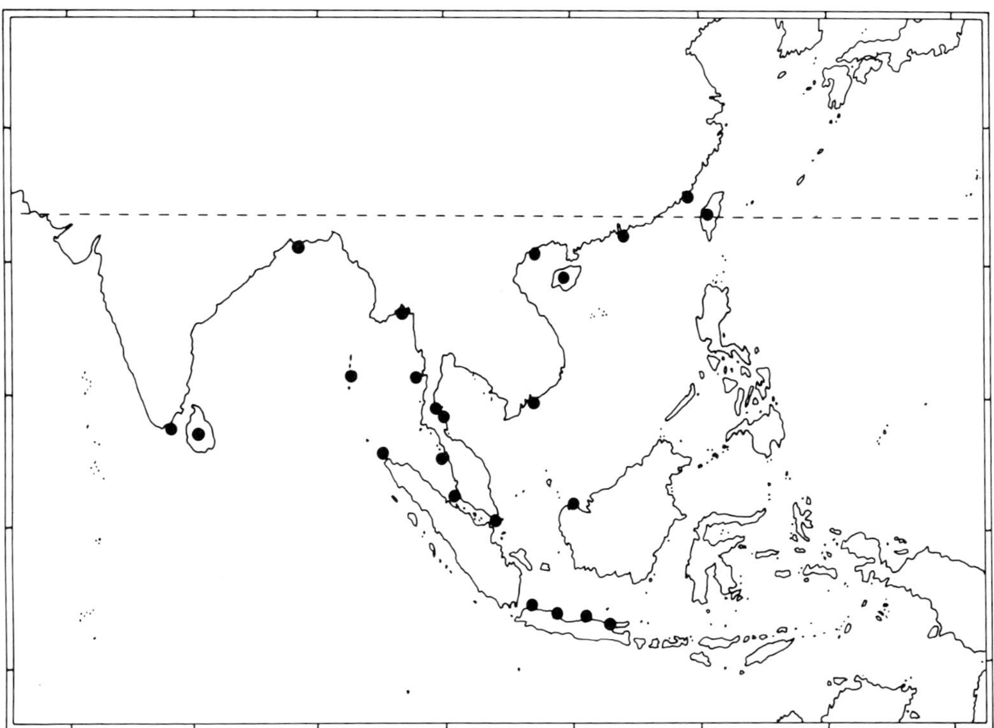

Fig. 76· Distribution of *Littoraria* (*Palustorina*) *melanostoma*.

*Records*. **India:** Manali I., off Mandapam (MCZ); Port Canning, Bengal (BMNH, RNHL); Calcutta (BMNH, AMS); **Andaman Is:** Port Blair (BMNH); **Sri Lanka** (BMNH); **Burma:** Pegu Estuary (BMNH); Mergui (BMNH); **Thailand:** Chonburi (ANSP, RNHL); Kanchanadit, near Surat Thani (DGR); Pakpun, Nakhon Si Thammarat (USNM); **Malaysia: Peninsula:** Batu Maung, Penang (DGR); Kuala Selangor (ANSP); 10 km N. of Port Kelang (DGR); **Sarawak:** Santubong (DGR); **Singapore:** Jelutong, Ubin I. (DGR); Kranji (BMNH); **Indonesia: Sumatra:** Weh I. (RNHL); **Java:** Tandjong Priok, Djakarta (RNHL); Tjirebon (RNHL); Pandjang I., near Djepara (RNHL); Surabaja (RNHL); **Vietnam:** Chilins, Vung Tau district (ANSP); Do Son, Tonkin (RNHL); **China:** Hainandao I. (ANSP); Amoy (ANSP, RNHL); **Hong Kong:** Three Fathoms Cove, New Territories (BMNH); Tai Tam Harbour (AMS); **Taiwan** (USNM).

REMARKS. This is a well known and distinctive species. As the type of the new subgenus *Palustorina*, the anatomy, especially that of the female reproductive tract, is discussed in some detail in the introductory sections. The radula is remarkable for the reduction in numbers of effective cusps, a feature shared with *L. conica*, but not found to such an extent in other members of the genus. Shell shape is constant, but degree of pigmentation shows some variation. The occurrence of a very few shells with pale pink ground colour is noteworthy, since this colour form is extremely rare in the subgenus, recorded elsewhere only in *L. articulata*. A possibly homologous pink morph is more frequent in the polymorphic species of the subgenus *Littorinopsis*. The relationship of *L. melanostoma* with *L. flammea* is discussed in the remarks upon the latter species.

SIMILAR SPECIES. *L. melanostoma* is so distinctive that confusion with any other species should not arise. *L. flammea* is the most similar, but the shell of *L. melanostoma* is more solid, the sides of the spire are straight, and the axial stripes or rows of aligned dots are more numerous.

## ***Littoraria (Palustorina) flammea*** (Philippi, 1847)

*Litorina flammea*—Philippi, 1847, vol. 3: 16, *Litorina* pl. 6, fig. 21 [China; lectotype here designated MNHNP, 19.5 mm]; Weinkauff, 1882: 56–57, pl. 7, figs 9,12
*Littorina flammea*—Reeve, 1857: *Littorina* pl. 9, figs 46a,b
*Littorina (Melaraphe) flammea*—Tryon, 1887: 245, pl. 43, figs 34, 35 [in part]
*Littorina (Malaraphe) flammea*—Casto de Elera, 1896: 311 [in part]
*Melarhaphe flammea*—Yen, 1942: 196
*Littorina fortunei* Reeve, 1857: *Littorina* pl. 9, figs 42a,b [China; lectotype (Rosewater, 1970) BMNH 1968309]
*Littorina (Littorinopsis) scabra scabra*—Rosewater, 1970: 456–461, pl. 352, figs 3, 4 [in part; not Linnaeus, 1758]

NOMENCLATURE. Rosewater (1970) designated the lectotype of *Litorina flammea* from amongst a lot of four specimens in the Cuming Collection (BMNH), supposing that the shell was that figured by Philippi (1847). The resemblance of the shell to the figure is indeed close. However, there is no documentary evidence to support the designation since, regarding the origin of the shell, Philippi merely notes the locality as China and that it was received from Largilliert. An old label accompanying the specimens and another by Tomlin record the locality of the shells as the Philippines. Nevertheless, Philippi did not in every case mention the origin of figured specimens when these were from Cuming (e.g. *L. intermedia*) and some mixing of labels could have occurred. The lectotype of *Littorina fortunei* Reeve was selected by Rosewater from the same lot, but does not bear close resemblance to Reeve's figure, although shell and figure are undoubtedly of the same species. Evidently Reeve himself did not recognize these shells as syntypes of Philippi's species, for he figured a broader shell from another lot as *Littorina flammea* Philippi. In the MNHNP are three shells of *L. flammea* from the Largilliert Collection housed in the Natural History Museum in Rouen (P. Bouchet, pers. comm.). It seems appropriate to redesignate as lectotype one of these which closely resembles Philippi's figure.

DIAGNOSIS. Shell: rather thin; spire tall, outline convex, whorls lightly rounded; aperture elongate; varices may be present; columella narrow, rounded; primary grooves 7–9; secondary sculpture absent; total of 16–20 low, rounded ribs on last whorl, with narrow grooves between; microsculpture of spiral striae on ribs, axial striae or pits in grooves; colour white to cream, pattern of narrow, brown, oblique axial stripes from suture to periphery, numbering 9–10 on last whorl; columella and parietal callus purple brown. Animal: unknown.

SHELL (Fig. 77). *Shape*. Height 16–20 mm. Teleoconch of about 7 whorls. Shell rather thin. Spire tall, outline gently convex; whorls only slightly rounded; sutures a little impressed. Peripheral keel absent. Aperture elongate; adult lip a little flared, not thickened. Varices 0–1(4). Columella narrow, rounded, not excavated; pillar concave. Sexual dimorphism: unknown.

*Dimensions*: Table 23.

*Sculpture* (Fig. 74e,f). Protoconch not seen. Primary grooves 7–9, equally spaced, or posterior 2 ribs a little narrower. Secondary sculpture absent. Ribs remain low and rounded, of similar width, but slightly narrower at suture and periphery; total of 16–20 on last whorl. Grooves narrow, $\frac{1}{6}$–$\frac{1}{3}$ rib width on last whorl; widest groove is that just below peripheral rib, which is $\frac{1}{2}$–1 width of this rib. Microsculpture of spiral striae on ribs, sometimes indistinct; on last whorl central stria of each rib is pronounced, but does not develop into secondary sculpture, seeming to be only a projection of the periostracum; irregular axial growth striae cover surface, stronger in grooves, which may appear pitted on last whorl.

*Colour*. Ground colour whitish or cream; pattern of red brown, oblique axial stripes from suture to periphery; stripes are narrow and distant, usually interrupted by primary grooves, and number 9–10 on last whorl; pattern becomes faint on base. External pattern visible within aperture. Columella and parietal callus purple brown or pink.

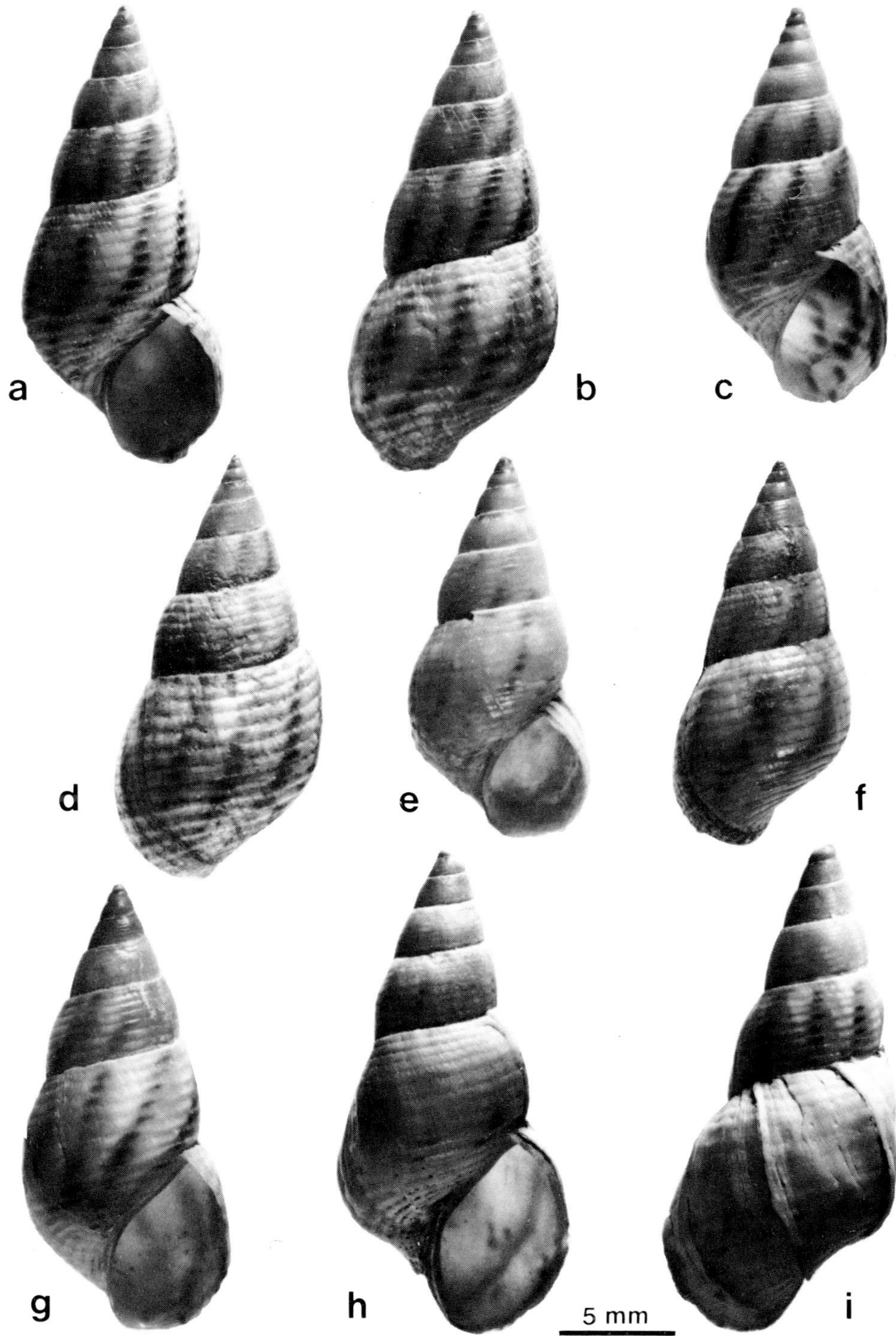


**Fig. 77** *Littoraria* (*Palustorina*) *flammea*, China: **(a,b)** lectotype of *Litorina flammea* Philippi (MNHNP); **(c)** specimen designated lectotype of *Litorina flammea* Philippi by Rosewater (1970) (BMNH 1968310); **(d)** (BMNH); **(e)** lectotype of *Littorina fortunei* Reeve (BMNH 1968309); **(f)** (BMNH); **(g)** paralectotype (MNHNP); **(h,i)** paralectotype (MNHNP).

**Table 23** Dimensions of *Littoraria* (*Palustorina*) *flammea*.

| Specimen | Locality | Sex | Primary grooves | H (mm) | B (mm) | LA (mm) | WA (mm) | C (mm) | P | S | SH |
|---|---|---|---|---|---|---|---|---|---|---|---|
| *Litorina flammea* lectotype, MNHNP | China | | 8 | 19.5 | 9.2 | 8.0 | 5.1 | 0.6 | 2.12 | 0.64 | 2.44 |
| paralectotype, MNHNP | China | | 9 | 20.5 | 9.9 | 9.4 | 6.3 | 0.8 | 2.07 | 0.67 | 2.18 |
| *Littorina fortunei* lectotype, BMNH 1968309 | [China] | | 8 | 16.1 | 8.5 | 6.5 | 4.6 | | 1.89 | 0.71 | 2.48 |
| BMNH 1968310 | [China] | | 8 | 16.7 | 7.4 | 7.9 | 4.8 | 0.6 | 2.26 | 0.61 | 2.11 |
| BMNH acc. 1829 | China | | 8 | 16.1 | 8.9 | 7.8 | 4.8 | 0.5 | 1.81 | 0.62 | 2.06 |
| BMNH acc. 1829 | China | | 7 | 15.9 | 8.0 | 7.3 | 4.6 | 0.4 | 1.99 | 0.63 | 2.18 |
| BMNH acc. 1829 | China | | 8 | 17.8 | 9.1 | 8.5 | 4.9 | 0.5 | 1.96 | 0.58 | 2.09 |
| BMNH acc. 1829 | China | | 8 | 15.4 | 8.0 | 7.0 | 4.4 | 0.4 | 1.93 | 0.63 | 2.20 |

Animal. Unknown.

Distribution. *Habitat.* Unknown.

*Range.* Recorded from 'China' in literature and on museum labels. A doubtful record from Mindoro, Philippines (Casto de Elera, 1896).

*Records.* **China** (BMNH, MNHNP, MCZ).

Remarks. This species is very rare in museum collections; all material seen dates from the early nineteenth century and none is localized more precisely than 'China'. Since the species is not known from Indochina, Hong Kong or Japan, it is possible that it occurs on the shores of the East China Sea or the Yellow Sea. Despite the record by Casto de Elera (1896) and the label accompanying one collection in the BMNH, occurrence in the Philippines appears most unlikely, for large collections of other littorinids are available from the area.

The animal of *L. flammea* is unknown, but the species is placed in the subgenus *Palustorina* on the basis of the resemblance of the shell to that of *L. melanostoma*. The most significant similarity is the type of microsculpture, but other points of resemblance include the dark coloration of the parietal callus, narrow columella, elongate spire and pronounced axial alignment of the colour pattern. Despite these similarities, it does not at present seem possible that *L. flammea* could be a form of *L. melanostoma*. The latter is a species of constant characters throughout its rather wide range; even specimens from Hong Kong and Amoy on the Chinese coast are entirely typical and do not approach *L. flammea*. Furthermore, the few shells of *L. flammea* available are closely similar to each other.

Similar species. *L. flammea* could be confused only with *L. melanostoma*, but is distinguished from that species by its thinner shell, more convex spire outline and by the pattern of fewer axial stripes. Authors have compared this species with *L. luteola*, but in the latter the shell sculpture and coloration are quite different, as, if the subgeneric placement of *L. flammea* is correct, the animal may be also.

## ***Littoraria* (*Palustorina*) *conica*** (Philippi, 1846)

*Littorina conica* Philippi, 1846: 141 [Java; lectotype (Rosewater, 1970) BMNH 1968225]; Reeve, 1857: *Littorina* pl. 8, figs 36a,b; Nevill, 1885: 149 [in part]
*Litorina conica*—Philippi, 1847, vol. 3: 9, *Litorina* pl. 6, figs 1,2; Weinkauff, 1882: 54, pl. 7, figs 1,4
*Littorina* (*Melaraphis*) *conica*—Tapparone-Canefri, 1874: 48
*Littorina* (*Littorinopsis*) *conica*—von Martens, 1897: 198; Prashad, 1921: 485; Oostingh, 1927: 3

Fig. 78 *Littoraria (Palustorina) conica:* **(a)** lectotype of *Littorina conica* Philippi, Java (BMNH 1968225); **(b,c)** ♂, Santubong, Sarawak (DGR); **(d–g)** Sungei Merbok estuary, Malaysia (DGR); **(d,g)** ♂; **(e,f)** ♀; **(h,i)** ♀, Santubong, Sarawak (DGR).

*Littorinopsis conica*—Kuroda & Habe, 1952: 64; Brandt, 1974: 55, pl. 4, fig. 64
*Littoraria conica*—Higo, 1973: 46
*Littorina* (*Melaraphe*) *undulata*—Tryon, 1887: 244, pl. 43, fig. 41 [in part; not Gray, 1839]
*Littorina* (*Littorinopsis*) *carinifera*—Rosewater, 1970: 464–465, pl. 355, figs 10,11 [in part; not Menke, 1830]

NOMENCLATURE. Philippi described *Littorina conica* from specimens in the Cuming Collection, giving the type locality as Java. Rosewater (1970) designated as lectotype a shell in a lot from this collection which bears some resemblance to the figure of Philippi (1847). However, no original label exists and labels in Tomlin's hand give the locality as 'Japan', possibly misled by Reeve (1857) who also gave this locality for his figure. This might perhaps be an error for 'Java'; no specimens with reliable data have been seen from Japan, although the species appears in several Japanese faunal lists, possibly on the basis of Reeve's record. Philippi (1847) did not mention the owner of the figured shell in this case, but did not invariably do so (e.g. *L. intermedia* and its 'varieties'). From the available evidence the lectotype designation seems acceptable. Sowerby (1825) described *Turbo conicus*, a species of Upper Albian age from the Blackdown Greensand and subsequently (1840) referred it to the genus *Littorina*. This is a problematical species, but does not belong to either of the genera *Littorina* or *Littoraria* (N.J. Morris & R. Cleevely, pers. comm.).

DIAGNOSIS. Shell: apex blunt with papillose protoconch; spire outline convex, whorls flattened, sutures scarcely impressed; peripheral keel prominent; columella excavated, pillar convex; primary grooves 10–12; total of 50–70 flat ribs on last whorl, with narrow grooves between; microsculpture indistinct; colour variable, cream, usually with dense pattern of brown flecks, aligned into axial stripes only on spire whorls. Animal: penis not bifurcate, filament long, tapering; oviparous.

SHELL (Fig. 78). *Shape*. Height 16–25 mm. Teleoconch 5.5–6.5 whorls. Shell of moderate thickness, solid. Spire outline convex; first whorls of teleoconch are broad, so that apex appears blunt with a papillose protoconch; first 3 whorls rounded, subsequent whorls become flattened; sutures scarcely impressed. Peripheral keel prominent, becoming obsolete at end of last whorl. Adult lip only slightly thickened, not usually flared. Varices rare, usually merely strong growth lines behind lip. Columella of moderate width, excavated; pillar noticeably convex, narrowed at base. Sexual dimorphism: males smaller, relatively lower spire and larger aperture, aperture more elongate.

*Dimensions*: Table 24.

**Table 24** Dimensions of *Littoraria* (*Palustorina*) *conica*.

| Specimen | Locality | Sex | Primary grooves | H (mm) | B (mm) | LA (mm) | WA (mm) | C (mm) | P | S | SH |
|---|---|---|---|---|---|---|---|---|---|---|---|
| *Littorina conica* lectotype, BMNH 1968225 | Java | | 11 | 22.3 | 15.5 | 12.8 | 9.5 | 2.1 | 1.44 | 0.74 | 1.74 |
| DGR | Sungei Merbok, Malaysia | ♂ | 11 | 18.2 | 11.9 | 10.3 | 7.5 | 1.5 | 1.53 | 0.73 | 1.77 |
| DGR | Sungei Merbok, Malaysia | ♀ | 10 | 21.5 | 14.3 | 11.7 | 9.2 | 2.0 | 1.50 | 0.79 | 1.84 |
| DGR | Santubong, Sarawak | ♂ | 11 | 19.7 | 14.2 | 12.0 | 8.6 | 1.4 | 1.39 | 0.72 | 1.64 |
| DGR | Santubong, Sarawak | ♂ | 10 | 18.9 | 13.1 | 11.1 | 8.2 | 1.7 | 1.44 | 0.74 | 1.70 |
| DGR | Santubong, Sarawak | ♀ | 10 | 22.6 | 15.9 | 13.0 | 10.2 | 2.5 | 1.42 | 0.78 | 1.74 |
| DGR | Santubong, Sarawak | ♀ | 10 | 21.3 | 14.6 | 12.4 | 9.5 | 2.3 | 1.46 | 0.77 | 1.72 |
| DGR, mean of 10 | Santubong, Sarawak | ♂ | | 18.12 | | | | | 1.428 | 0.734 | 1.687 |
| standard error | | | | 0.48 | | | | | 0.011 | 0.004 | 0.013 |
| DGR, mean of 10 | Santubong, Sarawak | ♀ | | 20.59 | | | | | 1.431 | 0.761 | 1.730 |
| standard error | | | | 0.67 | | | | | 0.011 | 0.004 | 0.010 |
| statistic t or U | | | | 2.999 | | | | | 55.5 | 94 | 79 |
| probability | | | | 0.008 | | | | | 0.712 | <.001 | 0.028 |

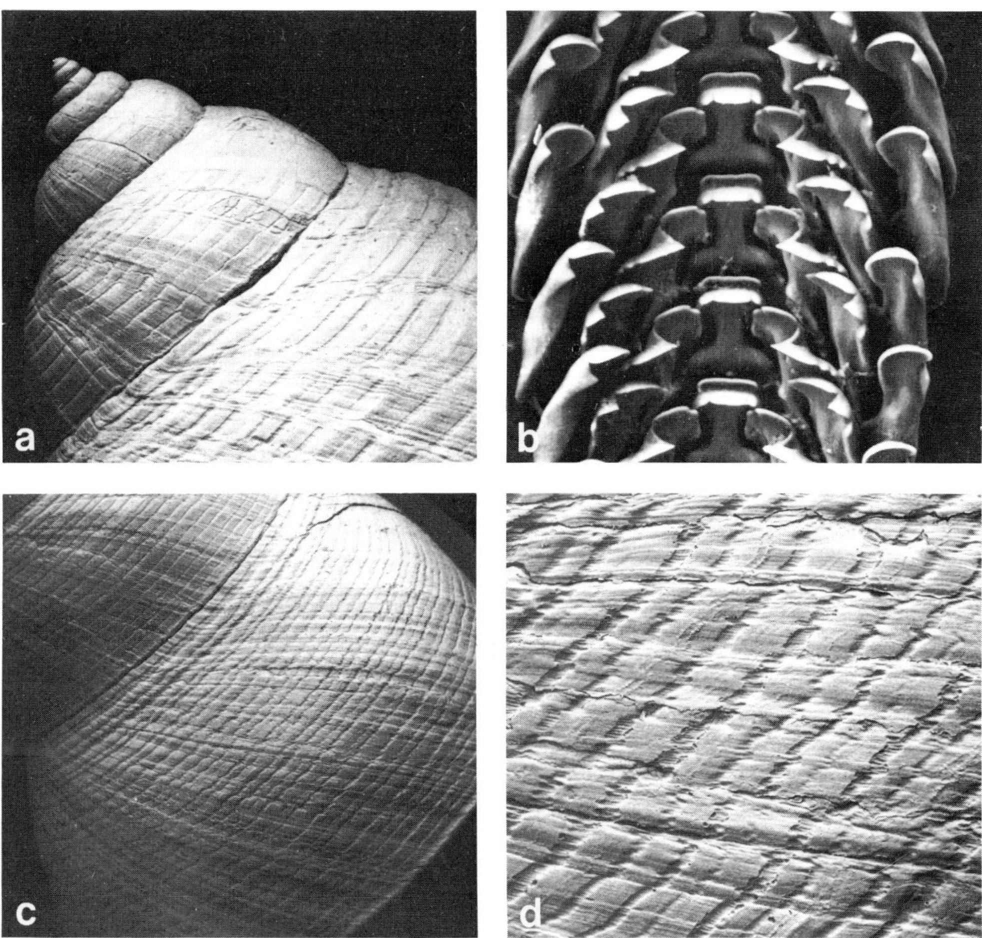


**Fig. 79** *Littoraria* (*Palustorina*) *conica:* **(a)** spire, Santubong, Sarawak ( × 18); **(b)** radula, Penang, Malaysia ( × 220); **(c,d)** Santubong, Sarawak; **(c)** last whorl ( × 7); **(d)** detail ( × 27).

*Sculpture* (Fig. 79a,c,d). Protoconch of normal shape (sculpture not seen). First whorl of teleoconch probably smooth. Primary grooves (9)10–12(13) by whorl 4; posterior 4 ribs about half width of the equally spaced ribs over rest of whorl, having been formed by division of 2 larger ribs on earlier whorls. On whorls 4–5 primary ribs become divided by secondary grooves at anterior $\frac{1}{3}$–$\frac{1}{2}$ width; posterior 4 ribs divide later. Tertiary grooves appear on whorl 6. Grooves remain narrow, less than $\frac{1}{5}$ width of widest ribs on last whorl; all grooves are of similar size, spaced rather unevenly across surface of last whorl to produce 50–70 ribs. All ribs remain low and flat, that at periphery often a little more prominent owing to slightly greater width and to sharp angulation of shell. Microsculpture of irregular axial growth striae covering surface, stronger in grooves, which may appear pitted on last whorl; faint spiral striae sometimes visible.

*Colour.* Variable. Ground colour cream or pale brown, usually with pattern of dark brown flecks and dashes densely scattered over surface. Pattern often roughly aligned on spire to form axial stripes numbering 11–13 per whorl, but last whorl is uniformly speckled and flecked with dark pigment. Degree of pigmentation varies; uniform cream shells entirely lacking pigment are not infrequent. Purple brown flecks and dashes corresponding to external pattern are visible within aperture, clouded by a thin whitish callus. Columella and parietal callus dark purple, occasionally white in palest shells; columellar pillar whitish in all individuals.

ANIMAL. *Colour*. Pigmentation grey to black; sides of foot darkly mottled; head darkest; tentacles banded, unpigmented stripe each side of base.

*Penis* (Fig. 80a–c). Length to 6.1 mm. Base simple, incorporating glandular disc. Filament long, tapering. Sperm groove open. Base and filament fawn, glandular disc pale brown.

*Sperm*. Eupyrene sperm 308–344 $\mu$m. Nurse cells (Fig. 80g–i) 35–51 $\mu$m, elongate oval or fusiform, usually with mucronate or papillose apex; basal flagellum to 250 $\mu$m; rods 1(2), basal, lozenge shaped to rounded oval; yolk granules small, becoming larger towards base.

*Pallial oviduct* (Fig. 80d–f). Length to 7.6 mm. Spiral section to 5.7 mm diam., $5\frac{1}{2}$ or $6\frac{1}{2}$ whorls; opaque albumen gland $\frac{1}{4}$ whorl, white; translucent albumen gland off white; opaque capsule gland 1 whorl, pale pink; translucent capsule gland pink to red brown; spiral usually distinct externally; egg groove darkly pigmented. Straight section to 4.0 mm, pale brown; no terminal papilla. Bursa posterior, extending into spiral section. Development assumed oviparous.

*Radula* (Fig. 79b). Length to 28 mm; relative length 1.21–1.53. Chisel-toothed type; central rachidian cusp wide, same length as 2 small flanking cusps, giving almost continuous cutting edge; cusps of paired teeth extremely obliquely triangular, minor cusps reduced; lateral with gap anterior to main cusp.

*Alimentary system*. Anterior pair of oesophageal pouches large, to 1.9 mm diam., dark red and glandular. Mid-oesophagus also dark red and glandular.

DISTRIBUTION. *Habitat*. Occasional in *Bruguiera* and *Ceriops* zones, but typically found in the landward fringe, on trunks and leaves of *Avicennia, Excoecaria, Acanthus, Nypa* and others, 0.2–2.0 m above the ground. A continental species.

*Range* (Fig. 81). Southern Burma, west coast of Malay Peninsula, north-eastern Sumatra, Java, Borneo and southern Vietnam.

*Records*. **Burma:** King I., Mergui Arch. (BMNH); **Malaysia: Peninsula:** Batu Maung, Penang (DGR); 10 km N. of Port Kelang (DGR); **Sarawak:** Santubong (DGR): **Sabah:** Po Bui I., Sandakan (USNM); **Singapore:** Ubin I. (DGR); Kranji (ANSP); **Indonesia: Sumatra:** Belawan R., Deli (RNHL); **Java** (Philippi, 1846); **Kalimantan:** Boeloengan (MCZ); **Vietnam:** Chilins, Vung Tau district (ANSP).

REMARKS. *L. conica* is a distinctive species which has in the past been well known, despite its restricted range and infrequent occurrence in museum collections. The species is a typical member of the subgenus *Palustorina*. Noteworthy anatomical features include the chisel-toothed radula, like that of *L. melanostoma*, and the anterior pair of oesophageal pouches which, like those of *L. carinifera*, are red and glandular. The form and sculpture of the shell are constant, but coloration shows considerable variation. Shells from leaves and trunks of *Avicennia* and *Excoecaria* are often the palest in colour. Together with *L. carinifera*, this species is characteristically found in the landward fringe of the mangrove forest.

SIMILAR SPECIES. *L. carinifera* shares the strong peripheral keel of this species, but colour pattern and shell sculpture are at once diagnostic of each. Small specimens may be confused with *L. delicatula*, but in *L. conica* the protoconch projects as a papilla from the large first whorl of the teleoconch, the shell is broader, the columella wider and varices are absent. The animal of *L. delicatula* is unknown, but if placement of the species in the subgenus *Littorinopsis* is correct, then anatomical features should readily separate it from *L. conica*. Despite Tryon's synonymy, confusion with L. *undulata* should never occur.

## *Littoraria (Palustorina) carinifera* (Menke, 1830)

*Phasianella carinifera* Menke, 1830: 51, 141 [lectotype figure (Rosewater, 1970) Philippi, 1847, vol. 2: *Litorina* pl. 5, fig. 22; type locality Negros Occidental, Philippines]

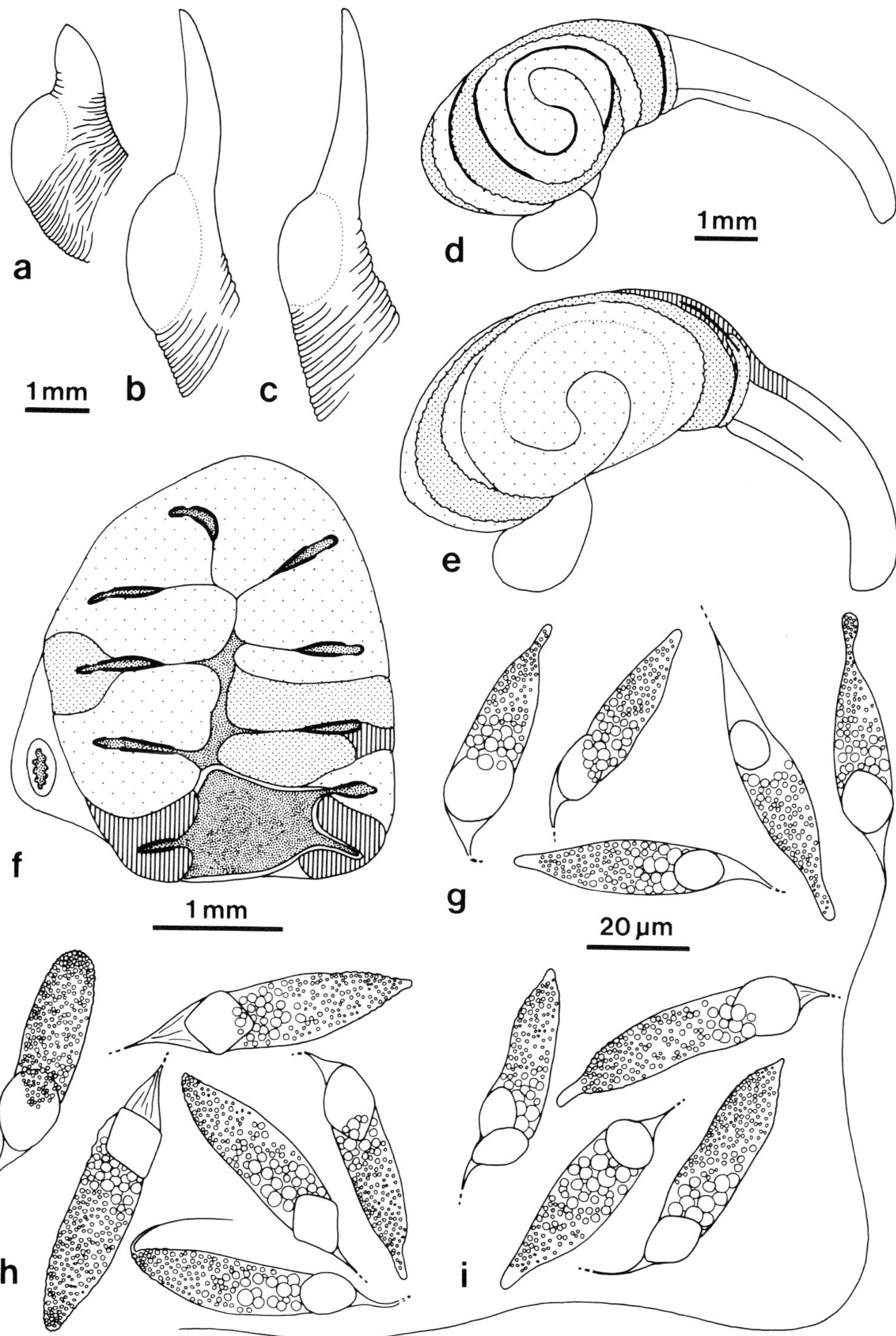


**Fig. 80** *Littoraria* (*Palustorina*) *conica:* **(a–c)** penes, Santubong, Sarawak; **(a)** contracted; **(b,c)** relaxed; **(d–f)** pallial oviducts, with transverse section; **(d)** Santubong, Sarawak; **(e,f)** Penang, Malaysia; **(g–i)** sperm nurse cells, flagellum shown on one only; **(g)** Penang, Malaysia; **(h,i)** Santubong, Sarawak.

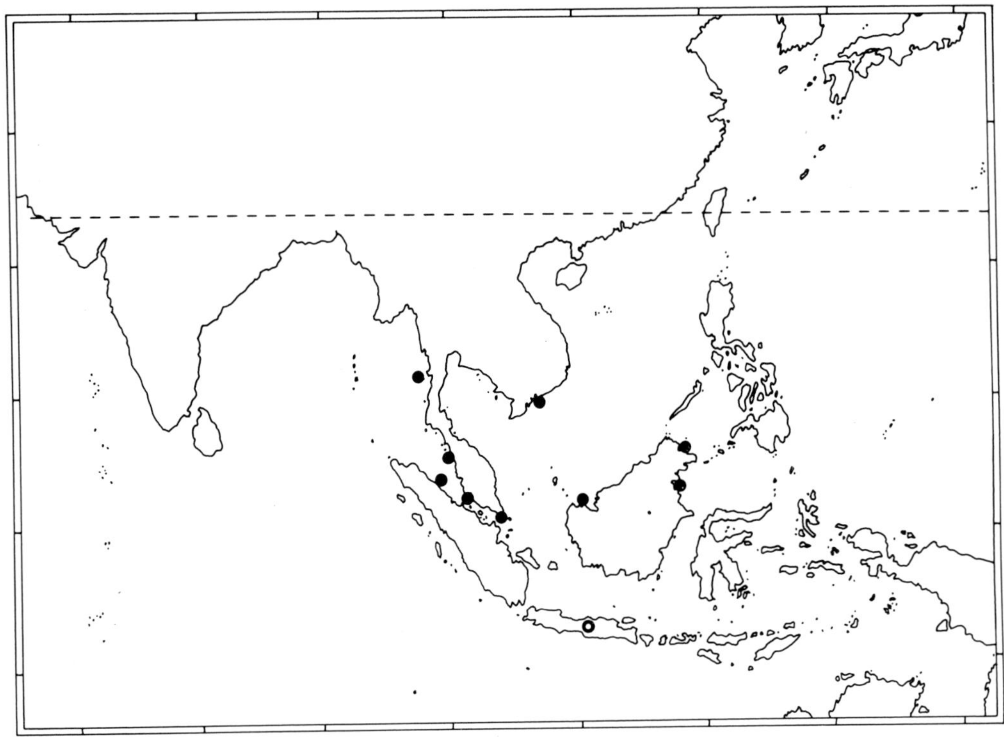

**Fig. 81** Distribution of *Littoraria* (*Palustorina*) *conica*.

*Litorina carinifera*—Philippi, 1847, vol. 2: 227, *Litorina* pl. 5, figs 22–24; Weinkauff, 1882: 48–49, pl. 6, figs 2, 3
*Littorina carinifera*—Reeve, 1857: *Littorina* pl. 6, figs 29a–c; Nevill, 1885: 151; Frith *et al.*, 1976 [zonation]; Nielsen, 1976 [note on habitat]
*Littorina* (*Littorinopsis*) *carinifera*—von Martens, 1897: 198 [in part]; Prashad, 1921: 484; Oostingh, 1927: 2–3; Rosewater, 1970: 464–465, pl. 325, fig. 5, pl. 355, figs 5–9, 12, 13 [in part]
*Littorina* (*Melaraphe*) scabra var. *carinifera*—Melvill & Standen, 1901: 363
*Littorinopsis carinifera*—Kuroda & Habe, 1952: 64; Brandt, 1974: 55–56
*Littoraria carinifera*—Higo, 1973: 46
*Littorina carinifera* subvar. *pyramidalis* Nevill, 1885: 151 [*nomen nudum*]
*Littorina carinifera* var. *laevior* Nevill, 1885: 151 [*nomen nudum*]
*Littorina rubropicta* von Martens, 1887: 170, pl. 16, figs 2a,b [King I., Mergui [Burma]; lectotype (Rosewater, 1970) BMNH 1887.3.10.140–4]
*Littorina* (*Melaraphe*) *scabra* var. *filosa*—Tryon, 1887: 244, pl. 42, fig. 28 [in part; not Sowerby, 1832]
*Littorina scabra* var. *filosa*—Pilsbry, 1895: 62 [not Sowerby, 1832]
*Littorina* (*Malaraphe*) *filosa*—Casto de Elera, 1896: 310 [in part; not Sowerby, 1832]
*Littorinopsis scabra*—Brandt, 1974: 53–54, pl. 4, fig. 61 [in part; not Linnaeus, 1758]

NOMENCLATURE. This well known species has usually been correctly determined in the literature, although occasionally it has been united with other carinate species such as *L. filosa*. Several authors have listed *Littorina perdix* King & Broderip (1832) in the synonymy of *L. carinifera*, apparently following Philippi (1847). No type specimen or figure of *Littorina perdix* is known to exist, but the original description, although brief, could not apply to *L. carinifera*; no locality was given.

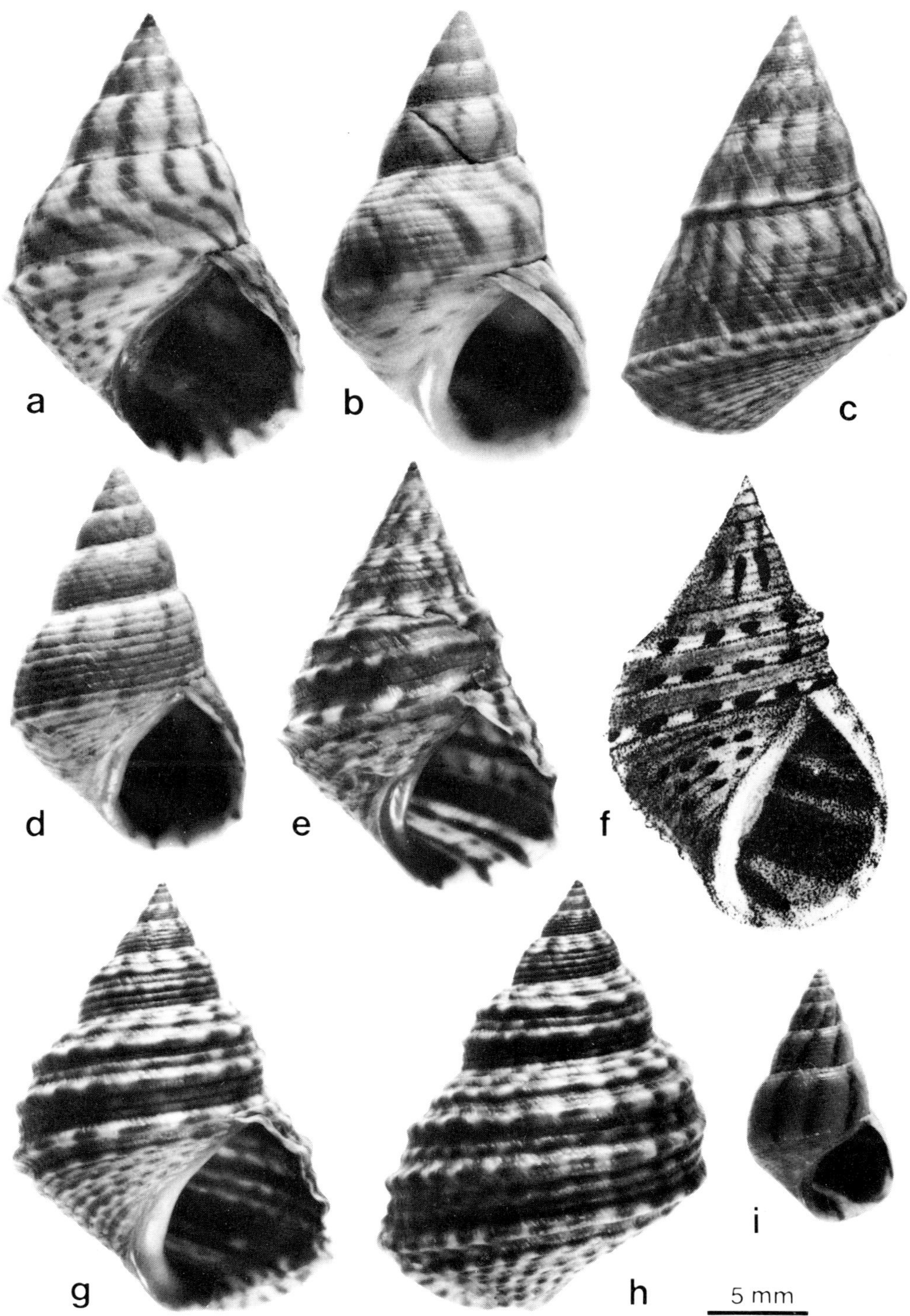


**Fig. 82** *Littoraria* (*Palustorina*) *carinifera:* **(a)** ♀, Santubong, Sarawak (DGR); **(b)** Bombay, India (BMNH); **(c)** ♀, Sungei Merbok estuary, Malaysia (DGR); **(d)** lectotype of *Littorina rubropicta* von Martens, King I., Mergui Arch., Burma (BMNH 1887.3.10.140–4); **(e)** ♀, Santubong, Sarawak (DGR); **(f)** lectotype figure of *Phasianella carinifera* Menke (Philippi, 1847, *Litorina* pl. 5, fig. 22); **(g,h)** Johore Strait (BMNH); **(i)** Tg. Rhu, Langkawi Is, Malaysia (WAM).

DIAGNOSIS. Shell: thick; spire whorls almost flat; peripheral keel strong, carinate; columella wide, excavated; primary grooves 7–9; secondary sculpture weak or absent; 15–30 ribs on last whorl, of which up to 9 may become carinate, or only peripheral rib may be carinate while others remain flat; grooves narrow; microsculpture of spiral striae on ribs, axial striae in grooves; colour grey with narrow orange or brown axial stripes, numbering 13–18 on last whorl; aperture pale with brown bands corresponding to external carinae. Animal: penis not bifurcate, filament small, tapering; oviparous.

SHELL (Fig. 82). *Shape*. Height 11–24 mm. Teleoconch 6–7 whorls. Shell thick, solid. Spire outline slightly convex; whorls almost flat, sometimes slightly turreted; sutures indistinct. Peripheral keel strong, carinate. Adult lip not thickened or flared; varices absent. Columella wide, shallowly excavated; pillar straight or with a slight convex flexure at base. No sexual dimorphism.

*Dimensions:* Table 25.

*Sculpture* (Fig. 83a–d). Protoconch normal. All teleoconch whorls sculptured. Primary grooves (6)7–9(10); spacing subequal, usually posterior 1–3 ribs narrowest, the following 1–2 ribs on shoulder widest. Secondary sculpture often absent, or confined to intercalation of a narrow riblet on the shoulder, sometimes as early as whorl 3, or division of a few primary ribs by faint secondary grooves on last whorl. In the most strongly sculptured carinate shells each primary rib may be divided at anterior $\frac{1}{3}$–$\frac{1}{2}$ of width on whorl 6, and on whorl 7 tertiary grooves may appear. Rib development on last whorl is variable: in smooth shells all 15–30 ribs remain low and flat except the prominent sharp peripheral carina; in carinate shells alternate ribs become more prominently rounded on whorl 6, developing into carinae on whorl 7, numbering up to 4 above and 4 below the large peripheral carina. Grooves remain narrow, less than $\frac{1}{4}$ rib width. Microsculpture of regular, fine, spiral striae on ribs; surface covered with fine, irregular, axial growth striae, strong and regular in grooves.

*Colour*. Rather constant. Ground colour blue grey, whitish or pale yellow, palest near suture and on carinae. Pattern of rusty orange or dark brown vertical, oblique or zigzag axial stripes, usually

**Table 25** Dimensions of *Littoraria* (*Palustorina*) *carinifera*.

| Specimen | Locality | Sex | Primary grooves | H (mm) | B (mm) | LA (mm) | WA (mm) | C (mm) | P | S | SH |
|---|---|---|---|---|---|---|---|---|---|---|---|
| *Phasianella carinifera* lectotype figure (Philippi, 1847) | unknown | | | 21.5 | 14.5 | 11.4 | 8.3 | | 1.48 | 0.73 | 1.89 |
| *Littorina rubropicta* lectotype, BMNH 1887.3.10.140–4 | King I., Mergui, Burma | | 9 | 18.3 | 12.0 | 9.2 | 6.9 | | 1.53 | 0.75 | 1.99 |
| DGR | Santubong, Sarawak | ♂ | 9 | 19.8 | 13.4 | 11.0 | 8.7 | 1.8 | 1.48 | 0.79 | 1.80 |
| DGR | Santubong, Sarawak | ♂ | 6 | 17.7 | 13.8 | 10.5 | 8.4 | 1.8 | 1.28 | 0.80 | 1.69 |
| DGR | Santubong, Sarawak | ♀ | 8 | 22.7 | 16.1 | 12.4 | 9.5 | 2.5 | 1.41 | 0.77 | 1.83 |
| DGR | Santubong, Sarawak | ♀ | 8 | 19.4 | 16.1 | 12.1 | 10.2 | 2.6 | 1.20 | 0.84 | 1.60 |
| BMNH acc. 1838 | Johore Strait | | 7 | 18.3 | 15.6 | 11.3 | 9.2 | 2.5 | 1.17 | 0.81 | 1.62 |
| BMNH acc. 2176 | Bombay, India | | 8 | 22.1 | 14.5 | 11.5 | 9.1 | 2.2 | 1.52 | 0.79 | 1.92 |
| BMNH acc. 2176 | Bombay, India | | 8 | 20.1 | 12.9 | 10.1 | 8.4 | 2.4 | 1.56 | 0.83 | 1.99 |
| WAM 1037–81 | Langkawi Is, Malaysia | | 8 | 12.1 | 7.8 | 6.0 | 4.6 | 0.8 | 1.55 | 0.77 | 2.02 |
| WAM 1037–81 | Langkawi Is, Malaysia | | 8 | 11.4 | 7.5 | 6.3 | 4.5 | 0.9 | 1.52 | 0.71 | 1.81 |
| DGR, mean of 10 | Sungei Merbok, Malaysia | ♂ | | 19.15 | | | | | 1.412 | 0.752 | 1.753 |
| standard error | | | | 0.54 | | | | | 0.018 | 0.009 | 0.023 |
| DGR, mean of 10 | Sungei Merbok, Malaysia | ♀ | | 19.29 | | | | | 1.378 | 0.768 | 1.815 |
| standard error | | | | 0.50 | | | | | 0.024 | 0.012 | 0.024 |
| statistic t or U | | | | 0.189 | | | | | 64 | 69 | 69 |
| probability | | | | 0.852 | | | | | 0.314 | 0.166 | 0.166 |

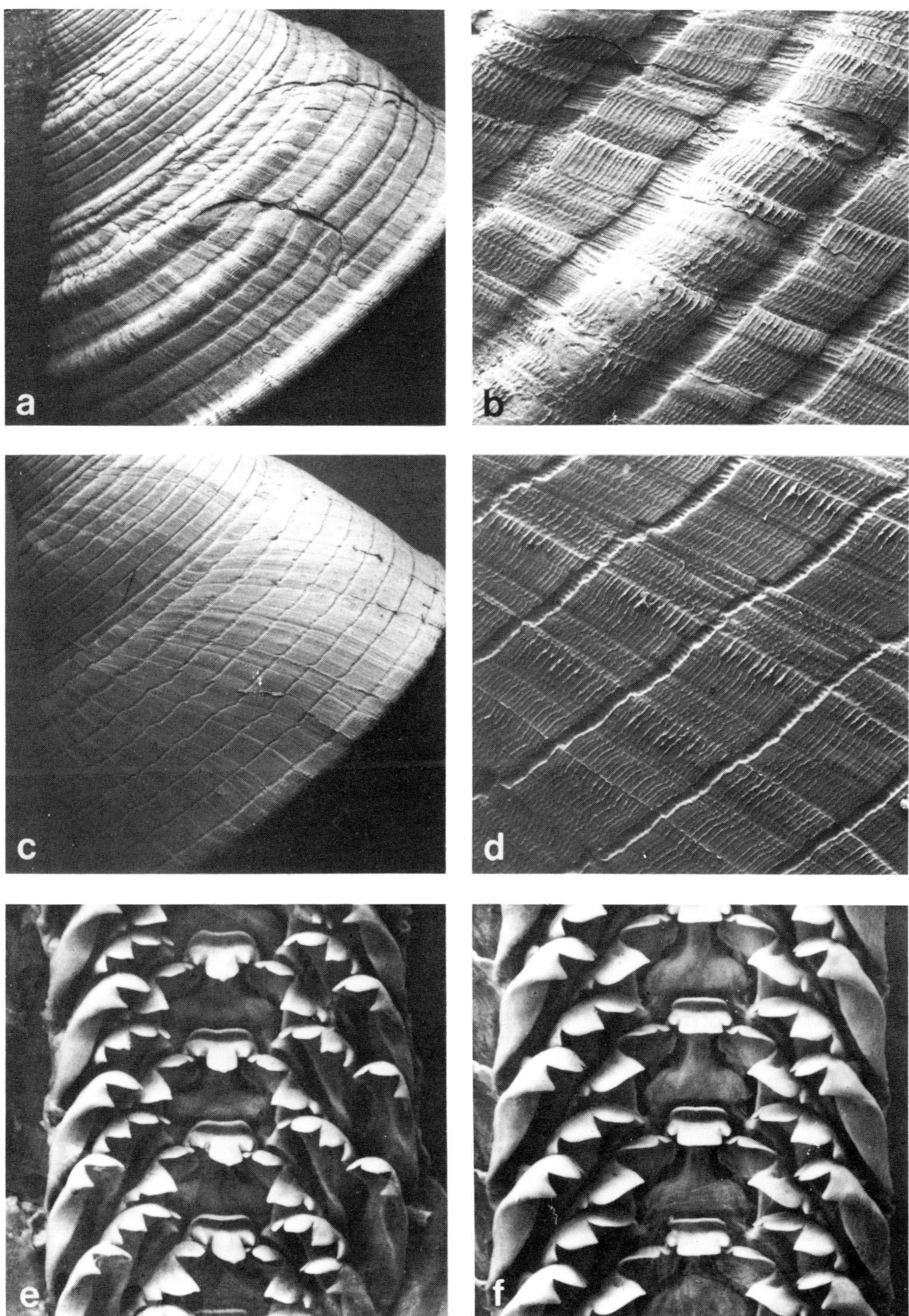


**Fig. 83** *Littoraria* (*Palustorina*) *carinifera:* **(a,b)** Santubong, Sarawak; **(a)** last whorl ( × 10); **(b)** detail ( × 40); **(c,d)** Sungei Merbok estuary, Malaysia; **(c)** last whorl ( × 10); **(d)** detail ( × 40); **(e)** radula, Phuket I., Thailand ( × 275); **(f)** radula, Santubong, Sarawak ( × 220).

continuous over grooves, from suture to periphery, but on base broken into spots confined to ribs. Stripes narrow, numbering (10)13–18(25) on last whorl. Aperture with blacks or dark purple brown spiral bands corresponding to major carinae; usually the 3 bands at suture, periphery and around columella are widest, with lines and spots between; sometimes aperture is uniformly dark with a single pale band anteriorly. Columella pale pink, purple brown, grey or white.

ANIMAL. *Colour.* Ground colour ochraceous to pale brown; pigmentation grey to black; sides of foot darkly mottled and speckled with opaque white flecks; head and tentacles darkly mottled, red buccal mass and oesophageal pouches visible within; tentacles sometimes conspicuously yellow ochre if dark pigment not too dense, pale stripes at base are indistinct.

*Penis* (Fig. 84a–e). Small, to 5.0 mm. Base simple, incorporating glandular disc. Filament small, tapering. Sperm groove open. Base and filament ochre, darkest around glandular disc; disc off white to cream.

*Sperm.* Eupyrene sperm 270–350 μm. Nurse cells (Fig. 84h–j) 30–46 μm; elongate oval, tapering to basal flagellum up to 220 μm long; rods 1(2–4), basal, shape variable, usually oval, occasionally narrow and elongate; yolk granules small, becoming larger towards base.

*Pallial oviduct* (Fig. 84f,g). Length to 8.0 mm. Spiral section to 5.7 mm diam., $5\frac{1}{2}$ whorls; opaque albumen gland $\frac{1}{4}$ whorl, cream yellow; translucent albumen gland ochre to fawn; opaque capsule gland $1–1\frac{1}{2}$ whorls, cream to pale pink; translucent capsule gland orange brown to red; spiral distinct externally, but dark pigmentation of egg groove only visible in section. Straight section long, to 3.1 mm, orange brown, reddish translucent capsule gland extends around pigmented egg groove into straight section; no terminal papilla. Bursa posterior, extending into spiral section. Development assumed oviparous.

*Radula* (Fig. 83e,f). Length to 25 mm; relative length 0.99–1.36. Intermediate between saw- and chisel-toothed types; central rachidian cusp wide, usually with a straight edge; cusps of paired teeth obliquely triangular; lateral with gap anterior to main cusp.

*Alimentary tract.* Anterior pair of oesophageal pouches large (to 2.0 mm diam.), glandular, pink to dark red.

*Pallial complex.* Hypobranchial gland large, 1.5–2.2 mm wide, brown.

DISTRIBUTION. *Habitat.* At back of mangrove forests; in landward fringe on trunks and dead wood, also leaves of *Nypa*, 0–0.5 m above ground; extends into zones of *Bruguiera* and *Ceriops*; in marsh grass (von Martens, 1887).

*Range* (Fig. 85). Pakistan to Malaysia, Indonesia and Philippines.

*Records.* **Pakistan:** China Creek, Karachi (MCZ); **India:** Bandra, N. of Bombay (USNM); Bombay (BMNH, RNHL, MCZ); Vingurla, N. of Goa (USNM); Netravati R., Mangalore (USNM, ANSP); Beypore estuary (BMNH); **Burma:** King I., Mergui Arch. (BMNH); **Thailand:** Ranong (MCZ); Ao Nam-Bor, Phuket I. (DGR); Songkla (MCZ); Chonburi (USNM, ANSP); Ko Kut (USNM); **Malaysia: Peninsula:** Tg. Rhu, Langkawi Is (WAM); Sungei Merbok estuary, Kedah (DGR); Endau (AMS); Kuantan (NUS); **Sarawak:** Santubong (DGR); **Sabah:** Labuan I. (RNHL); Tandjong Aru, Kota Kinabalu (ANSP); Po Bui I., Sandakan (USNM); **Singapore:** Jelutong, Ubin I. (DGR); Loyang Besar (WAM); **Indonesia: Sumatra:** Weh I. (RNHL); Belawan R., Deli (RNHL); Belinju, Bangka I. (RNHL); **Java:** Tandjong Priok, Djakarta (RNHL); Madura I. (RNHL); **Sulawesi:** Makassar (RNHL); **Moluccas:** Ternate I. (RNHL); **Irian Jaya:** Skroe, NW. of Fakfak (RNHL); **Vietnam:** Chilins, Vung Tau district (ANSP); **Philippines:** Pancol, Palawan I. (USNM, ANSP); Busuanga I. (RNHL); Viejo Victorias, Negros Occidental (USNM); Iloilo, Panay (USNM); Silanga, Samar (USNM); Medio I., Galera Bay, Mindoro (USNM); Alfonso XIII I., near Manila (ANSP).

REMARKS. Although the colour of the shell is rather constant, *L. carinifera* shows geographical variation in shape and sculpture. The form described by von Martens (1887) as *Littorina rubropicta*, with a rather smooth shell and single conspicuous peripheral carina, occurs in north-eastern Sumatra, the west coast of the Malayan Peninsula and India. At Singapore and to the south and east

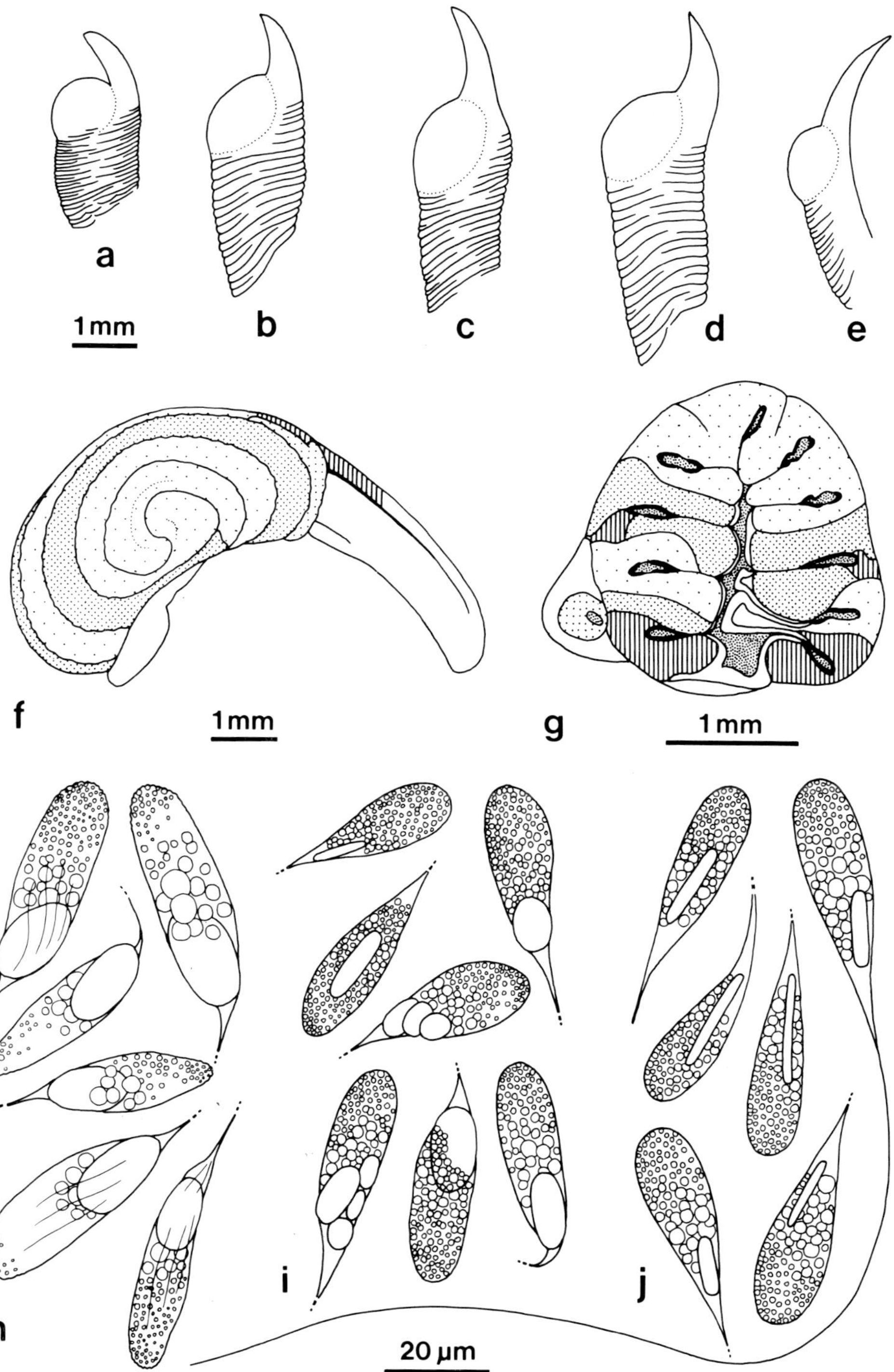


**Fig. 84** *Littoraria* (*Palustorina*) *carinifera*: **(a–e)** penes; **(a)** contracted; **(b–e)** relaxed; **(a–d)** Santubong, Sarawak; **(e)** Phuket I., Thailand; **(f,g)** pallial oviduct, with transverse section, Santubong, Sarawak; **(h–j)** sperm nurse cells, flagellum shown on one only; **(h)** Santubong, Sarawak; **(i,j)** Phuket I., Thailand.

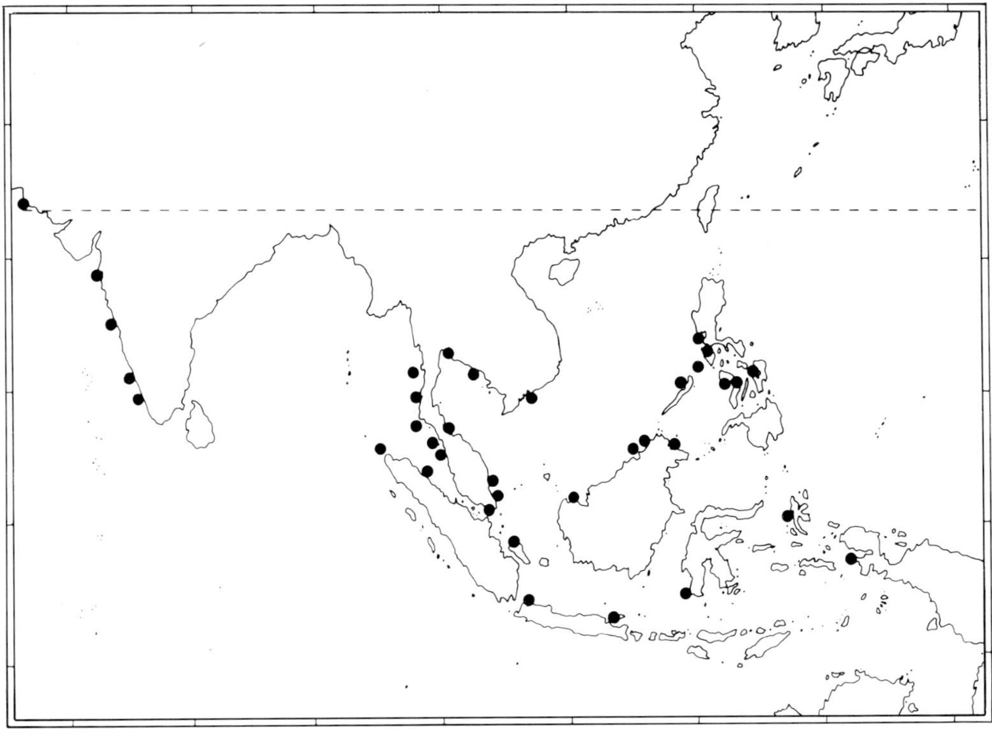

**Fig. 85** Distribution of *Littoraria* (*Palustorina*) *carinifera*.

the shells are of the typical, broad, strongly carinate form. Shells from Borneo and the Philippines are variable and include intermediates, so that taxonomic recognition of the two main forms is not justified. In a collection from the Langkawi Islands, western Malaysia (WAM) the shells are small, less than 14 mm in height, the periphery is merely angular, not carinate, the spiral grooves are shallow and occasionally obsolete, and the axial colour stripes are narrow (Fig. 82i). Intermediates link this unusual form with the more common '*rubropicta*' variety. The animal of *L. carinifera* is unusual in the large, red, glandular oesophageal pouches, found also in *L. conica*.

SIMILAR SPECIES. *L. carinifera* is highly distinctive, the combination of carinate shell, red brown axial stripes, banded aperture and wide columella distinguishing it immediately from the strongly ribbed or carinate species *L. filosa*, *L. cingulata*, *L. pallescens* and *L. sulculosa*. Only *L. conica* shows such a sharp angulation at the periphery, but is a thinner shell with more numerous ribs.

## ***Littoraria* (*Palustorina*) *sulculosa*** (Philippi, 1846)

*Littorina sulculosa* Philippi, 1846: 142 [north coast of Australia; lectotype (Rosewater, 1970) BMNH 1968279]; Reeve, 1857: *Littorina* pl. 8, figs 39a,b; von Martens, 1889: 193

*Litorina sulculosa*—Philippi, 1847, vol. 3: 18, *Litorina* pl. 6, fig. 10; Weinkauff, 1882: 55, pl. 7, figs 5, 8

*Littorina* (*Melaraphe*) *sulculosa*—Tryon, 1887: 247, pl. 43, fig. 52

*Melaraphe sulculosa*—Hedley, 1916: 38

*Melarhaphe sulculosa*—Hedley, 1918*b*: 275

*Littorina filosa*—Nevill, 1885: 148 [in part; not Sowerby, 1832]

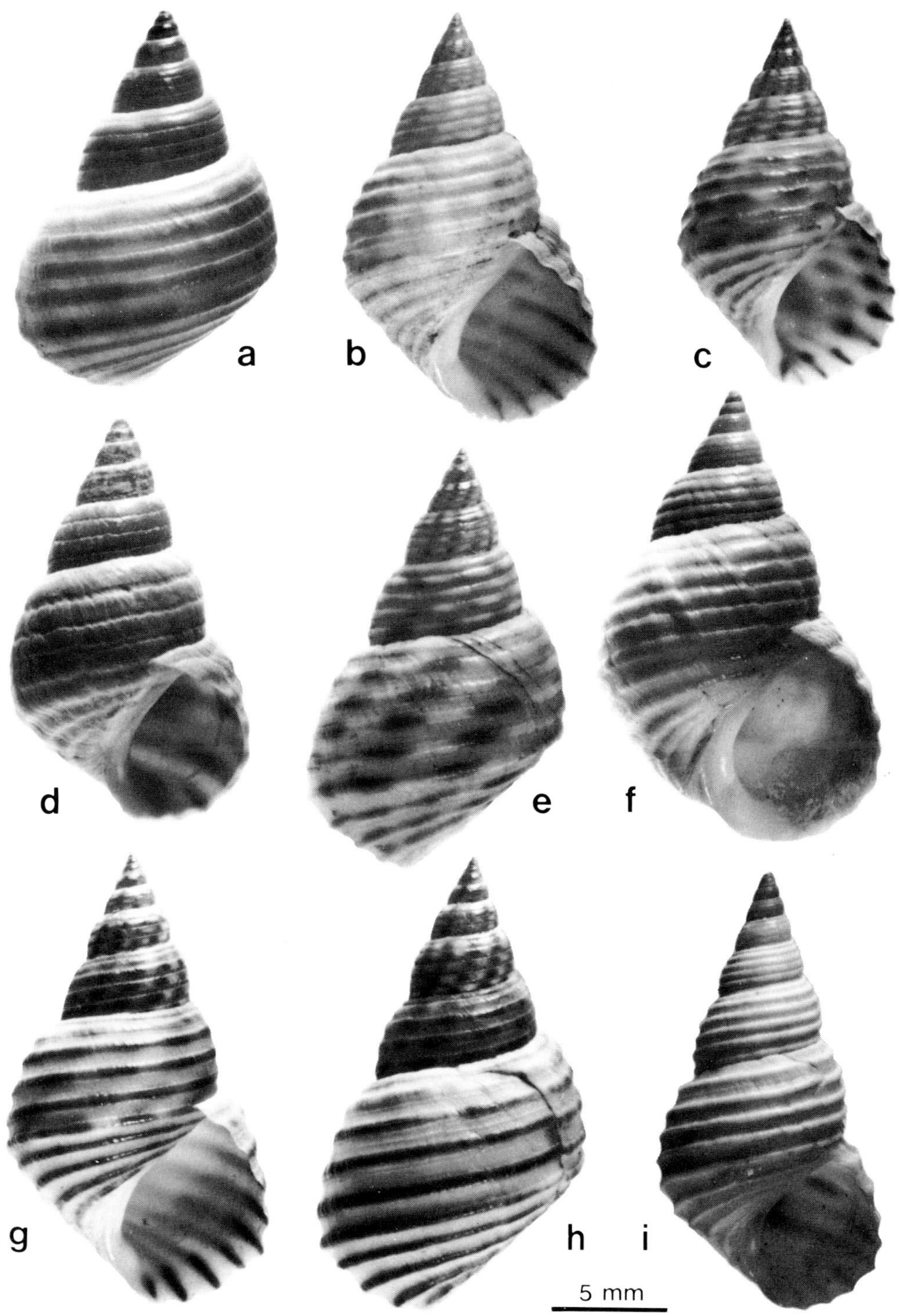


**Fig. 86** *Littoraria* (*Palustorina*) *sulculosa:* **(a)** Broome, W.A. (NMV); **(b,c)** Broome, W.A. (DGR); **(b)** ♂; **(c)** ♂; **(d)** lectotype of *Littorina sulculosa* Philippi, N. Australia (BMNH 1968279); **(e)** ♀, Broome, W.A. (DGR); **(f)** Broome, W.A. (NMV); **(g,h)** ♀, Broome, W.A. (DGR); **(i)** Derby, W.A. (AMS).

*Littorina* (*Littorinopsis*) *scabra scabra*—Rosewater, 1970: 456–461, pl. 352, figs 17, 18 [in part; not Linnaeus, 1758]

NOMENCLATURE. The species was described by Philippi (1846) from material in the Cuming Collection. Only one lot in the BMNH corresponds with Philippi's description and Rosewater (1970) designated as lectotype the shell closest in appearance to Philippi's (1847) figure. No original label exists and a label written by Tomlin erroneously records the type locality as Java.

DIAGNOSIS. Shell: very thick; whorls rounded or turreted; columella wide, scarcely excavated, white; primary grooves 5–6; secondary sculpture usually absent; ribs rounded and prominent, numbering 9–11 on last whorl, with wide grooves between; microsculpture of faint spiral striae on ribs, strong axial striae in grooves; colour cream, ribs pale brown, unicolorous or marked with long dashes. Animal: penis not bifurcate, filament long, tapering; oviparous.

SHELL (Fig. 86). *Shape*. Height 14–23 mm. Teleoconch 7–8 whorls. Shell of relatively great thickness (to 1.5 mm) when large. Spire outline straight; whorls gently rounded, often somewhat turreted; sutures impressed. Peripheral keel moderately developed, marked by the most prominent of the primary ribs. Adult lip sharp, not flared; varices absent. Columella wide, scarcely excavated; pillar concave, with small constriction at base, producing a low, rounded knob. Sexual dimorphism: males smaller.

*Dimensions*: Table 26.

*Sculpture* (Fig. 87a–d). Protoconch normal. First 0.5–1 whorl of teleoconch smooth. Primary grooves (4)5–6(7), almost equidistant, but posterior rib usually narrowest; occasionally posterior groove does not form until whorl 4–5, then posterior rib is the widest on earlier whorls. Secondary sculpture usually absent, but on whorl 8 of largest shells 1–3 narrow riblets may be intercalated in each primary groove. Primary ribs become rounded and prominent on whorls 7–8, numbering 9–11 on last whorl. Primary grooves at first narrow, almost equal to rib width on whorl 7, but 1–3 times rib width on whorl 8. Microsculpture of faint spiral striae on ribs, sometimes indistinct; surface covered by irregular axial growth lines; grooves sculptured by fine, strong, axial striae.

*Colour*. Rather constant. Ground colour cream, whitish or pale fawn. Ribs fawn, pale orange pink or grey brown, often unicolorous, or marked with long, indistinct dashes. Spire whorls coloured like rest of shell, but usually with dashes discrete and aligned into axial series; spire whorls sometimes grey, paler at sutures. Aperture white, with dark lines and dashes corresponding to external pattern, glazed within by a cream callus. Columella and parietal callus usually white, sometimes tinged pink.

**Table 26** Dimensions of *Littoraria* (*Palustorina*) *sulculosa*.

| Specimen | Locality | Sex | Primary grooves | H (mm) | B (mm) | LA (mm) | WA (mm) | C (mm) | P | S | SH |
|---|---|---|---|---|---|---|---|---|---|---|---|
| *Littorina sulculosa* lectotype, BMNH 1968279 | North coast of Australia | | 4 | 17.2 | 10.4 | 8.7 | 6.2 | 1.6 | 1.65 | 0.71 | 1.98 |
| DGR | Broome, W. A. | ♂ | 5 | 17.4 | 11.4 | 9.2 | 7.0 | 1.6 | 1.53 | 0.76 | 1.89 |
| DGR | Broome, W. A. | ♂ | 6 | 15.9 | 10.1 | 8.1 | 5.9 | 1.2 | 1.57 | 0.73 | 1.96 |
| DGR | Broome, W. A. | ♀ | 5 | 22.5 | 15.3 | 12.5 | 8.6 | 1.9 | 1.47 | 0.69 | 1.80 |
| DGR | Broome, W. A. | ♀ | 6 | 18.5 | 11.4 | 9.4 | 7.0 | 1.2 | 1.62 | 0.74 | 1.97 |
| AMS C.131743 | Derby, W. A. | | 5 | 18.0 | 9.9 | 7.7 | 5.7 | 0.8 | 1.82 | 0.74 | 2.34 |
| AMS C.131743 | Derby, W. A. | | 5 | 15.0 | 8.2 | 6.6 | 4.7 | 0.8 | 1.83 | 0.71 | 2.27 |
| DGR, mean of 10 | Broome, W. A. | ♂ | | 14.01 | | | | | 1.531 | 0.715 | 1.837 |
| standard error | | | | 0.58 | | | | | 0.008 | 0.010 | 0.018 |
| DGR, mean of 10 | Broome, W. A. | ♀ | | 17.75 | | | | | 1.560 | 0.733 | 1.874 |
| standard error | | | | 0.41 | | | | | 0.011 | 0.008 | 0.018 |
| statistic t or U | | | | 5.250 | | | | | 75 | 67 | 73 |
| probability | | | | <.001 | | | | | 0.064 | 0.218 | 0.090 |

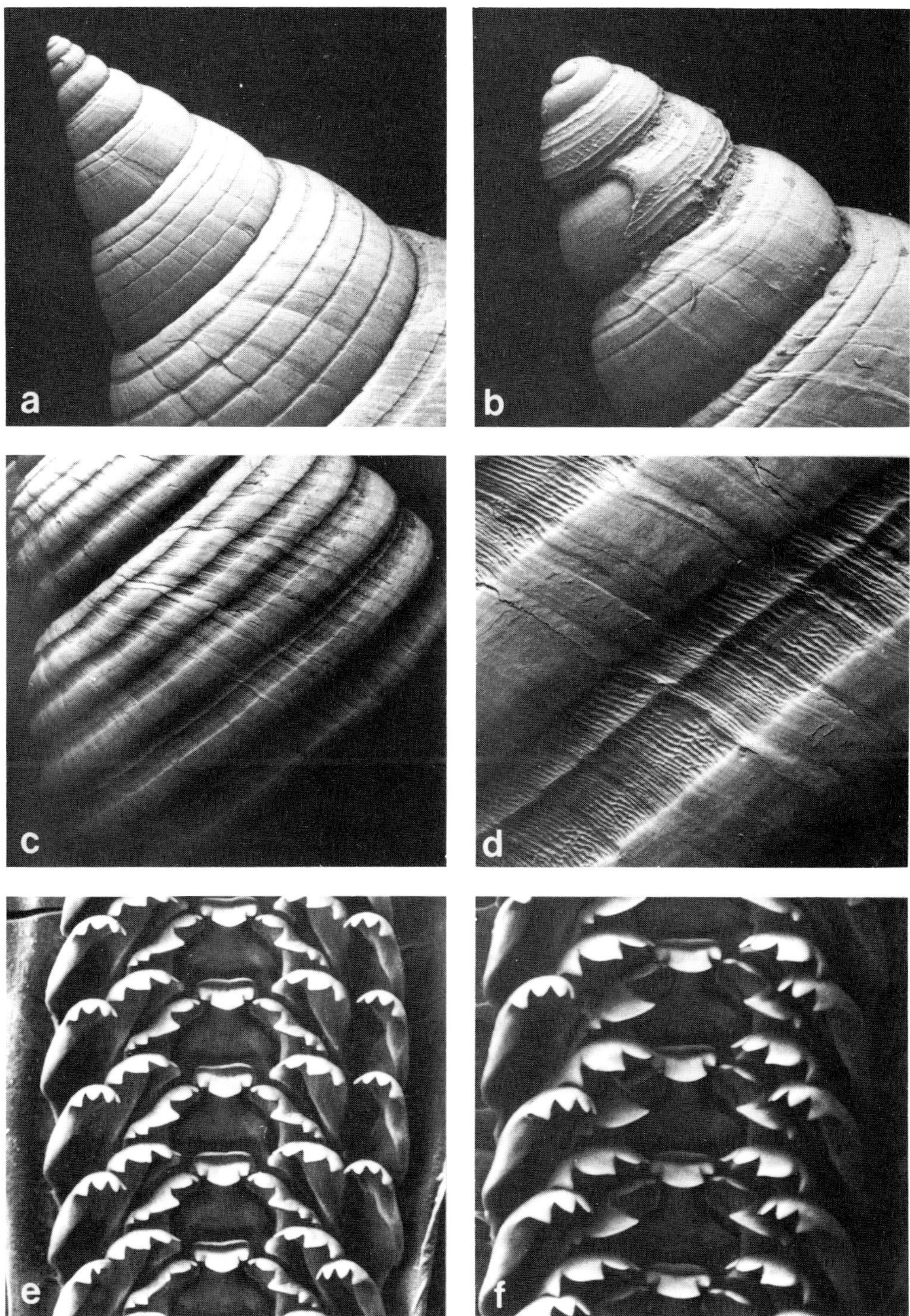


**Fig. 87** **(a–e)** *Littoraria* (*Palustorina*) *sulculosa*, Broome, W.A.: **(a)** spire ( × 19); **(b)** protoconch ( × 90); **(c)** last whorl ( × 9); **(d)** detail ( × 35); **(e)** radula ( × 210). **(f)** *Littoraria* (*Palustorina*) *articulata*: radula, Kanchanadit, Thailand ( × 325).

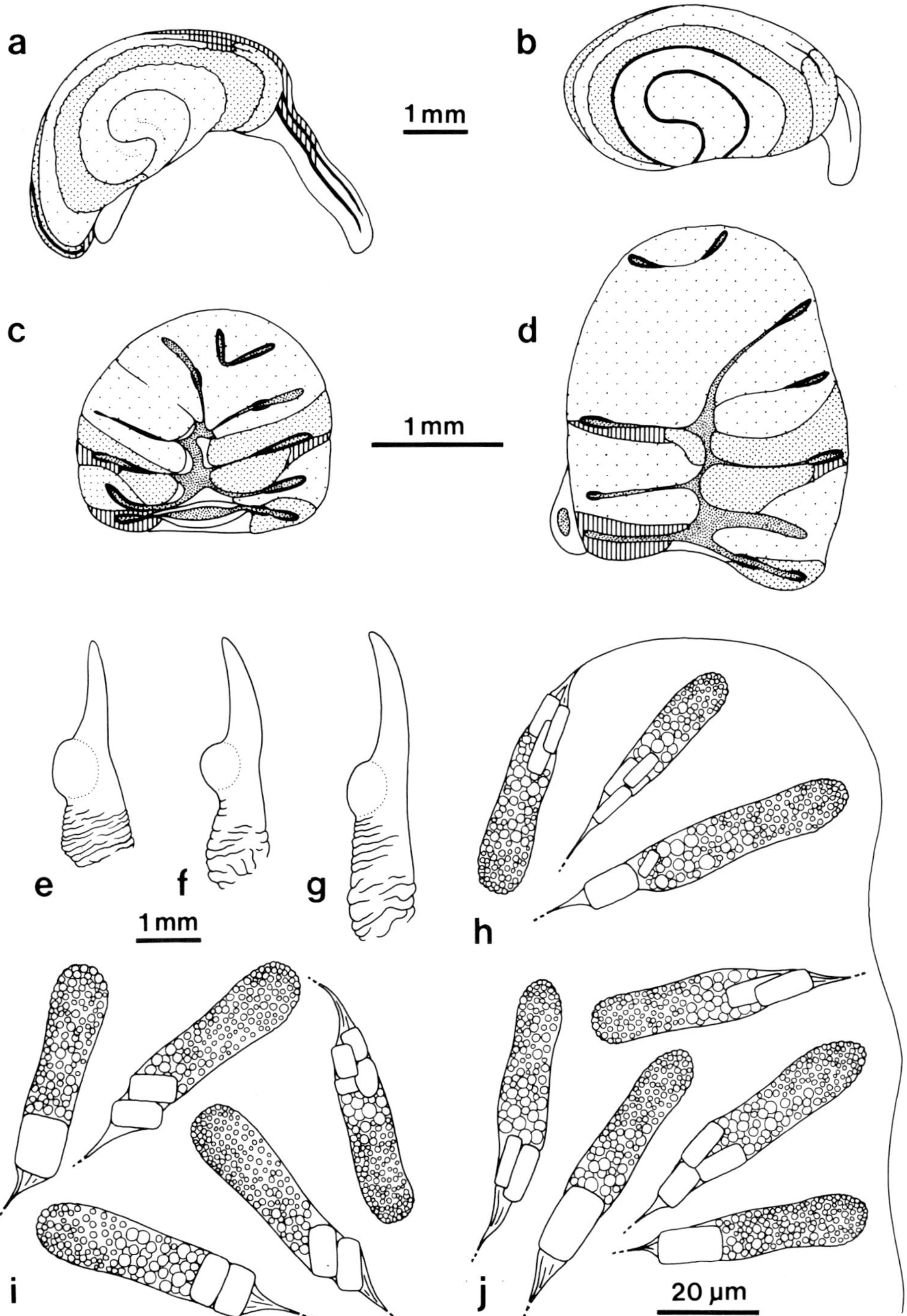


**Fig. 88** *Littoraria* (*Palustorina*) *sulculosa*, Broome, W.A.: **(a–d)** pallial oviducts, with transverse sections; **(e–g)** penes; **(h–j)** sperm nurse cells, flagellum shown on one only.

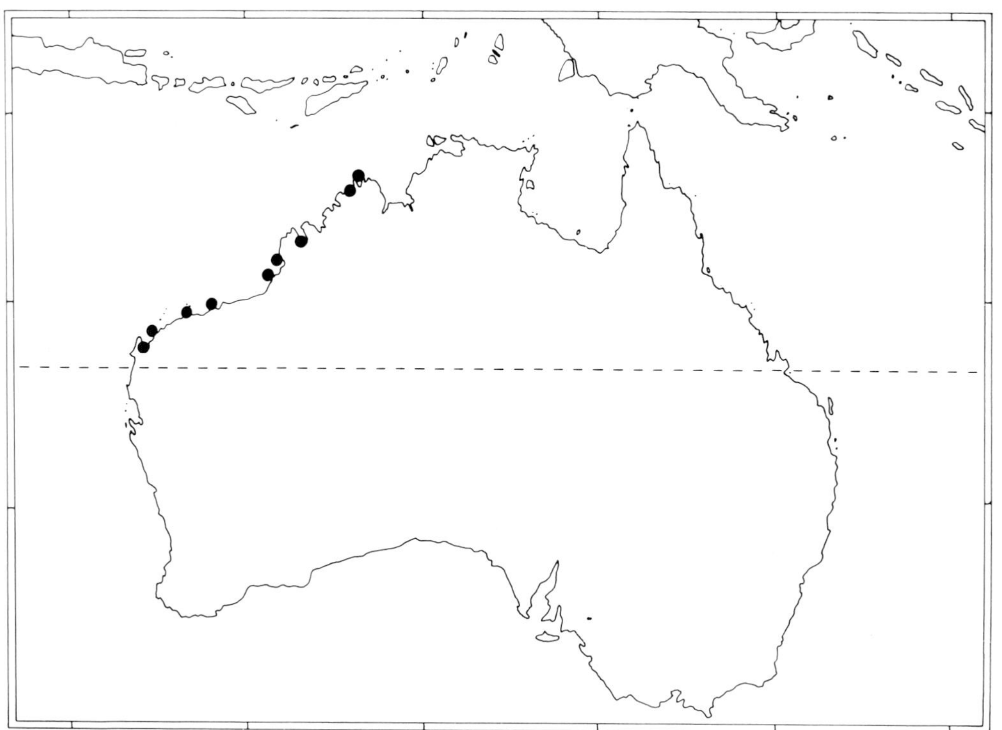

Fig. 89 Distribution of *Littoraria* (*Palustorina*) *sulculosa*.

ANIMAL. *Colour.* Pigmentation pale grey; sides of foot pale, mottled and lined; head pale grey, red buccal mass visible internally; tentacles banded, unpigmented stripe each side of base.

*Penis* (Fig. 88e–g). Length to 4.4 mm. Base simple, incorporating small glandular disc. Filament long, tapering. Sperm groove open. Penis cream.

*Sperm.* Eupyrene sperm 205–240 $\mu$m. Nurse cells (Fig. 88h–j) 36–55 $\mu$m, elongate oval, tapering at base, slightly constricted at mid-length; basal flagellum to 205 $\mu$m; rods 1–4, basal, rectangular; yolk granules small, becoming larger towards base.

*Pallial oviduct* (Fig. 88a–d). Length to 5.7 mm. Spiral section to 5.2 mm diam., $5\frac{1}{2}$ whorls; opaque albumen gland $\frac{1}{4}$ whorl, white; translucent albumen gland white to pale grey; opaque capsule gland $1\frac{1}{2}$ whorls, white or pale pink; translucent capsule gland orange brown to red; spiral distinct externally, egg groove usually darkly pigmented. Straight section to 2.4 mm, grey to pale brown, brown translucent capsule gland visible posteriorly; no terminal papilla. Bursa posterior, extending into spiral section. Development assumed oviparous.

*Radula* (Fig. 87e). Length to 17 mm; relative length 0.78–0.89. Intermediate between saw- and chisel-toothed types; central rachidian cusp broad, edge slightly pointed; cusps of paired teeth rather obliquely triangular; lateral with gap anterior to main cusp.

DISTRIBUTION. *Habitat.* At Broome, Western Australia, occurs from middle almost to landward side of a forest entirely of *Avicennia* trees, on trunks and also on leaves, 0–1.4 m above ground; sometimes under logs at back of forest. Also seen occasionally on *Rhizophora* roots. Fairly frequent on and under rocks on sheltered shores.

*Range* (Fig. 89). North-western Australia, from Exmouth Gulf to Vansittart Bay.

*Records.* **Australia: W.A.:** Bay of Rest, Exmouth Gulf (WAM); Onslow (AMS); Antonni Mia, Point

Sampson (WAM); Cossack (BMNH); Port Hedland (AMS); False Cape Creek, La Grange Bay (AMS, ANSP); Broome (DGR, AMS, NMV); Derby (AMS, WAM); Port Warrender, Admiralty Gulf (WAM); Vansittart Bay (AMS, WAM).

REMARKS. Because of the restricted and remote distribution of *L. sulculosa*, it is rare in museum collections outside Australia. Anatomical features of this species are typical of the genus *Palustorina*, but the shell bears a superficial resemblance to that of the sympatric *L. (Littorinopsis) cingulata cingulata*, from which separation may be difficult in the field. In mangrove forests the habitats of these two species overlap, but *L. sulculosa* occurs at lower levels and is less common towards the landward side of the forest. The shell of *L. sulculosa* is rather constant in form and colour, although in a collection from Derby, Western Australia (AMS) all specimens have unusually tall spires (Fig. 86i). Typical shells of this species are remarkable for their extreme thickness (1.5 mm at the outer apertural lip in a shell 18.7 mm in height) and are probably thereby adapted to resist crushing by crab predators, which are more common at the lower levels on the trees.

SIMILAR SPECIES. *L. sulculosa* is unlike any other member of the subgenus *Palustorina*. Although the penis and oviduct are immediately diagnostic, the shells of *L. sulculosa* and *L. cingulata cingulata* are similar in form and colour. *L. sulculosa* is most reliably distinguished by the strong axial microsculpture in the spiral grooves, but in addition there are fewer primary ribs on the last whorl, the colour pattern is of long dashes or continuous spiral lines, and the columella is usually white.

## *Littoraria (Palustorina) articulata* (Philippi, 1846)

*Littorina intermedia* var. *articulata* Philippi, 1846: 141 [Swan Point [W.A., Australia]; neotype here designated BMNH 198348, just S. of Lookout Hill, Broome, W.A., Australia; additional specimens from neotype lot AMS, WAM, USNM; not *Litorina scabra* var. *articulata* Philippi, 1847]
*Litorina intermedia* var. *articulata*—Philippi, 1847, vol. 2: 223
? *Litorina intermedia* var. *strigata*—Philippi, 1847, vol. 2: 223, *Litorina* pl. 5, fig. 7 [in part; not Philippi, 1846; original citation of figures of var. *strigata* corrected to pl. 5, figs 7, 8, 9]
*Littoraria strigata* ('Philippi')—Habe, 1964: 29, pl. 9, fig. 31 [not Philippi, 1846]
*Littorina (Melaraphe) scabra* var. *intermedia*—Tryon, 1887: 224, pl. 42, fig. 23 [in part; not Philippi, 1846]
*Littorina (Malaraphe) intermedia*—Casto de Elera, 1896: 309–310 [in part; not Philippi, 1846]
*Littorina (Littorinopsis) intermedia*—von Martens, 1897: 197 [in part; not Philippi, 1846]
*Littorina intermedia*—Yen, 1933: 93; Fischer, 1970: 99 [both not Philippi, 1846]
*Melarhaphe intermedia*—Yen, 1942: 196 [not Philippi, 1846]
*Litorina sinensis* Philippi, 1847, vol. 3: 16–17, *Litorina* pl. 6, fig. 23 [China; lectotype here designated MNHP, 13.4 mm]; Lischke, 1871*b*: 71–72; Weinkauff, 1882: 83–84, pl. 11, figs 9, 12
*Littorina intermedia* var. *sinensis*—Nevill, 1885: 147
*Littorina sinensis*—Pilsbury, 1895: 62
*Melaraphe (Litt.) blanfordi* Dunker, 1871: 150 [Rockhampton [Qld., Australia]; location of type unknown, not in Museum für Naturkunde, E. Berlin, R. Kilias, pers. comm.]
*Melarapha blanfordi*—Iredale & McMichael, 1962: 38
*Litorina strigata* Lischke, 1871*a*: 148–149 [Nagasaki, Japan; type not in Zoological Museum, Leningrad, A.N. Golikov, pers. comm.]; Lischke. 1871*b*: 73, pl. 5, fig. 22 [both not *Littorina intermedia* var. *strigata* Philippi, 1846]
*Littorina strigata* Lischke—Pilsbry, 1895: 62 [not Philippi, 1846]
*Littorina (Melaraphe) strigata* Lischke—Tryon, 1887: 245, pl. 43, fig. 33 [not Philippi, 1846]
*Littorinopsis strigata* (Lischke)—Kuroda & Habe, 1952: 64; Oyama & Takemura, 1961: fig. 10 [both not Philippi, 1846]
*Littoraria strigata* (Lischke)—Kojima, 1958*c* [egg capsule]; Azuma, 1960: 10; Higo, 1973: 46 [all not Philippi, 1846]
*Littoraria strigata* ('Dunker')—Yoo, 1976: 56, pl. 7, figs 18, 19 [not Philippi, 1846]
*Melaraphe scabra*—Hedley, 1916: 37 [in part; not Linnaeus, 1758]

*Littorina* (*Littorinopsis*) *scabra scabra*—Rosewater, 1970: 456–461, pl. 352, figs 28, 29 [in part; not Linnaeus, 1758]
*Littorina undulata*—Berry, 1963 [in part; not Gray, 1839]
*Littorinopsis undulata*—Brandt, 1974: 54 [in part; not Gray, 1839]

NOMENCLATURE. Philippi (1846) described *Littorina intermedia* var. *articulata* in a work based upon material in the Cuming Collection, now in the BMNH. Although the types of other species described by Philippi in the same article are located in the BMNH, that of var. *atticulata* is lost. From the original descriptions and from the figures published by Philippi in the following year, it is evident that his concept of *Littorina intermedia* embraced the species here defined as *L. strigata* (Philippi; not Lischke), *L. articulata* and possibly also *L. subvittata*, in addition to *L. intermedia s. s.* Of these four, only *L. articulata* occurs at Swan Point, Western Australia, the type locality of var. *articulata*, and shells from north-western Australia fit the original description: '*testa interstitiis sulcorum regularito albido et fusco articulatis* ... only 6 lin. [13.2 mm] high'. The var. *articulata* was not figured as such by Philippi, but his *Litorina* pl. 5, fig. 7, cited in the text as *Litorina intermedia* var. *strigata* (see p. 125 for a discussion of the error in the original citation) resembles specimens of *L. articulata* from north-western Australia. This resemblance is also seen in one of the paralectotypes of var. *strigata*, which was probably the original of the figure. In view of several known errors in the citations of the figures in Philippi's text, it is not impossible that the figure does indeed represent var. *articulata*, and that the specimen is its type, which has become mixed with the type collection of var. *strigata*. However, in the absence of an original label and because of the variability of *L. strigata*, there is insufficient evidence for the designation of shell or figure as lectotype of var. *articulata*. Although there is no reasonable doubt as to the identity of var. *articulata*, a neotype with preserved animal has been designated. This is necessary in view of confusion in the literature between this species and *L. intermedia*, and because of its close relationship with *L. strigata*, from which unequivocal separation is only possible on the basis of anatomical features. No material is available from Swan Point, but the neotype lot was collected from Broome, only 200 km to the south.

In 1847 Philippi described and figured *Litorina sinensis* from a collection from China received from Largilliert. In the MNHNP there are three shells from the Largilliert Collection housed in the Natural History Museum of Rouen (P. Bouchet, pers. comm.) and one of these is here designated as lectotype. Although Lischke (1871b) distinguished his new species *Litorina strigata* from *L. sinensis* on the basis of shell shape, sculpture and pattern, the form seems to fall within the range of specimens of *L. articulata* from Japan. Since *L. strigata* (Philippi) is not known from China or Japan, *Litorina sinensis* and *Litorina strigata* Lischke are placed in the synonymy of *L. articulata*. *Melaraphe blanfordi* was described by Dunker (1871) from Rockhampton, Queensland, but even in the absence of a type specimen or a figure, can be determined as *L. articulata* since *L. strigata* (Philippi) is not recorded from Australia.

*L. articulata* (and probably also *L. strigata*) has sometimes been misidentified as *L. undulata*, particularly in recent ecological studies in South East Asia (p. 58) and also in museum collections. *L. undulata* does not occur in mangrove habitats.

DIAGNOSIS. Shell: small (less than 20 mm); solid; spire relatively low, whorls rounded; peripheral keel absent; columella wide, excavated; primary grooves 8–10; secondary sculpture often weak or absent; total of 20–33 flat ribs on last whorl; grooves usually impressed lines only; microsculpture indistinct; colour variable, cream yellow to white, with pattern of brown or black dashes on ribs, dashes usually aligned at suture and periphery to form short axial stripes numbering 8–11 on last whorl. Animal: penis not bifurcate, base broad, filament less than half length of base; oviparous.

SHELL (Fig. 90). *Shape*. Height 6–20 mm. Teleoconch 5–6 whorls. Shell of moderate thickness, solid. Spire relatively low, outline slightly convex; whorls rounded, sutures impressed. Peripheral keel absent, but 1–2 ribs at periphery may be slightly enlarged, and pattern also emphasizes periphery. Adult lip not thickened or flared; varices absent. Columella wide, deeply excavated; pillar straight, strongly pinched at base. Sexual dimorphism: males smaller.

*Dimensions*: Table 27.

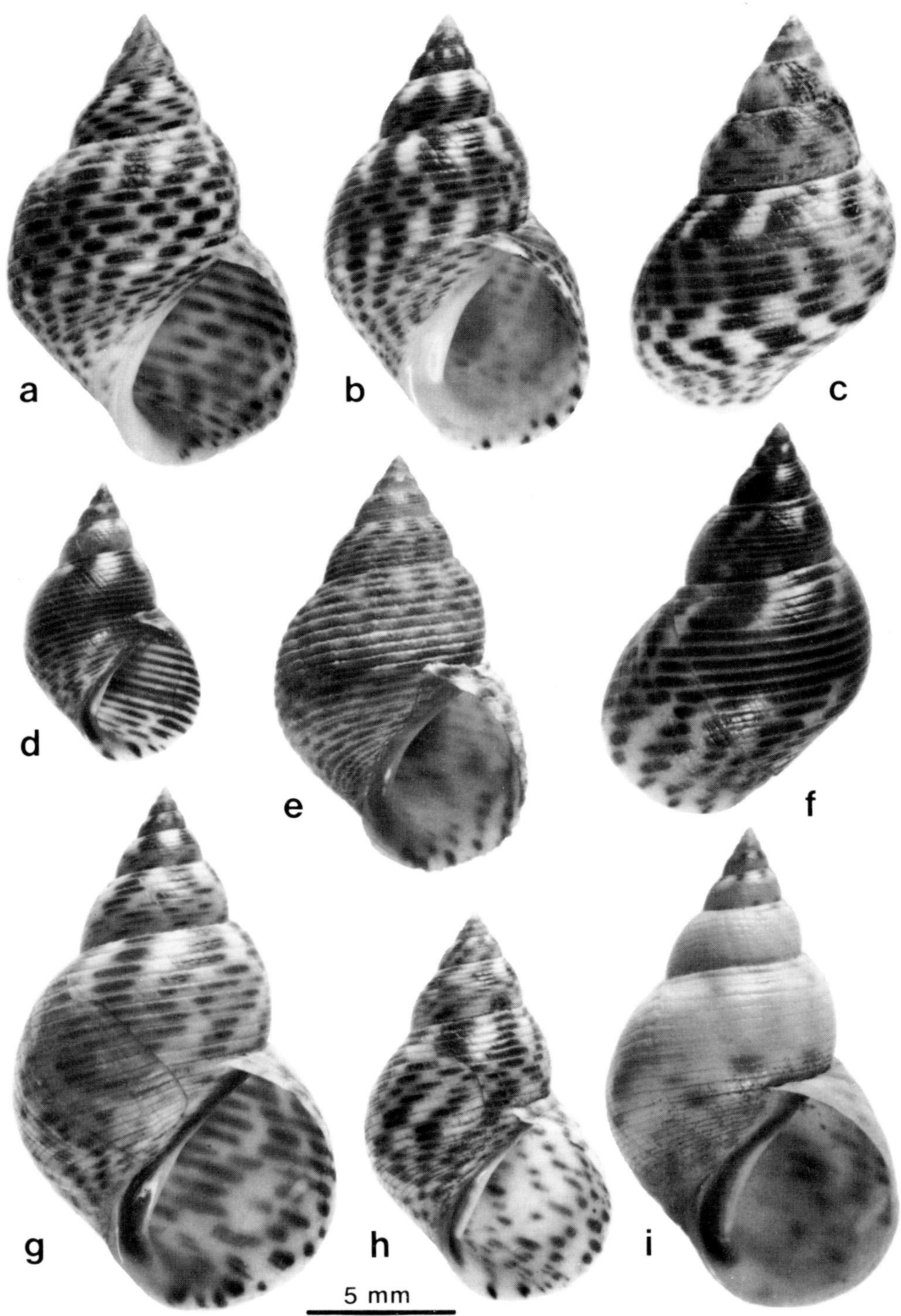


**Fig. 90** *Littoraria* (*Palustorina*) *articulata:* **(a)** ♀, Broome, W.A. (DGR); **(b)** neotype of *Littorina intermedia* var. *articulata* Philippi, ♂, Broome, W.A. (BMNH 198348); **(c)** lectotype of *Litorina sinensis* Philippi, China (MNHNP); **(d)** ♂, Magnetic I., Qld. (DGR); **(e)** ♂, Tolo Harbour, Hong Kong (AMS); **(f)** ♀, Magnetic I., Qld. (DGR); **(g)** ♀, Karumba, Qld. (BMNH); **(h)** ♂, Kanchanadit, Thailand (DGR); **(i)** ♀, Magnetic I., Qld. (DGR).

**Table 27** Dimensions of *Littoraria* (*Palustorina*) *articulata*.

| Specimen | Locality | Sex | Primary grooves | H (mm) | B (mm) | LA (mm) | WA (mm) | C (mm) | P | S | SH |
|---|---|---|---|---|---|---|---|---|---|---|---|
| *Littorina intermedia* var. *articulata* neotype, BMNH 198348 | Broome, W. A. | ♂ | 9 | 14.4 | 9.6 | 8.3 | 6.3 | 2.0 | 1.50 | 0.80 | 1.73 |
| *Litorina sinensis* lectotype, MNHNP | China | | 8 | 13.4 | 9.5 | 7.4 | 5.5 | 1.1 | 1.41 | 0.74 | 1.81 |
| DGR | Broome, W. A. | ♂ | 10 | 13.9 | 9.8 | 8.2 | 6.1 | 1.5 | 1.42 | 0.74 | 1.70 |
| DGR | Broome, W. A. | ♂ | 9 | 14.5 | 10.1 | 8.9 | 6.9 | 1.9 | 1.44 | 0.78 | 1.63 |
| DGR | Broome, W. A. | ♀ | 9 | 15.6 | 11.1 | 9.1 | 6.9 | 1.8 | 1.41 | 0.76 | 1.71 |
| DGR | Broome, W. A. | ♀ | 9 | 15.9 | 10.4 | 9.2 | 6.8 | 1.9 | 1.53 | 0.74 | 1.73 |
| DGR | Magnetic I., Qld. | ♂ | 8 | 12.3 | 8.9 | 7.4 | 5.3 | 1.2 | 1.38 | 0.72 | 1.66 |
| DGR | Magnetic I., Qld. | ♀ | 9 | 15.5 | 11.2 | 9.4 | 7.5 | 1.6 | 1.38 | 0.80 | 1.65 |
| DGR | Ubin I., Singapore | ♂ | 9 | 10.8 | 7.2 | 6.4 | 4.9 | 1.2 | 1.50 | 0.77 | 1.69 |
| DGR | Ubin I., Singapore | ♀ | 9 | 12.4 | 8.6 | 7.3 | 5.6 | 1.3 | 1.44 | 0.77 | 1.70 |
| DGR, mean of 10 | Broome, W. A. | ♂ | | 12.92 | | | | | 1.430 | 0.756 | 1.677 |
| standard error | | | | 0.39 | | | | | 0.007 | 0.009 | 0.014 |
| DGR, mean of 10 | Broome, W. A. | ♀ | | 14.77 | | | | | 1.419 | 0.768 | 1.715 |
| standard error | | | | 0.20 | | | | | 0.007 | 0.009 | 0.013 |
| statistic t or U | | | | 4.189 | | | | | 62 | 61.5 | 72 |
| probability | | | | 0.001 | | | | | 0.394 | 0.415 | 0.106 |

*Sculpture* (Fig. 91). Protoconch normal, usually eroded smooth, first whorl of teleoconch relatively large. All teleoconch whorls sculptured by spiral grooves. Primary grooves 8–10 by whorl 4, usually equidistant, but posterior 1–3 ribs may be a little wider and posterior groove deepest. Secondary sculpture often absent; on whorl 6 some ribs, or all but posterior rib, may be divided by a small secondary groove at $\frac{1}{2}$ width. Tertiary sculpture seldom developed. Total of primary and secondary ribs on last whorl (16)20–33, up to 50 ribs if tertiary sculpture is present; ribs of approximately equal width, low and flattened but for posterior rib which is usually rounded and prominent; in strongly sculptured shells 1–2 peripheral ribs may also be somewhat raised and rounded. Grooves usually impressed lines only, but up to $\frac{1}{4}$ rib width in shells with strongest sculpture; primary grooves occasionally become obsolete on shoulder of last whorl. Microsculpture of faint, regular, spiral striae confined to ribs, but sometimes indistinct or absent; irregular axial growth striae cover surface; strong, regular, axial striae are visible in wider grooves.

*Colour*. Variable. Ground colour usually cream yellow, sometimes pale grey or white (very rarely orange pink), with pattern of black or dark brown dashes on ribs. Typically, dashes aligned across the 3–4 posterior ribs to form axial stripes, numbering 8–11 at suture of body whorl, dashes aligned also at peripheral 4 ribs; these zones of alignment separated by a band of dense non-aligned dashes, which may merge to form continuous spiral stripes; on the base dashes are sparse and not conspicuously aligned. Degree of pigmentation is variable: shells may be black with pale grooves and yellow spots at suture and periphery; pattern may be pale and confined to suture and periphery, but is never absent; dashes may be roughly aligned over whole whorl from suture to base, but remain discrete, separated by pale grooves. Aperture yellow or cream, with purple black dashes corresponding to external pattern, glazed by whitish callus within. Columellar pillar white, parietal callus and columellar excavation dark purple, but occasionally merely tinged pink or entirely white.

ANIMAL. *Colour*. Pigmentation dark grey to black; sides of foot darkly mottled; head black, sometimes with pale central streak; tentacles darkly banded, unpigmented stripe each side of base.

*Penis* (Fig. 92a–k). Length to 5.0 mm. Base simple, incorporating glandular disc. Filament small, less than half length of base, tapering. Sperm groove open. Base and filament off white to pale ochre; glandular disc opaque white to cream.

*Sperm* (Fig. 5e,f). Eupyrene sperm 236–252 μm. Nurse cells (Fig. 93d–f) 30–42 μm, elongate oval

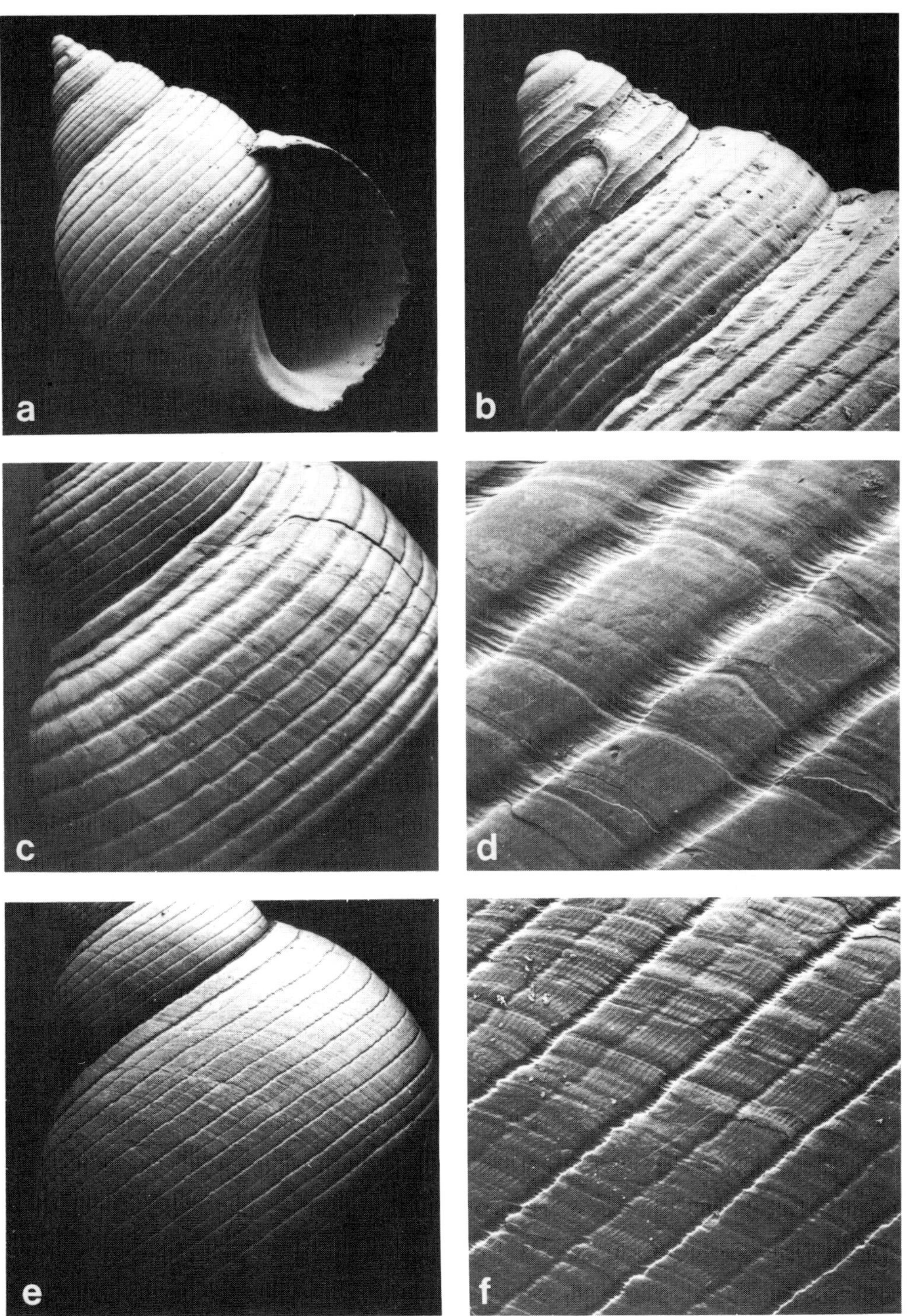


**Fig. 91** *Littoraria* (*Palustorina*) *articulata:* **(a,b)** Magnetic I., Qld.; **(a)** young shell ( × 18); **(b)** protoconch ( × 85); **(c,d)** Broome, W.A.; **(c)** last whorl ( × 9); **(d)** detail ( × 36); **(e,f,)** Magnetic I., Qld.; **(e)** last whorl ( × 10); **(f)** detail ( × 42).

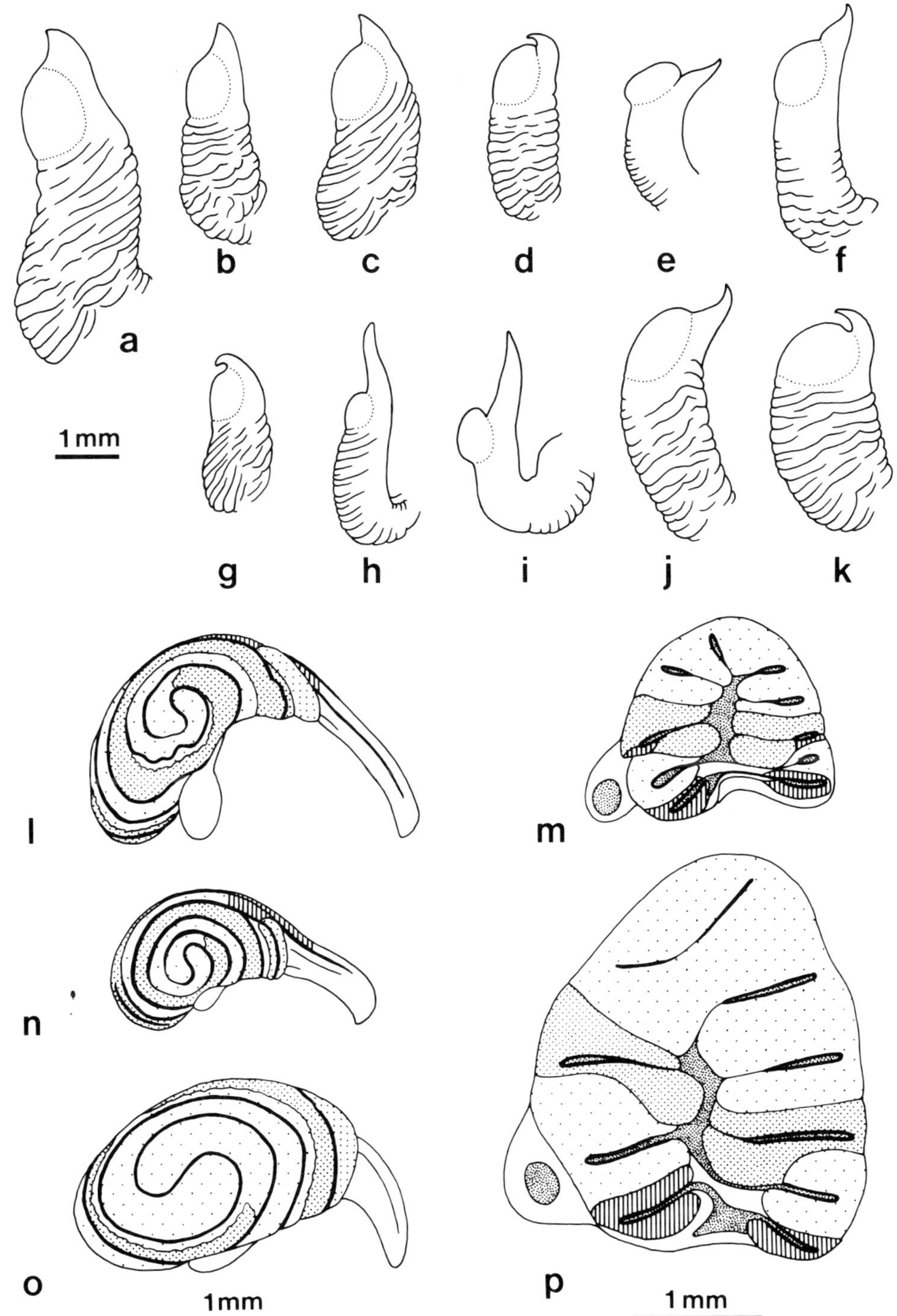


**Fig. 92** *Littoraria* (*Palustorina*) *articulata:* **(a–k)** penes; **(a–c)** Broome, W.A.; **(d–f)** Magnetic I., Qld.; **(g)** Tolo Harbour, Hong Kong; **(h)** Iso, Kagoshima Bay, Japan; **(i)** Sanchun-Kun, Kyongsang-Namdo, S. Korea; **(j,k)** Kanchanadit, Thailand; **(l-p)** pallial oviducts, with transverse sections; **(l,m)** Broome, W.A.; **(n)** Magnetic I., Qld.; **(o,p)** Kanchanadit, Thailand.

to fusiform; basal flagellum to 170 μm; rods 1–2(4), basal, elongate oval; yolk granules small, becoming larger towards base.

*Pallial oviduct* (Fig. 92l–p). Length to 5.7 mm. Spiral section to 4.8 mm diam., 5½ whorls, opaque albumen gland ⅓ whorl, white; translucent albumen gland white to grey or fawn; opaque capsule gland 1–1½ whorls, cream, ochre or pale pink; translucent capsule gland red brown; spiral distinct externally; egg groove darkly pigmented. Straight section to 2.0 mm, fawn; no terminal papilla. Bursa posterior, extending into spiral section. Development oviparous.

*Egg capsules* (Fig. 93a–c). Capsules 248–268 μm diam., transparent, biconvex discs, convexity usually more pronounced on one side, with broad circumferential flange; containing single ova; egg covering 76–82 μm diam. (pers. obs. at Magnetic I., Qld.). At Amakusa, Kyushu, Japan, Kojima (1958*c*) reported symmetrically biconvex capsules 350 μm in diam., lacking circumferential flange, and with egg covering 100 μm diam. (Fig. 93c).

*Radula* (Fig. 87f). Length to 14 mm; relative length 0.85–1.18. Intermediate between saw- and chisel-toothed types; central rachidian cusp square to broadly rectangular, edge rounded or slightly pointed; cusps of paired teeth obliquely triangular; lateral with gap anterior to main cusp.

Distribution. *Habitat.* Most abundant in *Avicennia* fringe, less so in *Sonneratia* fringe, common also on outermost *Rhizophora* trees. On trunks at outer edge of forest, 0.3–1.2 m above ground, up to 2.0 m if tidal range is great. Frequent at lower levels on roots in *Rhizophora* forests, but does not extend further than the back of this zone. Sometimes abundant on sheltered rocks and wooden pilings, at and above water level. A continental species.

*Range* (Fig. 94). Known range (determination based on anatomy, indicated by 'A' in list of records below) extends from Moreton Bay, Queensland, around northern coast of Australia to Exmouth Gulf, Western Australia; elsewhere Singapore, Sarawak, Gulf of Thailand, Vietnam, Hong Kong, southern Japan. Based on identification of shells alone, the range is extended to central Indonesia; occurence in India is doubtful.

*Records.* **India:** Bombay (BMNH); Mandapam (ANSP); Madras (BMNH); Waitair, Vishakhapatnam (ANSP); Diamond Harbour, Bengal (BMNH); **Sri Lanka:** Trincomalee (BMNH); **Burma:** near Cape Negrais, Arakan (BMNH); Salween R. (BMNH); King I., Mergui Arch. (BMNH); **Andaman Is** (BMNH); **Thailand:** Songkhla (MCZ); Kanchanadit, near Surat Thani (DGR, A); Hua Hin (WAM, A); Bang Saen, 14 km SW. of Chon Buri (ANSP, A); **Malaysia: Peninsula:** Batu Maung, Penang (DGR, A); Sungei Merbok estuary, Kedah (DGR, A); 10 km N. of Port Kelang (DGR, A); Kuantan (AMS); **Sarawak:** Santubong (DGR, A); **Sabah:** Po Bui I., Sandakan (USNM); **Singapore:** Jelutong, Ubin I. (DGR, A); Loyang Besar (WAM); **Indonesia: Sumatra:** Weh I. (RNHL); Bangka I. (RNHL); **Java:** Tandjong Priok, Djakarta (RNHL, MCZ); Rembang (RNHL); S. coast Madura I. (RNHL); Paternoster I. (RNHL); **Kalimantan:** Balikpapan (RNHL); **Sulawesi:** Makassar (RNHL); **Irian Jaya:** Merauke (AMS, RNHL); **Vietnam:** Ben Dinh, Vung Tau district (ANSP, A); Ile de la Table, Tonkin (MCZ); **China:** Hainandao I. (ANSP); Amoy (USNM, RNHL); mouth of Yangtze R. (BMNH); Tsingtao, Shantung (Academia Sinica; BMNH); **Hong Kong:** Tolo Channel (BMNH, A); Deep Bay, New Territories (AMS, A); **S. Korea:** Sachun-Kun, Kyongsang-Namdo (ANSP, A); **Japan:** Ryukyu Is (AMS); Iso, Kagoshima Bay, Kyushu (NSMT, A); Hirado, Hizen, Kyushu (AMS); Amakusa, Kyushu (USNM; NSMT, A); Ehime Pref., Shikoku (USNM); **Australia: W.A.:** Bay of Rest, Exmouth Gulf (WAM); Onslow (AMS); Broome (DGR, A; AMS; NMV); Derby (AMS); Port Warrender, Admiralty Gulf (WAM, A); Wyndham (AMS); **N.T.:** Ludmilla Creek, Darwin (DGR, A); Port Essington (AMS, A); Woolen R., Arnhemland (AMS); **Qld.:** Sweers I. (AMS); Karumba (AMS); Thursday I. (DGR, A): Boigu I. (WAM); Quintell Beach, Iron Range (QM); Low Is (QM); Missionary Bay, Hinchinbrook I. (DGR, A; USNM); Magnetic I. (DGR, A); Bowen (WAM); Mackay (WAM); N. Keppel I. (AMS); Yeppoon (AMS); Gladstone (AMS); Urangan (AMS); Sandgate, Moreton Bay (QM, NMV).

Remarks. Shells of *L. articulata* and *L. strigata* (Philippi) are often indistinguishable and anatomical characters are required for their separation. The close relationship between these two species is discussed in the remarks upon *L. strigata*. The distribution map (Fig. 94) is compiled largely from museum collections of shells so that determinations are often doubtful. In particular, occurrence of

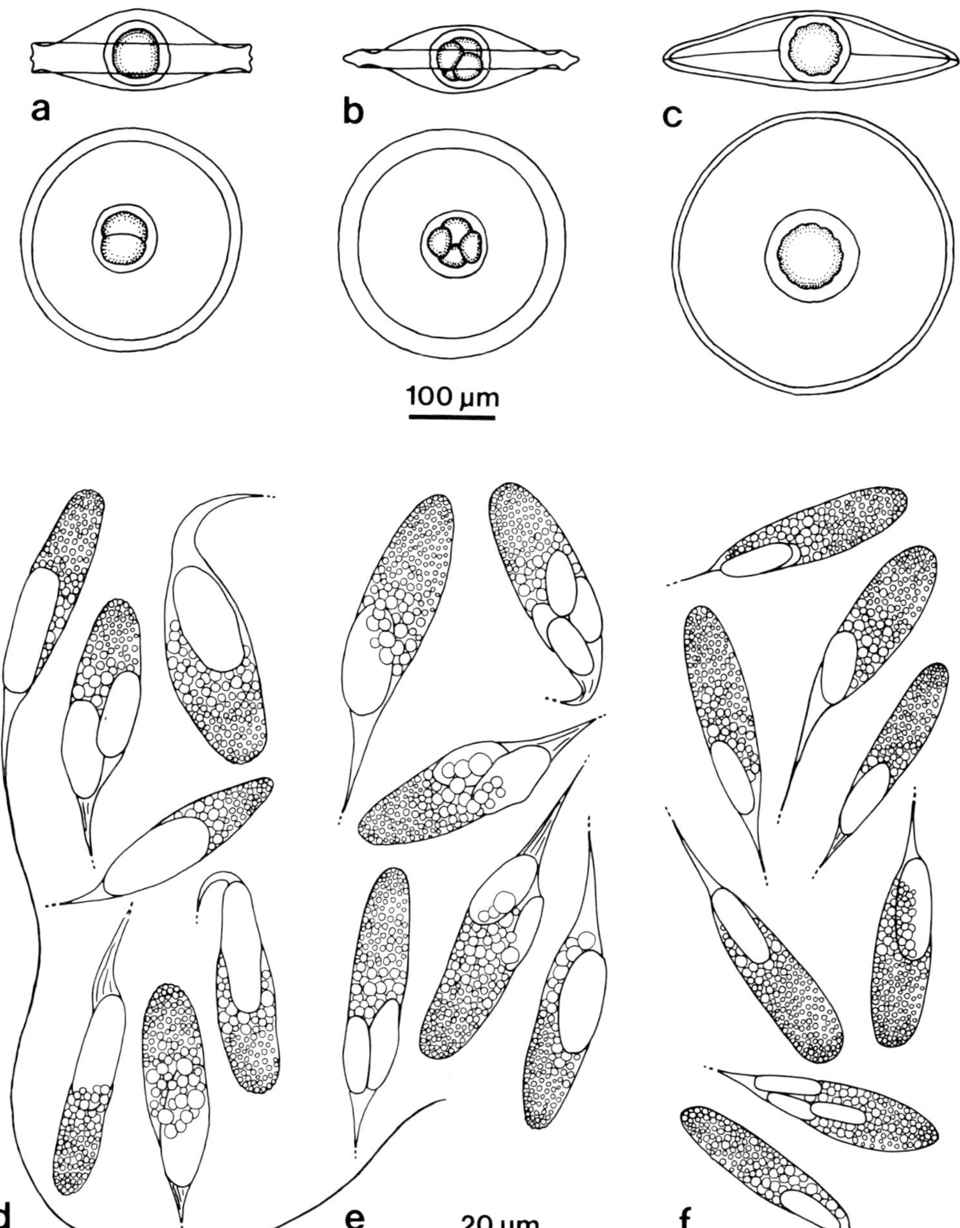


**Fig. 93** *Littoraria* (*Palustorina*) *articulata:* **(a–c)** egg capsules; **(a)** normal form, Magnetic I., Qld.; **(b)** abnormal form, Magnetic I., Qld.; **(c)** Amakusa, Kyushu, Japan (after Kojima, 1958c); **(d–f)** sperm nurse cells, flagellum shown on one only; **(d,e)** Magnetic I., Qld.; **(f)** Kanchanadit, Thailand.

the species in India has not been confirmed by preserved material and it is possible that only *L. strigata* occurs there.

Amongst preserved specimens, shell characters are fairly constant, varying only in prominence of the spiral grooves and in details of colour pattern. A rather well defined geographical form

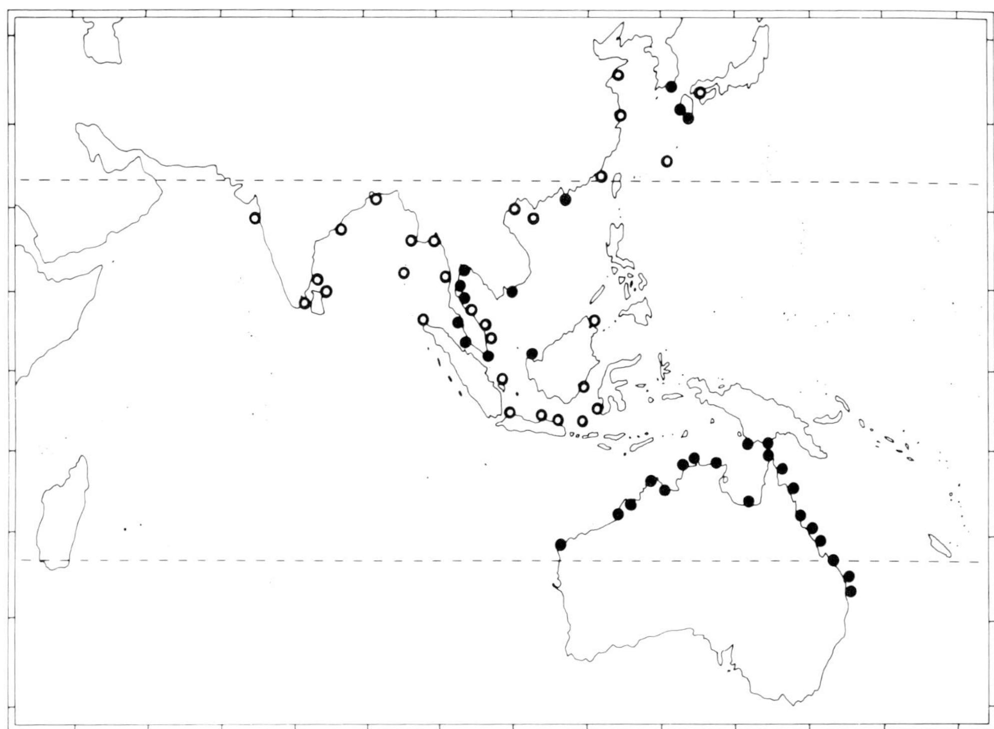

**Fig. 94** Distribution of *Littoraria* (*Palustorina*) *articulata*. Closed circles, records confirmed by anatomical data; open circles, records based on shells alone.

(Fig. 90a,b) occurs in north-western Australia, from Admiralty Gulf to Exmouth Gulf; the neotype is of this form, as, presumably, was the specimen on which Philippi based this species. In this region the shells are more strongly sculptured, with wider grooves, the columella is white as opposed to purple, and the shell coloration is characteristic, the ground colour white with a pattern of discrete black or brown dashes. There are no anatomical differences, however, and the form is not considered worthy of taxonomic recognition.

In specimens from China and Japan, the colour pattern is sometimes diffuse and the sculpture may be strong (Fig. 90c,e). In addition the penial filament of such shells is sometimes relatively longer than in specimens from South East Asia and Australia. The egg capsule of the Japanese form has been illustrated by Kojima (1958*c*; reproduced Fig. 93c herein), and is larger than that of the species at Magnetic I., Queensland (Fig. 93a), and lacks the circumferential flange. However, even at the latter locality there is some variation in capsule shape, and an unusual form (Fig. 93b) is more similar to that drawn by Kojima. The eastern Asian form is tentatively assigned to *L. articulata*. Further work is necessary on the forms of *L. articulata* and also on their relationship with the form here recognized as a separate species, *L. strigata*.

Of the many thousands of specimens seen, only 4 have been found in which the ground colour of the shell is orange pink. This colour form is known elsewhere in the subgenus only in *L. melanostoma* and is rare in that species also, although pink shells are not infrequent in the subgenus *Littorinopsis*. The species occupies a wide range of habitats and zones. At the seaward edge of continental mangrove forests it occurs in great numbers, often closely packed on trunks at and just above the water level.

SIMILAR SPECIES. Differentiation of *L. articulata* from the two similar species *L. strigata* and *L. vespacea* is discussed in the remarks upon these two (pp. 88, 216). Anatomically, *L. intermedia* and *L.*

*articulata* are perfectly distinct, belonging to different subgenera. The shells are usually easily separated, but in doubtful cases the following characters can be used: that of *L. articulata* is smaller, the whorls more rounded and the spire lower; spiral sculpture is usually less pronounced; the ground colour is white or yellow, not grey; the pattern of dark dashes (except in the NW. Australian form) is less discrete and regular than in *L. intermedia*, and there are fewer axial colour stripes at the suture of the last whorl. The shell of *L. undulata* (Fig. 99a–c) is rarely at all similar to that of *L. articulata;* in the former the spiral sculpture is usually obsolete, the columellar pillar relatively longer, straight and lilac in colour, and if darkly coloured the shell pattern is of uninterrupted axial zigzag stripes. Anatomically, *L. undulata* is distinguished by a penial glandular disc of unusually large size (Fig. 4e). The West African *L. cingulifera* (Fig. 99h,i) can be extremely similar to *L. articulata*, but the spire is usually lower and the aperture more patulous, the columellar pillar relatively longer, straight and not so sharply pinched at the base, the shell pattern more often of spiral lines than dashes; anatomically the species are separated by the form of the penis, for in *L. cingulifera* the glandular disc is borne on a small projection of the base and the filament is broad and blunt (Fig. 4g).

## *Littoraria (Palustorina) strigata* (Philippi, 1846)

*Littorina intermedia* var. *strigata* Philippi, 1846: 141 [Jimamailan [Himamaylan], Negros I., Philippines; lectotype here designated BMNH 1968353]; Nevill, 1885: 146–147 [in part]
*Litorina intermedia* var. *strig*ata—Philippi, 1847, vol. 2: 223, *Litorina* pl. 5, fig. 8 [in part; original citation corrected to figs 7, 8, 9]
*Littorina intermedia*—von Martens, 1887: 169–170; Hidalgo, 1904–5: 206 [both in part; not Philippi, 1846]
*Littorina* (*Melaraphe*) *scabra* var. *intermedia*—Tryon, 1887: 244 [in part; not Philippi, 1846]
*Littorina* (*Malaraphe*) *intermedia*—Casto de Elera, 1896: 309–310 [in part; not Philippi, 1846]
*Littorina* (*Littorinopsis*) *intermedia*—von Martens, 1897: 197 [in part; not Philippi, 1846]
*Littorinopsis intermedia*—Brandt, 1974: 54 [in part; not Philippi, 1846]
*Littorina* (*Littorinopsis*) *scabra scabra*—Rosewater, 1970: 456–461, pl. 352, fig. 8 [in part; not Linnaeus, 1758]
*Littorina undulata*—Berry, 1972 [in part; not Gray, 1839]
*Littorinopsis undulata*—Brandt, 1974: 54 [in part; not Gray, 1839]

Nomenclature. Since separation of *L. strigata* from *L. articulata* is dependent largely upon anatomical characters, type specimens, figures and descriptions based on shells alone are difficult to interpret. *Littorina intermedia* var. *strigata* was described by Philippi in 1846 from material in the Cuming Collection, and the variety was figured by the author the following year. Of the three figures referred to in the text as var. *strigata* (see p. 125 for a discussion of the error in the original citation), figures 7 and 9 may represent respectively the species here recognized as *L. articulata* and *L. subvittata*. Only figure 8 is typical of the species to which the name *strigata* is here restricted. The origin of the figured specimens was not mentioned in the text, but there is little doubt that a lot from the type locality in the BMNH contains the original specimen of Philippi's figure 8, although no label in Cuming's hand remains. This figured specimen was recognized by Rosewater (1970), who designated it as lectotype of *Littorina intermedia*. However, this designation is not acceptable (p. 125) and the same specimen is here designated lectotype of *Littorina intermedia* var. *strigata*. The determination of this specimen is not in doubt, for the shell (Fig. 95c) is of the type most characteristic of *L. strigata* and not seen in *L. articulata*; furthermore, of these two species only the former is known to occur in the Philippines. It should be noted that *Litorina strigata* Lischke is apparently a synonym of *L. articulata*, and not of *L. strigata* (Philippi).

Diagnosis. Shell: small (less than 21 mm); solid; spire whorls rounded; peripheral keel absent; columella of moderate width, excavated; primary grooves 8–10; secondary sculpture often weak or absent; total of 20–33 flat ribs on last whorl; grooves narrow; microsculpture indistinct; colour variable, cream yellow to white with black or brown pattern typically forming 6–8 broad, oblique,

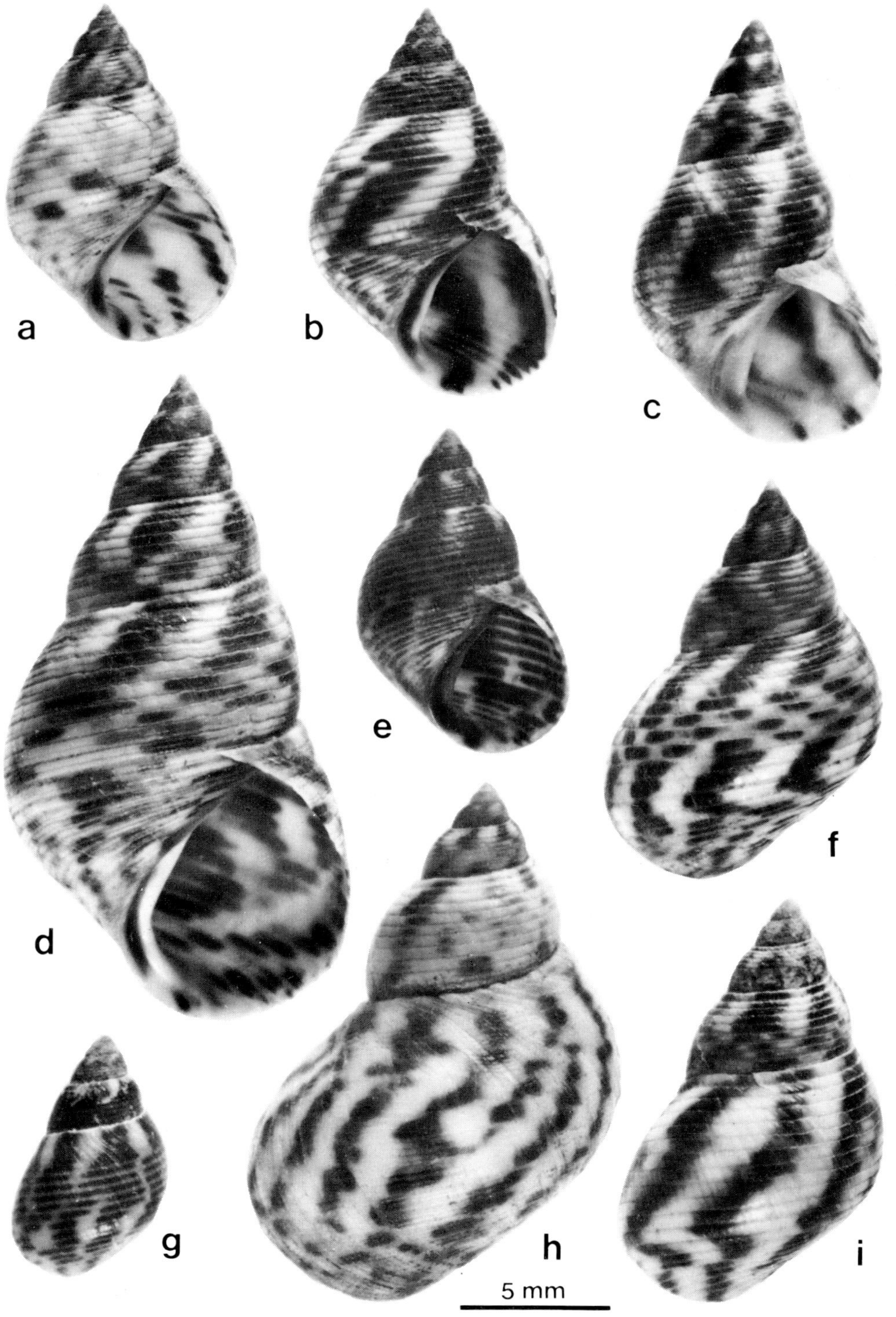


**Fig. 95** *Littoraria* (*Palustorina*) *strigata:* **(a)** ♂, Penang, Malaysia (DGR); **(b)** ♂, Kanchanadit, Thailand (DGR); **(c)** lectotype of *Littorina intermedia* var. *strigata* Philippi, Himamaylan, Negros I., Philippines (BMNH 1968353); **(d)** ♀, Kanchanadit, Thailand (DGR); **(e,f)** Penang, Malaysia (DGR); **(e)** ♂; **(f)** ♀; **(g)** ♂, Manila Bay, Philippines (AMS); **(h)** ♀, Penang, Malaysia (DGR); **(i)** ♀, Kanchanadit, Thailand (DGR).

axial stripes, but pattern may be broken up into poorly aligned dashes on ribs. Animal: penis not bifurcate, base narrow, filament long, greater than length of base; oviparous.

SHELL (Fig. 95). *Shape*. Height 6–21 mm. Teleoconch 5–7 whorls. Shell of moderate thickness, solid. Spire outline slightly convex, whorls rounded; sutures impressed. Peripheral keel absent, but pattern may sometimes emphasize periphery. Adult lip not thickened or flared; varices absent. Columella of moderate width, excavated; pillar straight, pinched at base. Sexual dimorphism: males smaller, aperture narrower.

*Dimensions*: Table 28.

*Sculpture* (Fig. 96a,b). Protoconch not seen, usually eroded smooth; first whorl of teleoconch relatively large. All teleoconch whorls sculptured by spiral grooves. Primary grooves number 8–10 by whorl 4, approximately equidistant, posterior groove usually deepest. Secondary sculpture often absent, but on whorl 6 some ribs, or all ribs but the most posterior, may be divided by a small secondary groove at about $\frac{1}{2}$ width. A few tertiary grooves are formed in largest shells only. Ribs usually remain flattened, but for the rounded and prominent posterior rib; in the most strongly sculptured forms all ribs may be somewhat rounded. Total of primary and secondary ribs on body whorl 20–33, up to 60 on largest shells with tertiary sculpture. Grooves usually about $\frac{1}{5}$ rib width, but sometimes impressed lines only, and rarely up to $\frac{1}{2}$ rib width. Microsculpture of faint, regular spiral striae confined to ribs, occasionally indistinct or absent; surface covered by irregular axial growth striae; strong and regular striae visible in wider grooves.

*Colour*. Variable. Ground colour cream yellow or white, with a black or dark brown pattern. Typically, pattern is conspicuously aligned from suture to base, to form 6–8 broad, oblique or zigzag, axial stripes, separated by an approximately equal width without pigment; stripes are continuous across grooves and may divide and anastomose. Pattern sometimes broken up into separate spiral dashes on ribs, aligned only at suture and periphery; shells may be sparsely or densely patterned. Aperture yellow or cream, with black lines and stripes corresponding to external pattern; in largest shells glazed by a thin whitish callus within. Parietal callus and columellar excavation dark purple, occasionally only tinged pink; columellar pillar white.

**Table 28** Dimensions of *Littoraria* (*Palustorina*) *strigata*.

| Specimen | Locality | Sex | Primary grooves | H (mm) | B (mm) | LA (mm) | WA (mm) | C (mm) | P | S | SH |
|---|---|---|---|---|---|---|---|---|---|---|---|
| *Littorina intermedia* var. *strigata* lectotype, BMNH 1968353 | Himamaylan, Negros I., Philippines | | 10 | 14.1 | 9.0 | 7.5 | 5.5 | 1.4 | 1.57 | 0.73 | 1.88 |
| DGR | Kanchanadit, Thailand | ♂ | 10 | 12.1 | 8.0 | 6.6 | 4.8 | 0.9 | 1.51 | 0.86 | 1.83 |
| DGR | Kanchanadit, Thailand | ♂ | 9 | 12.5 | 8.9 | 7.2 | 5.3 | 1.1 | 1.40 | 0.74 | 1.74 |
| DGR | Kanchanadit, Thailand | ♀ | 9 | 21.0 | 12.5 | 10.6 | 7.7 | 1.7 | 1.68 | 0.73 | 1.98 |
| DGR | Kanchanadit, Thailand | ♀ | 9 | 13.7 | 9.0 | 7.6 | 5.5 | 1.2 | 1.52 | 0.73 | 1.80 |
| DGR | Penang, Malaysia | ♂ | 9 | 13.4 | 8.4 | 7.5 | 5.4 | 1.0 | 1.60 | 0.72 | 1.79 |
| DGR | Penang, Malaysia | ♀ | 9 | 14.9 | 10.2 | 8.6 | 6.7 | 2.0 | 1.46 | 0.78 | 1.73 |
| AMS C.131827 | Manila, Philippines | ♂ | 8 | 7.0 | 5.3 | 4.6 | 3.3 | 0.5 | 1.32 | 0.72 | 1.52 |
| AMS C.131827 | Manila, Philippines | ♀ | 8 | 7.2 | 5.5 | 4.8 | 3.5 | 0.8 | 1.31 | 0.73 | 1.50 |
| DGR, mean of 10 | Penang, Malaysia | ♂ | | 11.20 | | | | | 1.422 | 0.737 | 1.655 |
| standard error | | | | 0.31 | | | | | 0.022 | 0.007 | 0.022 |
| DGR, mean of 10 | Penang, Malaysia | ♀ | | 12.94 | | | | | 1.419 | 0.773 | 1.681 |
| standard error | | | | 0.68 | | | | | 0.022 | 0.010 | 0.032 |
| statistic t or U | | | | 2.321 | | | | | 51 | 83 | 56 |
| probability | | | | 0.032 | | | | | 0.970 | 0.012 | 0.436 |

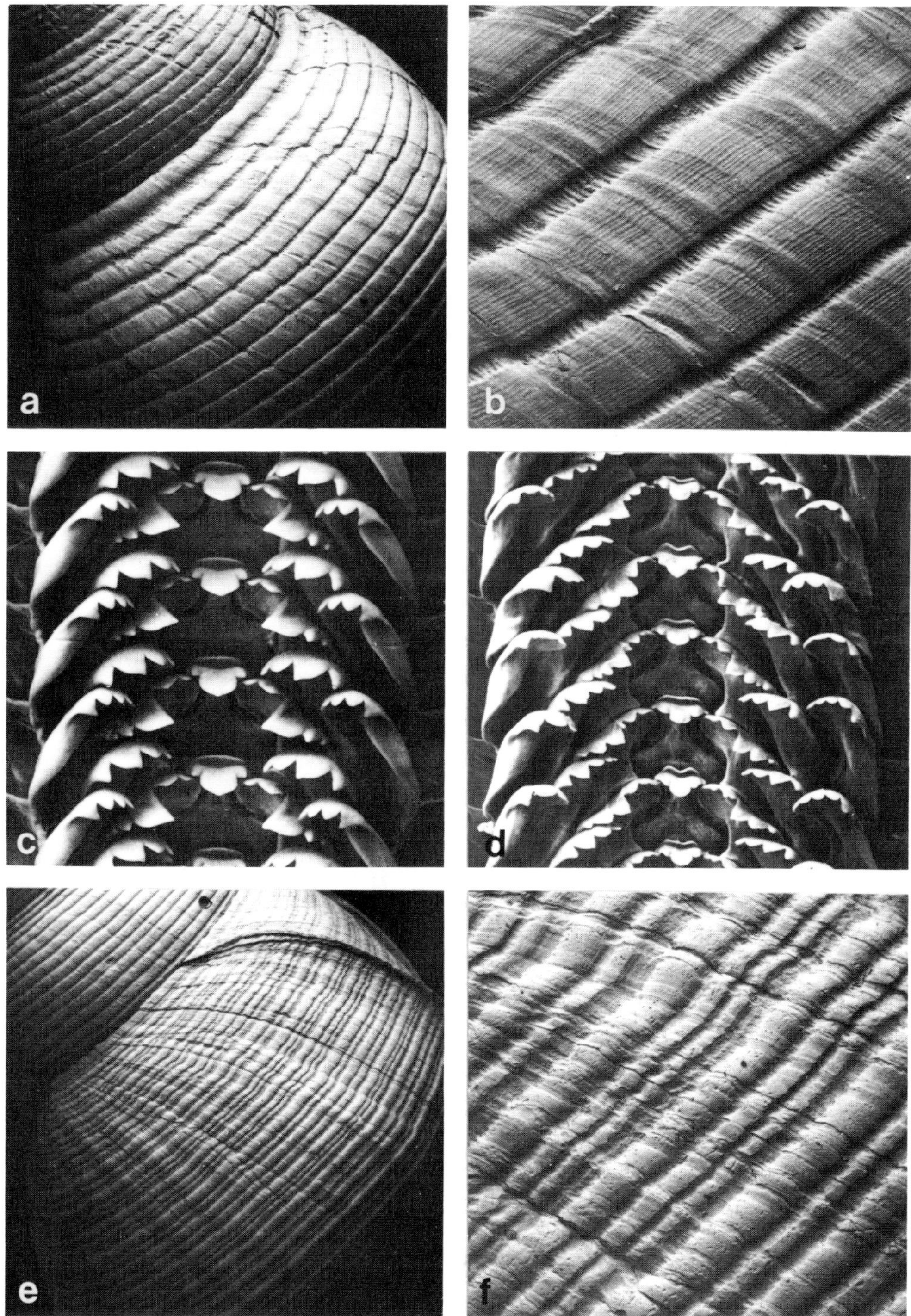


**Figs. 96** **(a–c)** *Littoraria* (*Palustorina*) *strigata:* **(a,b)** Kanchanadit, Thailand; **(a)** last whorl ( × 11); **(b)** detail ( × 43); **(c)** radula, Penang, Malaysia ( × 260). **(d–f)** *Littoraria* (*Littorinopsis*) *ardouiniana*, Three Fathoms Cove, Hong Kong: **(d)** radula ( × 170); **(e)** last whorl ( × 7); **(f)** detail ( × 26).

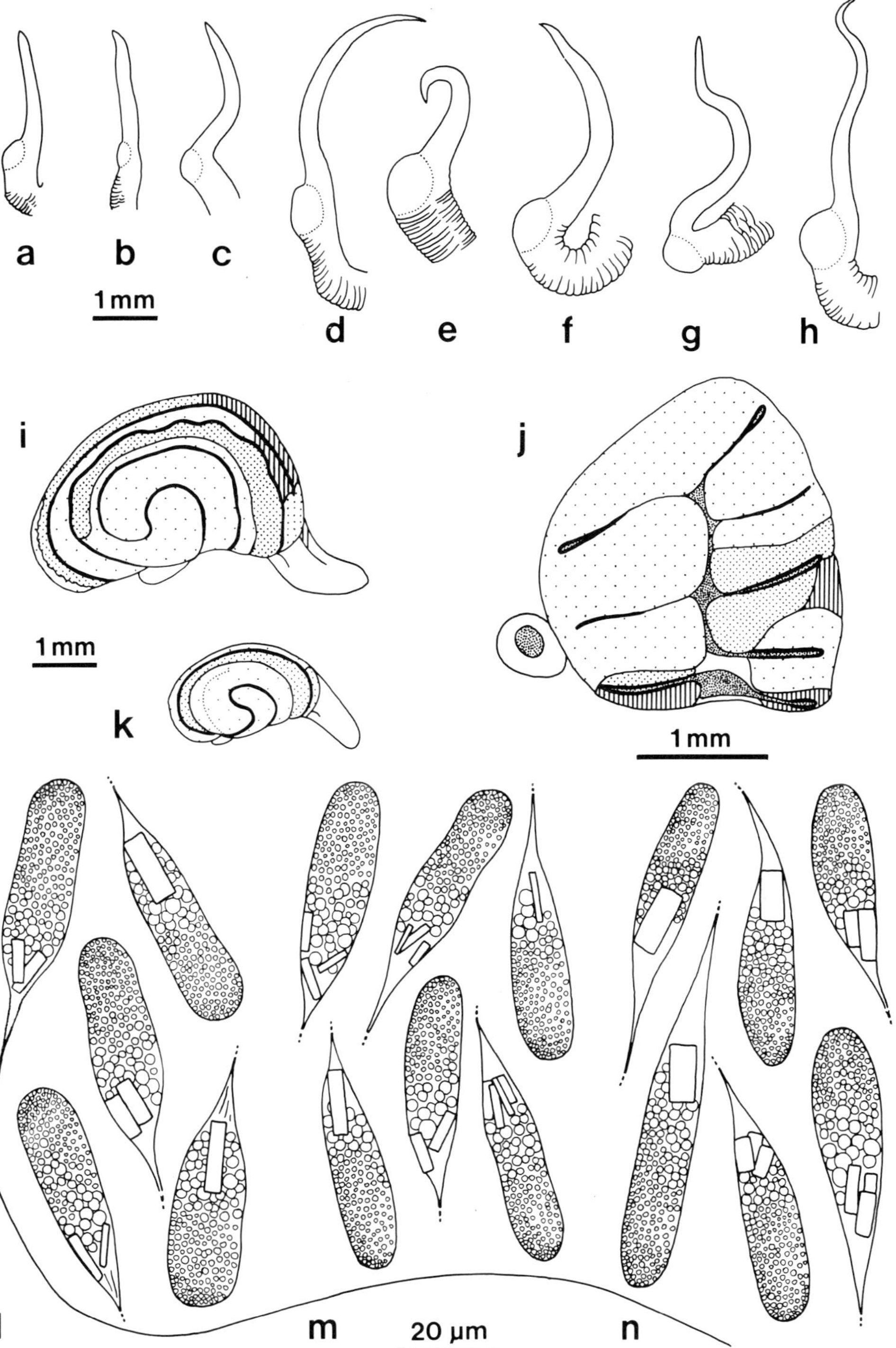


**Fig. 97** *Littoraria* (*Palustorina*) *strigata:* **(a–h)** penes; **(a,b)** Manila Bay, Philippines; **(c)** Phuket I., Thailand; **(d–f)** Kanchanadit, Thailand; **(g,h)** Penang, Malaysia; **(i–k)** pallial oviducts, with transverse section; **(i,j)** Kanchanadit, Thailand; **(k)** Manila Bay, Philippines; **(1–n)** sperm nurse cells, flagellum shown on one only; **(l)** Kanchanadit, Thailand; **(m)** Penang, Malaysia; **(n)** Phuket I., Thailand.

ANIMAL. *Colour.* Pigmentation black; sides of foot darkly mottled; head darkest; tentacles darkly banded, stripes at base often absent or indistinct.

*Penis* (Fig. 97a–h). Length to 6.0 mm. Base simple, rather narrow, incorporating small glandular disc. Filament long, greater than length of base, tapering. Sperm groove open. Penis white to cream.

*Sperm.* Eupyrene sperm 240 μm. Nurse cells (Fig. 97l–n) 34–47 μm; elongate oval, tapering to basal flagellum, sometimes slightly constricted at mid-length; basal flagellum to 175 μm; rods 1–4(6), small, basal, rectangular; yolk granules small, becoming larger towards base.

*Pallial oviduct* (Fig. 97i–k). Length to 5.0 mm. Spiral section to 4.2 mm diam., $4\frac{1}{2}$ whorls; opaque albumen gland $\frac{1}{4}$ whorl, white; translucent albumen gland white; opaque capsule gland $\frac{1}{2}$–$\frac{3}{4}$ whorl, pale pink or white; translucent capsule gland red brown; spiral distinct externally; egg groove darkly pigmented. Straight section short, to 1.4 mm, pale brown; no terminal papilla. Bursa posterior, extending into spiral section. Development assumed oviparous.

*Radula* (Fig. 96c). Length to 17 mm; relative length 1.07–1.32. Intermediate between saw- and chisel-toothed types; central rachidian cusp broad, edge straight or slightly pointed; cusps of paired teeth obliquely triangular; lateral with gap anterior to main cusp.

DISTRIBUTION. *Habitat.* Most common on trunks in *Avicennia* fringe, 0.3–1.8 m above ground; also on *Sonneratia* trunks and roots of outermost *Rhizophora* trees, but scarce in *Rhizophora* forest. Sometimes abundant on sheltered rocks and wooden pilings.

*Range* (Fig. 98). Known range (determination based on anatomy, indicated by 'A' in list of records below) south-western Thailand, Singapore, Gulf of Thailand, Java, Sarawak, Philippines. Based on identification of shells, range is extended west to India and Pakistan.

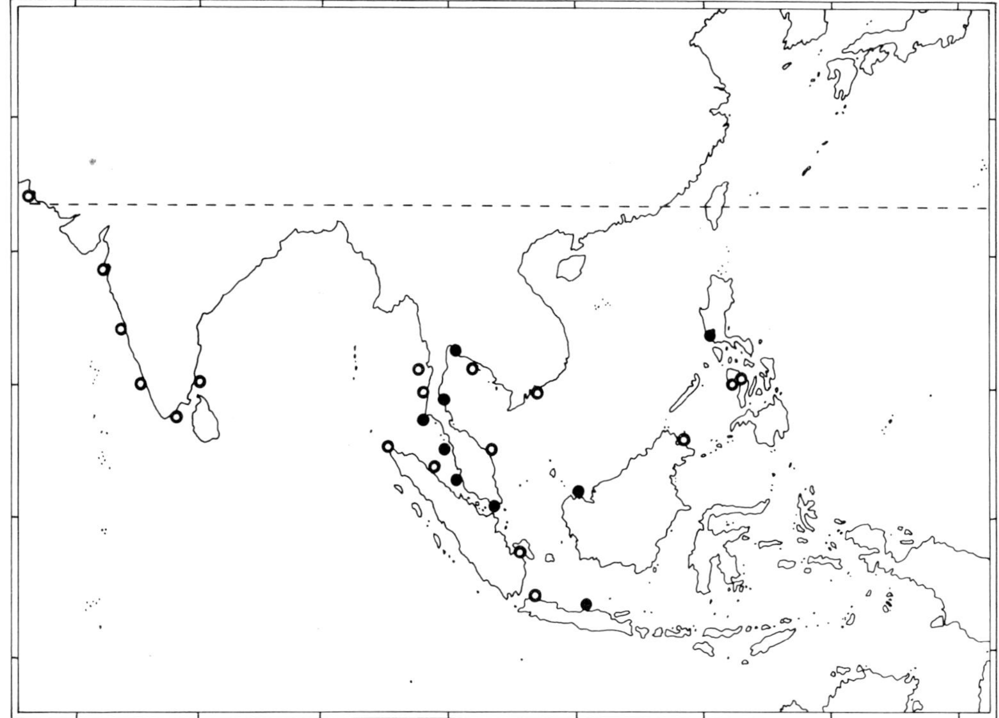

**Fig. 98** Distribution of *Littoraria* (*Palustorina*) *strigata*. Closed circles, records confirmed by anatomical data; open circles, records based on shells alone.

*Records*. **Pakistan:** west wharf, Karachi (MCZ); **India:** Bombay (BMNH, MCZ, USNM); Vingurla, N. of Goa (USNM); Netravati R., Mangalore (USNM); Gulf of Mannar (USNM); Adyar R. estuary, Madras (USNM); **Burma:** Elphinstone I., Mergui Arch. (BMNH); Victoria Point (MCZ); **Thailand:** Tanga I., Butang Group (USNM); Ao Nam-Bor, Phuket I. (DGR, A); Kanchanadit, near Surat Thani (DGR, A); Bang Saen, 14 km SW. of Chon Buri (ANSP, A); Ko Kut (USNM); **Malaysia: Peninsula:** Batu Maung, Penang (DGR, A); Sungei Merbok estuary, Kedah (DGR, A); 10 km N. of Port Kelang (DGR, A); Trengganu (ANSP); **Sarawak:** Santubong (DGR, A); **Sabah:** Po Bui I., Sandakan (USNM); **Singapore:** Jelutong, Ubin I. (DGR, A); **Vietnam:** Chilins, Vung Tau district (ANSP); **Indonesia: Sumatra:** Weh I. (RNHL); Belawan R., Deli (RNHL); Tandjong Njioer, Bangka I. (RNHL); **Java:** Tandjong Priok, Djakarta (RNHL); Djepara (USNM, A); **Philippines:** Viejo Victorias, Negros I. (USNM); Iloilo, Panay (USNM); SE. Bataan, Luzon (ANSP); Cavite, Manila Bay (AMS, A).

REMARKS. This species is clearly very closely related to *L. articulata* and shells alone are often indistinguishable. That the two are probably distinct was first suggested by the observation that in several South East Asian populations of supposed *L. 'articulata'* the penis was dimorphic. The dimorphism was not associated with reproductive condition, since all males examined were mature, nor with obvious parasitic infection. Furthermore, from observations on *L. articulata* throughout the year at Magnetic Island, Queensland, it was known that penial form did not show seasonal change. The narrow penial type with long filament was found to be correlated with a shell pattern of broad axial stripes, and all sperm nurse cells examined (from localities: Kanchanadit, SE. Thailand; Phuket I., SW. Thailand; Penang, Malaysia; Manila Bay, Philippines) contained small, rectangular rods, in contrast with the larger, oval rods of *L. articulata*. Males are immediately separable by penial and sperm characters, but determination of females is more difficult. Usually the pallial oviduct consists of $4\frac{1}{2}$ whorls and the opaque capsule gland of $\frac{1}{2}$–$\frac{3}{4}$ whorl ($5\frac{1}{2}$ and 1–$1\frac{1}{2}$ respectively in *L. articulata*), but the number of whorls visible depends upon the plane of the transverse section and the opaque capsule gland is somewhat variable in extent. At certain localities (e.g. those on the Malayan Peninsula listed above) shells of *L. strigata* can usually be distinguished by their stronger sculpture and by the fewer and uninterrupted axial black stripes. However, shells of *L. strigata* from Manila Bay (Fig. 95g) and some from Java, Singapore and the Gulf of Thailand, fall within the range of typical *L. articulata*. Other examples of separate species of littorinids with closely similar shells have been reported (Sacchi & Rastelli, 1967; Hannaford Ellis, 1979; Murray, 1979), but further research is necessary to verify that *L. articulata* and *L. strigata* are indeed distinct species.

There is no clear habitat difference between *L. strigata* and *L. articulata*. Both occur commonly at similar levels on the trunks of the outermost trees of mangrove forests. However, at the South East Asian localities where the ranges overlap, either one or other species is dominant (*L. strigata* at: Phuket I., SW Thailand; Batu Maung, Penang; Sungei Merbok estuary, Kedah; near Port Kelang, Malaysia; *L. articulata* at: Singapore; Santubong, Sarawak; all DGR). The only localities at which both species are abundant are in the Gulf of Thailand (Kanchanadit, SE Thailand, DGR, *L. strigata* 30%; and near Chon Buri, ANSP). There is a suggestion that *L. strigata* is dominant in the somewhat less turbid and more oceanic situations.

The geographical distribution of *L. strigata* is imperfectly known, since many records are based upon shells alone. Unfortunately no preserved material has been seen from India, but shells from this area often resemble typical *L. strigata* rather than *L. articulata*, and it is possible that only the former occurs there.

SIMILAR SPECIES. Separation from *L. articulata* is discussed above. Other similar species include *L. intermedia*, *L. undulata* and *L. cingulifera*, as discussed under *L. articulata* (p. 208). *L. vespacea* is superficially similar even in anatomical features; the penis of *L. strigata* lacks a constriction at the base of the filament and is more slender, the sperm nurse cells are flagellate, the bursa copulatrix is posterior and the oviduct of only $4\frac{1}{2}$ whorls; the shell of *L. strigata* is more narrow, the columella less deeply excavated and the axial stripes are fewer than in *L. vespacea*.

**Fig. 99** **(a–c)** *Littoraria* (*Littoraria*) *undulata:* **(a,b)** Green I., Qld.; **(a)** ♂; **(b)** ♀; **(c)** Sula Arch., Indonesia (BMNH). **(d–f)** *Littoraria* (*Littorinopsis*) *angulifera:* **(d)** Gabon (AMS); **(e)** Banana, Zaire (BMNH); **(f)** Colon, Panama (AMS). **(g)** *Littoraria* (*Littoraria*) *zebra*, Panama (BMNH). **(h,i)** *Littoraria* (*Littoraria*) *cingulifera*, Lagos, Nigeria (BMNH).

# REFERENCES


**Abbott, R. T. 1954.** Review of the Atlantic periwinkles, *Nodilittorina, Echininus*, and *Tectarius. Proceedings of the United States National Museum* **103:** 449–464.

**Abbott, R. T. 1960.** The genus *Strombus* in the Indo-Pacific. *Indo-Pacific Mollusca* **1:** 33–146.

**Abe, N. 1942.** Ecological observations on *Melaraphe* (*Littorinopsis*) *scabra* (Linnaeus) inhabiting the mangrove-tree. *Palao Tropical Biological Station Studies* **2:** 391–435.

**Adam, W. & Leloup, E. 1938.** Résultats scientifiques du voyage aux Indes Orientales Néerlandaises de LL. AA. RR. le Prince et la Princesse Léopold de Belgique. Prosobranchia et Opisthobranchia. *Mémoires du Musée Royal d'Histoire Naturelle de Belgique. Sér. 2* **2**(19): 1–209.

**Adams, H. & Adams, A. 1858.** *The Genera of Recent Mollusca* **1.** London.

**Allan, J. 1959.** *Australian Shells.* Ed. **2.** Melbourne.

**Allen, J. F. 1953.** The morphology of the radula of *Littorina irrorata* (Say). *Transactions of the American Microscopical Society* **71:** 139–145.

**Anderson, D. T. 1962.** The reproduction and early life histories of the gastropods *Bembicium auratum* (Quoy and Gaimard) (Fam. Littorinidae), *Cellana tramoserica* (Sower.) (Fam. Patellidae) and *Melanerita melanotragus* (Smith) (Fam. Neritidae). *Proceedings of the Linnean Society of New South Wales* **87:** 62–68.

**Anderson, D. T. 1966.** Further observations on the life-histories of littoral gastropods in New South Wales. *Proceedings of the Linnean Society of New South Wales* **90:** 242–251.

**Anderson, H. 1958.** The gastropod genus *Bembicium* Philippi. *Australian Journal of Marine and Freshwater Research* **9:** 546–568.

**Angas, G. F. 1871.** A list of additional species of marine Mollusca to be included in the fauna of Port Jackson and the adjacent coasts of New South Wales. *Proceedings of the Zoological Society of London* 1871: 87–101.

**Ankel, W. E. 1930.** Die atypische Spermatogenese von *Janthina. Zeitschrift für Zellforschung und Mikroskopische Anatomie* **11:** 491–604.

**Ankel, W. E. 1936.** Die Frasspuren von *Helcion* und *Littorina* und die Funktion der Radula. *Verhandlungen der Deutschen Zoologischen Gesellschaft. Supplementband, Zoologischer Anzeiger* **38:** 174–182.

**Ankel, W. E. 1938.** Wie frisst *Littorina?* I. Radula-Bewegung und Fresspur. *Senckenbergiana. Frankfurt a. M.* **19:** 317–333.

**Annandale, N. & Prashad, B. 1919.** Some gastropod molluscs from the Gangetic Delta. *Records of the Indian Museum* **16:** 241–257.

**Azuma, M. 1960.** *A Catalogue of the Shell-bearing Mollusca of Okinoshima, Kashiwajima and the Adjacent Area (Tosa Province), Shikoku, Japan.*

**Bakker, K. 1959.** Feeding habits and zonation in some intertidal snails. *Archives Néerlandaises de Zoologie* **13:** 230–257.

**Bandel, K. 1974.** Studies on Littorinidae from the Atlantic. *Veliger* **17:** 92–114.

**Bandel, K. 1975.** Das Embryonalgehäuse mariner Prosobranchier der Region von Banyuls-sur-Mer. Teil 1. *Vie et Milieu. Sér. A* **25:** 83–118.

**Bandel, K. & Kadolsky, D. 1982.** Western Atlantic species of *Nodilittorina* (Gastropoda: Prosobranchia): comparative morphology and its functional, ecological, phylogenetic and taxonomic implications. *Veliger* **25:** 1–42.

**Barkman, J. J. 1955.** On the distribution and ecology of *Littorina obtusata* (L.) and its subspecific units. *Archives Néerlandaises de Zoologie* **11:** 22–86.

**Barnard, K. H. 1951.** *A Beginner's Guide to South African Shells.* Cape Town.

**Bartsch, P. 1915.** Report on the Turton collection of South African marine mollusks, with additional notes on other South African shells contained in the United States National Museum, Smithsonian Institution. *Bulletin of the United States National Museum* **91:** 1–305.

**Battaglia, B. 1952.** Ricerche sulla spermatogenesi atipica dei Gasteropodi Prosobranchi. II. Le cellule nutrici nella spermatogenesi di *Littorina neritoides* L. (Gasteropodo Prosobranco). *Bollettino di Zoologia* **19:** 195–201.

**Bedford, L. 1965.** The histology and anatomy of the reproductive system of the littoral gastropod *Bembicium nanum* (Lamarck) (Family Littorinidae). *Proceedings of the Linnean Society of New South Wales* **90:** 95–105.

**Bequaert, J. C. 1943.** The genus *Littorina* in the Western Atlantic. *Johnsonia* **1**(7): 1–27.

**Berger, E. M. 1973.** Gene-enzyme variation in three sympatric species of *Littorina*. *Biological Bulletin. Marine Biological Laboratory Woods Hole* **145:** 83–90.

**Berry, A. J. 1961.** Some factors affecting the distribution of *Littorina saxatilis* (Olivi). *Journal of Animal Ecology* **30:** 27–45.

**Berry, A. J. 1963.** Faunal zonation in mangrove swamps. *Bulletin of the National Museum, State of Singapore* **32:** 90–98.

**Berry, A. J. 1972.** The natural history of the West Malaysian mangrove faunas. *Malayan Nature Journal* **25:** 135–162.

**Berry, A. J. & Chew, E. 1973.** Reproductive systems and cyclic release of eggs in *Littorina melanostoma* from Malayan mangrove swamps (Mollusca: Gastropoda). *Journal of Zoology. London* **171:** 333–344.

**Biggs, H. E. J. 1958.** Littoral collecting in the Persian Gulf. *Journal of Conchology. London* **24:** 270–275.

**Bingham, F. O. 1972*a*.** Shell growth in the gastropod *Littorina irrorata*. *Nautilus* **85:** 136–141.

**Bingham, F. O. 1972*b*.** Several aspects of the reproductive biology of *Littorina irrorata* (Gastropoda). *Nautilus* **86:** 8–10.

**Bingham, F. O. 1972*c*.** The mucus holdfast of *Littorina irrorata* and its relationship to relative humidity and salinity. *Veliger* **15:** 48–50.

**Borkowski, T. V. 1971.** Reproduction and reproductive periodicities of South Floridian Littorinidae (Gastropoda: Prosobranchia). *Bulletin of Marine Science* **21:** 826–840.

**Borkowski, T. V. 1974.** Growth, mortality, and productivity of south Floridian Littorinidae (Gastropoda: Prosobranchia). *Bulletin of Marine Science* **24:** 409–438.

**Borkowski, T. V. 1975.** Variability among Caribbean Littorinidae. *Veliger* **17:** 369–378.

**Borkowski, T. V. & Borkowski, M. R. 1969.** The *Littorina ziczac* species complex. *Veliger* **11:** 408–414.

**Boschma, H. 1948.** Thread spinning in *Littorina scabra*. *Proceedings of the Malacological Society. London* **27:** 223.

**Brandt, R. A. M. 1974.** The non-marine aquatic Mollusca of Thailand. *Archiv für Molluskenkunde* **105:** 1–423.

**Breton, J. le 1970.** Evolution et chute du pénis, étude de l'influence du jeûne, chez *Littorina littorea* L., mollusque, Gastropode, Prosobranche. *Compte Rendu Hebdomadaire des Séances de l'Académie des Sciences. Paris. Sér. D* **271:** 534–536.

**Buckland-Nicks, J. A. 1973.** The fine structure of the spermatozoon of *Littorina* (Gastropoda: Prosobranchia), with special reference to sperm motility. *Zeitschrift für Zellforschung und Mikroskopische Anatomie* **144:** 11–29.

**Buckland-Nicks, J. A. & Chia, F.-S. 1977.** On the nurse cell and the spermatozeugma in *Littorina sitkana*. *Cell and Tissue Research* **179:** 347–356.

**Buckland-Nicks, J. A., Chia, F.-S. & Behrens, S. 1973.** Oviposition and development of two intertidal snails, *Littorina sitkana* and *Littorina scutulata*. *Canadian Journal of Zoology* **51:** 359–365.

**Cason, J. E. 1950.** A rapid one-step Mallory Heidenhain stain for connective tissue. *Stain Technology* **25:** 225–226.

**Casto de Elera, R. P. 1896.** *Catálogo Sistemático de toda la Fauna de Filipinas* **3.** *Moluscos y Radiados*. Manila.

**Caugant, D. & Bergerard, J. 1980.** The sexual cycle and reproductive modality in *Littorina saxatilis* Olivi (Mollusca: Gastropoda). *Veliger* **23:** 107–112.

**Cernohorsky, W. O. 1972.** *Marine Shells of the Pacific* **2.** Sydney.

**Chapman, V. J. 1976.** *Mangrove Vegetation*. Vaduz.

**Chemnitz, J. H. 1795.** *Neues systematisches Conchylien-Cabinet* **11.** Nurnberg.

**Cook, L. M. 1983.** Polymorphism in a mangrove snail in Papua New Guinea. *Biological Journal of the Linnean Society of London* **20:** 167–173.

**Cossmann, M. 1916.** *Essais de Paléoconchologie Comparée* **10.** Paris.

**Daguzan, J. 1977.** Analyse biometrique du dimorphisme sexuel chez quelques Littorinidae (Mollusques, Gastéropodes, Prosobranches). *Haliotis* **6:** 17–40.

**Dance, S. P. 1967.** Report on the Linnaean shell collection. *Proceedings of the Linnean Society of London* **178:** 1–24.

**Dautzenberg, P. 1923.** Liste Préliminaire des Mollusques marins de Madagascar et description de deux espèces nouvelles. *Journal de Conchyliologie. Paris* **68:** 21–74.

**Dautzenberg, P. 1929.** Mollusques testacés marins de Madagascar. *Faune des Colonies françaises* **3.** Paris.

**Dautzenberg, P. & Fischer, H. 1905.** Liste des mollusques récoltés par M. le Capitaine de Frégate Blaise au Tonkin, et description d'espèces nouvelles. *Journal de Conchyliologie. Paris* **53:** 85–234.

**Dautzenberg, P. & Fischer, H. 1912.** Mollusques provenant des campagnes de l' "Hirondelle" et de la "Princesse Alice" dans les Mers du Nord. *Résultats des Campagnes Scientifiques accomplies par le Prince Albert I* **37:** 1–629.

**Deshayes, G. P. & Milne Edwards, H. 1843.** *Histoire Naturelle des Animaux sans Vertèbres par J. B. P. A. de Lamarck.* Ed. 2. **9.** Paris, London.

**Doutch, H. F. 1972.** The paleogeography of northern Australia and New Guinea and its relevance to the Torres Strait area. *In* D. Walker, *Bridge and Barrier: The Natural and Cultural History of Torres Strait*: 1–10. Canberra.

**Dunker, G. 1871.** Mollusca nova Musei Godeffroy Hamburgensis. *Malakozoologische Blätter* **18:** 150–175.

**Ekman, S. 1953.** *Zoogeography of the Sea.* London.

**Elner, R. W. & Raffaelli, D. G. 1980.** Interactions between two marine snails, *Littorina rudis* Maton and *Littorina nigrolineata* Gray, a predator, *Carcinus maenas* (L.) and a parasite, *Microphallus similis* Jagorskidd. *Journal of Experimental Marine Biology and Ecology* **43:** 151–160.

**Emerson, W. K. 1967.** Indo-Pacific faunal elements in the tropical eastern Pacific, with special reference to the mollusks. *Venus* **25:** 85–93.

**Emson, R. H. & Faller-Fritsch, R. J. 1976.** An experimental investigation into the effect of crevice availability on abundance and size-structure in a population of *Littorina rudis* (Maton): Gastropoda: Prosobranchia. *Journal of Experimental Marine Biology and Ecology* **23:** 285–297.

**Fischer, P. 1887.** *Manuel de Conchyliologie* **1.** Paris.

**Fischer, P. 1891.** Catalogue et distribution géographique des mollusques terrestres, fluviatiles et marins d'une partie de l'Indo-Chine. *Bulletin. Société d'Histoire Naturelle d'Autun* **4:** 87–276.

**Fischer, P. H. 1970.** Gastéropodes testacés marins du Golfe de Siam. *Journal de Conchyliologie. Paris* **108:** 93–121.

**Fischer-Piette, E. & Gaillard, J. M. 1971.** La variabilité (morphologique et physiologique) des *Littorina saxatilis* (Olivi) ibériques et ses rapport avec l'écologie. *Mémoires. Muséum National d'Histoire Naturelle. Paris. Sér. A. Zoologie* **70:** 1–90.

**Fish, J. D. & Fish, S. 1977.** The veliger larva of *Hydrobia ulvae* with observations on the veliger of *Littorina littorea* (Mollusca: Prosobranchia). *Journal of Zoology. London* **182:** 495–504.

**Ford, E. B. 1945.** Polymorphism. *Biological Reviews of the Cambridge Philosophical Society* **20:** 73–88.

**Fretter, V. 1980.** Observations on the gross anatomy of the female genital duct of British *Littorina* spp. *Journal of Molluscan Studies* **46:** 148–153.

**Fretter, V. & Graham, A. 1962.** *British Prosobranch Molluscs. Their Functional Anatomy and Ecology.* London.

**Fretter, V. & Manly, R. 1977.** The settlement and early benthic life of *Littorina neritoides* (L.) at Wembury, S. Devon. *Journal of Molluscan Studies* **43:** 255–262.

**Frith, D. W., Tantanasiriwong, R. & Bhatia, O. 1976.** Zonation of macrofauna on a mangrove shore, Phuket Island. *Research Bulletin of the Phuket Marine Biological Center* **10:** 1–37.

**Gaines, M. S., Caldwell, J. & Vivas, A. M. 1974.** Genetic variation in the mangrove periwinkle *Littorina angulifera. Marine Biology. Berlin* **27:** 327–332.

**Gallagher, S. B. & Reid, G. K. 1974.** Reproductive behaviour and early development in *Littorina scabra angulifera* and *Littorina irrorata* (Gastropoda: Prosobranchia) in the Tampa Bay region of Florida. *Malacological Review* **7:** 105–125.

**Gallagher, S. B. & Reid, G. K. 1979.** Population dynamics and zonation of the periwinkle snail, *Littorina angulifera*, of the Tampa Bay, Florida, Region. *Nautilus* **94:** 162–178.

**Gibson, D. G. 1964.** Mating behaviour in *Littorina planaxis* Philippi (Gastropoda: Prosobranchiata). *Veliger* 7: 134–139.

**Gmelin, J. F. 1791.** *Systema Naturae* **1**. Ed. 13. Holmiae.

**Goodwin, B. J. 1979.** The egg mass of *Littorina obtusata* and *Lacuna pallidula* (Gastropoda: Prosobranchia). *Journal of Molluscan Studies* **45**: 1–11.

**Goodwin, B. J. & Fish, J. D. 1977.** Inter- and intraspecific variation in *Littorina obtusata* and *L. mariae* (Gastropoda: Prosobranchia). *Journal of Molluscan Studies* **43**: 241–254.

**Grahame, J. 1969.** Shedding of the penis in *Littorina littorea*. *Nature. London* **221**: 976.

**Grahame, J. 1973.** Breeding energetics of *Littorina littorea* (L.) (Gastropoda: Prosobranchiata). *Journal of Animal Ecology* **42**: 391–404.

**Graus, R. R. 1974.** Latitudinal trends in the shell characteristics of marine gastropods. *Lethaia* **7**: 303–314.

**Gray, J. E. 1839.** Molluscous animals, and their shells. *In* The Admiralty, *The Zoology of Captain Beechey's Voyage ... performed in His Majesty's ship Blossom*: 103–155. London.

**Griffith, E. & Pidgeon, E. 1834.** *The Animal Kingdom by the Baron Cuvier* **12**. *The Mollusca and Radiata*. London.

**Guyomarc'h-Cousin, C. 1976.** Organogenèse descriptive de l'appareil génital chez *Littorina saxatilis* (Olivi), gastéropode prosobranche. *Bulletin de la Société Zoologique de France* **101**: 465–476.

**Habe, T. 1964.** *Shells of the Western Pacific in Colour* **2**. Osaka.

**Habe, T. & Kosuge, S. 1966.** *Shells of the World in Colour* **2**. *The Tropical Pacific*. Osaka.

**Hamilton, P. V. 1978.** Intertidal distribution and long-term movements of *Littorina irrorata* (Mollusca: Gastropoda). *Marine Biology. Berlin* **46**: 49–58.

**Hanley, S. C. T. 1855.** *Ipsa Linnaei Conchylia*. London.

**Hanley, S. C. T. 1860.** On the Linnaean manuscript of the 'Museum Ulricae'. *Journal of the Proceedings of the Linnean Society. London* **4**: 43–90.

**Hannaford Ellis, C. J. 1979.** Morphology of the oviparous rough winkle *Littorina arcana* Hannaford Ellis, 1978, with notes on the taxonomy of the *L. saxatilis* species-complex (Prosobranchia: Littorinidae). *Journal of Conchology. London* **30**: 43–56.

**Hannaford Ellis, C. J. 1983.** Patterns of reproduction in four *Littorina* species. *Journal of Molluscan Studies* **49**: 98–106.

**Hedley, C. 1916.** A preliminary index of the Mollusca of Western Australia. *Journal of the Royal Society of Western Australia* **1**: 3–77.

**Hedley, C. 1918*a*.** A check-list of the marine fauna of New South Wales. Part I. Mollusca. *Journal of the Proceedings of the Royal Society of New South Wales* **51**(suppl.): M1–M120.

**Hedley, C. 1918*b*.** Mollusca. *In* H. Basedow, Narrative of an expedition in North-Western Australia. *Proceedings of the Royal Geographical Society of Australasia, South Australian Branch* **18**: 263–283.

**Heller, J. 1975*a*.** The taxonomy of some British *Littorina* species with notes on their reproduction (Mollusca: Prosobranchia). *Zoological Journal of the Linnean Society of London* **56**: 131–151.

**Heller, J. 1975*b*.** Visual selection of shell colour in two littoral prosobranchs. *Zoological Journal of the Linnean Society of London* **56**: 153–170.

**Heller, J. 1976.** The effects of exposure and predation on the shell of two British winkles. *Journal of Zoology. London* **179**: 201–213.

**Hennig, W. 1966.** *Phylogenetic Systematics*. Urbana, Illinois.

**Heude, R. P. 1885.** Notes sur les mollusques terrestres de la Vallée du Fleuve Bleu. *Mémoires concernant l'Histoire Naturelle de l'empire Chinois* **1**, cah. 3.

**Hickman, C. S. 1977.** Integration of electron scan and light imagery in study of molluscan radulae. *Veliger* **20**: 1–8.

**Hidalgo, J. G. 1904–5.** *Catálogo de los Moluscos Testáceos de las Islas Filipinas, Joló y Marianas*. Madrid.

**Higo, S. 1973.** *A Catalogue of Molluscan Fauna of the Japanese Islands and the Adjacent Area*.

**Hirase, S. 1934.** *A Collection of Japanese Shells*. Tokyo.

**Hirase, S. & Taki, I.** [undated] *An Illustrated Handbook of Shells in Natural Colours from the Japanese Islands and Adjacent Territory*. Tokyo.

**Hughes, R. N. 1979*a*.** On the taxonomy of *Littorina africana* (Mollusca: Gastropoda). *Zoological Journal of the Linnean Society of London* **65**: 111–118.

**Hughes, R. N. 1979*b*.** South African populations of *Littorina rudis*. *Zoological Journal of the Linnean Society of London* **65:** 119–126.

**Iredale, T. & McMichael, D. F. 1962.** A reference list of the marine Mollusca of New South Wales. *Memoirs of the Australian Museum* **11:** 1–109.

**Issel, A. 1869.** *Malacologia del Mar Rosso*. Pisa.

**Jablonski, D. & Lutz, R. A. 1983.** Larval ecology of marine benthic invertebrates: paleobiological implications. *Biological Reviews of the Cambridge Philosophical Society* **58:** 21–89.

**James, B. L. 1968.** The characters and distribution of the subspecies and varieties of *Littorina saxatilis* (Olivi, 1792) in Britain. *Cahiers de Biologie Marine* **9:** 143–165.

**Janson, K. 1982*a*.** Genetic and environmental effects on the growth rate of *Littorina saxatilis*. *Marine Biology. Berlin* **69:** 73–78.

**Janson, K. 1982*b*.** Phenotypic differentiation in *Littorina saxatilis* Olivi (Mollusca, Prosobranchia) in a small area on the Swedish West coast. *Journal of Molluscan Studies* **48:** 167–173.

**Johansson, J. 1939.** Anatomische Studien über die Gastropodenfamilien Rissoidae und Littorinidae. *Zoologiska Bidrag fran Uppsala* **18:** 287–396.

**Jones, J. S., Leith, B. H. & Rawlings, R. 1977.** Polymorphism in *Cepaea*: a problem with too many solutions? *Annual Review of Ecology and Systematics* **8:** 109–143.

**Jones, M. L. 1972.** Comparisons of electrophoretic patterns of littorine snails of Panama: an attempt to define geminate species. *XVIII Congrès International de Zoologie. Monte Carlo*. Theme **3:** 1–10.

**Jones, W. T. 1971.** The field identification and distribution of mangroves in Eastern Australia. *Queensland Naturalist* **20:** 35–51.

**Jordan, J. & Ramorino, L. 1975.** Reproduccion de *Littorina* (*Austrolittorina*) *peruviana* (Lamarck, 1822) y *Littorina* (*Austrolittorina*) *araucana* Orbigny, 1840. *Revista de Biologia Marina. Valparaiso* **15:** 227–261.

**Kay, E. A. 1979.** *Hawaiian Marine Shells. Reef and Shore Fauna of Hawaii Section 4: Mollusca*. Honolulu.

**Kay, E. A. 1984.** Patterns of speciation in the Indo-West Pacific. *In* F. J. Radovsky, P. H. Raven & S. H. Sohmer, *Biogeography of the Tropical Pacific: Proceedings of a Symposium*: 15–31. Honolulu.

**Keen, A. M. 1971.** *Sea Shells of Tropical West America*. Ed. 2. Palo Alto, California.

**Kensley, B. 1973.** *Sea-Shells of Southern Africa. Gastropods*. Cape Town.

**Kilburn, R. N. 1972.** Taxonomic notes on South African marine Mollusca (2), with the description of new species and subspecies of *Conus*, *Nassarius*, *Vexillum* and *Demoulia*. *Annals of the Natal Museum* **21:** 391–437.

**Kilburn, R. & Rippey, E. 1982.** *Sea Shells of Southern Africa*. Johannesburg.

**Knorr, G. W. 1768.** *Vergnügen der Augen und des Gemüths, in Vorstellung einer allgemeinen Sammlung von Schnecken und Muscheln* **3**. Nurnberg.

**Kojima, Y. 1957.** On the breeding of a periwinkle, *Littorivaga brevicula* (Philippi). *Bulletin of the Biological Station of Asamushi* **8:** 59–62.

**Kojima, Y. 1958*a*.** On the breeding of a periwinkle, *Littorivaga atkana* (Dall). *Bulletin of the Biological Station of Asamushi* **9:** 35–37.

**Kojima, Y. 1958*b*.** A new type of the egg capsule of a periwinkle, *Littorina squalida* Broderip *et* Sowerby. *Bulletin of the Biological Station of Asamushi* **9:** 39–41.

**Kojima, Y. 1958*c*.** On the planktonic egg capsules of *Littorivaga mandschurica* (Schrenk) and *Littoraria strigata* (Lischke). *Venus* **20:** 81–86.

**Krauss, F. 1848.** *Die Südafrikanischen Mollusken*. Stuttgart.

**Kuroda, T. & Habe, T. 1952.** *Check List and Bibliography of the Recent Marine Mollusca of Japan*. Tokyo.

**Lamarck, J. B. P. A. de 1822.** *Histoire Naturelle des Animaux sans Vertèbres* **7**. Paris.

**Lear, R. & Turner, T. 1977.** *Mangroves of Australia*. St. Lucia, Qld.

**Lebour, M. V. 1935.** The breeding of *Littorina neritoides*. *Journal of the Marine Biological Association of the United Kingdom* **20:** 373–378.

**Lebour, M. V. 1945.** Eggs and larvae of some prosobranchs from Bermuda. *Proceedings of the Zoological Society of London* **114:** 462–489.

**Legeckis, R. & Cresswell, G. 1981.** Satellite observations of sea-surface temperature fronts off the coast of western and southern Australia. *Deep Sea Research* **28A:** 297–306.

**Leidy, J. 1845.** Anatomical description of the animal of *Littorina angulifera*, Lam. *Boston Journal of Natural History* **5:** 344–347.

**Lenderking, R.E. 1952.** Observations on *Littorina angulifera* Lam. from Biscayne, Florida. *Quarterly Journal. Florida Academy of Sciences* **14:** 247–250.

**Lenderking, R.E. 1954.** Some recent observations on the biology of *Littorina angulifera* Lam. of Biscayne and Virginia Keys, Florida. *Bulletin of Marine Science of the Gulf and Caribbean* **3:** 273–296.

**Lesson, M. 1831.** *Voyage autour du Monde, exécuté par ordre du Roi, sur la corvette de Sa Majesté, La Coquille, pendant les années 1822, 1823, 1824 et 1825. Zoologie* **2**(1). Paris.

**Linke, O. 1933.** Morphologie und Physiologie des Genitalapparatus der Nordsee Littorinen. *Wissenschaftliche Meeresuntersuchungen der Kommission zur Wissenschaften Untersuchung der Deutschen Meere. Abteilung Helgoland* **19**(5): 1–60.

**Linnaeus, C. 1758.** *Systema Naturae per Regna tria Naturae* **1.** *Regnum Animale.* Ed. 10. Holmiae.

**Linnaeus, C. 1764.** *Museum Ludovicae Ulricae Reginae.* Holmiae.

**Linnaeus, C. 1767.** *Systema Naturae per Regna tria Naturae* **1.** *Regnum Animale.* Ed. 12. Holmiae.

**Lischke, C.E. 1871*a*.** Diagnosen neuer Meeres-Conchylien von Japan. *Malakozoologische Blätter* **18:** 147–150.

**Lischke, C.E. 1871*b*.** Japanische Meeres-Conchylien **2.** *Novitates Conchologicae* Suppl. 4.

**Lysaght, A.M. 1941.** The biology and trematode parasites of the gastropod *Littorina neritoides* (L.) on the Plymouth breakwater. *Journal of the Marine Biological Association of the United Kingdom* **25:** 41–67.

**McCoy, E.D. & Heck, K.L. 1976.** Biogeography of corals, seagrasses and mangroves: an alternative to the centre of origin concept. *Systematic Zoology* **25:** 201–210.

**Macnae, W. 1968.** A general account of the fauna and flora of mangrove swamps and forests in the Indo-West-Pacific region. *Advances in Marine Biology* **6:** 74–270.

**McQuaid, C.D. 1981.** The establishment and maintenance of vertical size gradients in populations of *Littorina africana knysnaensis* (Philippi) on an exposed rocky shore. *Journal of Experimental Marine Biology and Ecology* **54:** 77–89.

**Marcus, E. & Marcus, E. 1963.** Mesogastropoden von der Kuste Säo Paulos. *Abhandlungen der Mathematisch-Naturwissenschaftlichen Klasse. Akademie der Wissenschaften und der Literatur. Mainz* 1963(1): 1–105.

**Martens, E. von 1871.** *In* E. von Martens & B. Langkavel, *Donum Bismarkianum. Eine Sammlung von Südsee-Conchylien.* Berlin.

**Martens, E. von 1880.** Mollusken. *In* K. Möbius, *Beiträge zur Meeresfauna der Insel Mauritius und der Seychellen:* 181–352. Berlin.

**Martens, E. von 1887.** List of the shells of Mergui and its Archipelago, collected for the trustees of the Indian Museum, Calcutta, by Dr. John Anderson, F.R.S., Superintendent of the Museum. *Journal of the Linnean Society of London. Zoology* **21:** 155–219.

**Martens, E. von 1897.** Süss- und Brackwasser-Mollusken das Indischen Archipels. *In* M. Weber, *Zoologische Ergebnisse einer Reise in Niederlandisch Ost-Indien* **4.** Leiden.

**Martens, E. von 1900.** Land and freshwater Mollusca. *In* F.D. Godman & O. Salvin, *Biologia Centrali-Americana*, livr. 60–76: 473–608. London.

**Mayr, E. 1969.** *Principles of Systematic Zoology.* New York.

**Melvill, J.C. & Standen, R. 1901.** The Mollusca of the Persian Gulf, Gulf of Oman, and Arabian Sea, as evidenced mainly through the collections of Mr. F.W. Townsend, 1893–1900; with descriptions of new species. Part I—Cephalopoda, Gastropoda, Scaphopoda. *Proceedings of the Zoological Society of London* 1901: 327–460.

**Melville, R.V. 1980.** Opinion 1159. *Littorina* (Mollusca, Gastropoda): author and date of this generic name, and type species of this nominal genus determined by use of the plenary powers. *Bulletin of Zoological Nomenclature* **37:** 103–106.

**Menke, C.T. 1830.** *Synopsis Methodica Molluscorum.* Ed. 2. Pyrmonti.

**Metcalfe, W. 1852.** An enumeration of species of recent shells, received by W.J. Hamilton, Esq., from Borneo, in November 1850, with descriptions of the new species. *Proceedings of the Zoological Society of London* 1851: 70–74.

**Mienis, H.K. 1973.** Notes on a small collection of Littorinidae from Somalia. *Basteria* **37:** 57–62.

**Mitchell, T.L. 1838.** *Three Expeditions into the Interior of Eastern Australia* **1.** London.

**Moore, H.B. 1937.** The biology of *Littorina littorea* (L.). Part 1. Growth of the shell and tissues, spawning, length of life and mortality. *Journal of the Marine Biological Association of the United Kingdom* **21**: 721–742.

**Mörch, O. A. L. 1876.** Synopsis molluscorum marinorum Indiarum occidentalium (contin.). *Malakozoologische Blätter* **23**: 87–143.

**Mowry, R. W. 1956.** Alcian blue techniques for the histochemical study of acidic carbohydrates. *Journal of Histochemistry and Cytochemistry* **4**: 409–410.

**Muggeridge, P. L. 1979.** *The Reproductive Biology of the Mangrove Littorinids* Bembicium auratum *(Quoy and Gaimard) and* Littorina scabra scabra *(Linné) (Gastropoda, Prosobranchiata), with Observations on the Reproductive Cycles of Rocky Shore Littorinids of New South Wales.* PhD Thesis, University of Sydney.

**Murray, T. 1979.** Evidence for an additional *Littorina* species and a summary of the reproductive biology of *Littorina* from California. *Veliger* **21**: 469–474.

**Naylor, R. & Begon, M. 1982.** Variation within and between populations of *Littorina nigrolineata* Gray on Holy Island, Anglesey. *Journal of Conchology. London* **31**: 17–30.

**Nevill, G. 1885.** *Hand List of Mollusca in the Indian Museum, Calcutta. Part 2.* Calcutta.

**Newkirk, G. F. & Doyle, R. W. 1975.** Genetic analysis of shell-shape variations in *Littorina saxatilis* on an environmental cline. *Marine Biology. Berlin* **30**: 227–237.

**Nielsen, C. 1976.** Notes on *Littorina* and *Murex* from the mangroves at Ao Nam-Bor, Phuket, Thailand. *Research Bulletin of the Phuket Marine Biological Center* **11**: 1–4.

**North, W. J. 1954.** Size distribution, erosive activities, and gross metabolic efficiency of the marine intertidal snails, *Littorina planaxis* and *L. scutulata*. *Biological Bulletin. Marine Biological Laboratory Woods Hole* **106**: 185–197.

**Odhner, N. 1953.** *Identification of the Linnean shells in Museum Ludovicae Ulricae.* (Mimeo.).

**Oostingh, C. H. 1923.** Recent shells from Java. Part 1. Gastropoda. *Mededeelingen van de Landbouwhoogeschool te Wageningen* **26**(3): 1–174.

**Oostingh, C. H. 1927.** Littorinidae and Naticidae from North East Sumatra. *Miscellanea Zoologica Sumatrana* **15**: 1–5.

**Orbigny, A. d' 1842.** Mollusques. *In* M. Ramon de la Sagra, *Histoire Physique, Politique et Naturelle de l'Ile de Cuba* **1**. Paris.

**Oyama, K. & Takemura, Y. 1961.** *The Molluscan Shells* **5**. Tokyo.

**Palant, B. & Fishelson, L. 1968.** *Littorina punctata* (Gmelin) and *Littorina neritoides* (L.), (Mollusca, Gastropoda) from Israel: ecology and annual cycle of genital system. *Israel Journal of Zoology* **17**: 145–160.

**Pease, W. H. 1868.** Descriptions of marine gasteropodae, inhabiting Polynesia. *American Journal of Conchology* **4**: 71–80, 91–132.

**Peile, A. J. 1937.** Some radula problems. *Journal of Conchology. London* **20**: 292–304.

**Pelseneer, P. 1926.** La proportion relative des sexes chez les animaux et particulièrement chez les mollusques. *Mémoires. Académie Royale de Belgique. Classes des Sciences. Collection in 4°* **8**: 1–258.

**Percival, M. & Womersley, J. S. 1975.** *Floristics and Ecology of the Mangrove Vegetation of Papua New Guinea. National Herbarium Botany Bulletin* **8**. Lae.

**Pettitt, C. W. 1974*a*.** An indexed bibliography of the family Littorinidae (Gastropoda: Mollusca) 1758–1973. *Manchester Museum Publications, N.S.* No. NS 4.74.

**Pettitt, C. W. 1974*b*.** An indexed bibliography of the family Littorinidae (Gastropoda: Mollusca) 1758–1973. First supplement: corrections and additions. (Duplicated).

**Pettitt, C. W. 1979.** Bibliography of Littorinidae (Supplement). *The Littorinid Tidings* 8(suppl.). *Manchester Museum Computer Produced Publication No. 4.*

**Philippi, R. A. 1846.** Descriptions of a new species of *Trochus*, and of eighteen new species of *Littorina*, in the collection of H. Cuming, Esq. *Proceedings of the Zoological Society of London* 1845: 138–143.

**Philippi, R. A. 1847–8.** *Abbildungen und Beschreibungen neuer oder wenig gekannter Conchylien* **2** and **3**. Cassel.

**Phillips, B. F., Brown, P. A., Rimmer, D. W. and Reid, D. D. 1979.** Distribution and dispersal of the phyllosoma larvae of the Western Rock Lobster, *Panulirus cygnus*, in the South-eastern Indian Ocean. *Australian Journal of Marine and Freshwater Research* **30**: 773–783.

**Picken, G. B. 1979.** Non-pelagic reproduction of some Antarctic prosobranch gastropods from Signy Island, South Orkney Islands. *Malacologia* **19**: 109–128.

**Pilkington, M. C. 1971.** Eggs, larvae and spawning in *Melaraphe cincta* (Quoy & Gaimard) and *M. oliveri* Finlay (Littorinidae, Gastropoda). *Australian Journal of Marine and Freshwater Research* **22:** 79–90.

**Pilkington, M. C. 1974.** The eggs and hatching stages of some New Zealand prosobranch molluscs. *Journal of the Royal Society of New Zealand* **4:** 411–432.

**Pilsbry, H. A. 1895.** *Catalogue of the Marine Mollusca of Japan.* Detroit.

**Ponder, W. F. 1966.** The New Zealand species previously known as *Zelaxitas* Finlay, 1927 (Mollusca, Gastropoda). *Records of the Dominion Museum. Wellington* **5:** 163–176.

**Ponder, W. F. 1968.** The morphology of some small New Zealand prosobranchs. *Records of the Dominion Museum. Wellington* **6:** 61–95.

**Ponder, W. F. 1976.** Three species of Littorinidae from Southern Australia. *Malacological Review* **9:** 105–114.

**Ponder, W. F. & Rosewater, J. 1979.** Rectifications in the nomenclature of some Indo-Pacific Littorinidae. *Proceedings of the Biological Society of Washington* **92:** 773–782.

**Prashad, B. 1921.** Report on a collection of Sumatran molluscs from fresh and brackish water. *Records of the Indian Museum* **22:** 461–507.

**Prashad, B. 1925.** Respiration of gastropod molluscs. *Proceedings of the Indian Science Congress* **12:** 126–143.

**Quoy, J. R. C. & Gaimard, J. P. 1833.** *Voyage de Découvertes de l'Astrolabe. Zoologie* **2**(2). Paris.

**Raffaelli, D. G. 1977.** Observations on the copulatory behaviour of *Littorina rudis* (Maton) and *Littorina nigrolineata* (Gray). *Veliger* **20:** 75–77.

**Raffaelli, D. G. 1979.** The taxonomy of the *Littorina saxatilis* species-complex, with particular regard to the systematic position of *Littorina patula* Jeffreys. *Zoological Journal of the Linnean Society of London* **65:** 219–232.

**Raffaelli, D. G. 1982.** Recent ecological research on some European species of *Littorina. Journal of Molluscan Studies* **48:** 342–354.

**Raffaelli, D. G. & Hughes, R. N. 1978.** The effects of crevice size and availability on populations of *Littorina rudis* and *Littorina neritoides. Journal of Animal Ecology* **47:** 71–83.

**Raup, D. M. 1966.** Geometrical analysis of shell coiling: general problems. *Journal of Paleontology* **40:** 1178–1190.

**Reeve, L. A. 1857–8.** *Conchologia Iconica* **10.** *Monograph of the genus* Littorina. Plates 1–16 (1857), 17–18 (1858). London.

**Reimchen, T. E. 1979.** Substrate heterogeneity, crypsis, and colour polymorphism in an intertidal snail (*Littorina mariae*). *Canadian Journal of Zoology* **57:** 1070–1085.

**Reimchen, T. E. 1981.** Microgeographical variation in *Littorina mariae* Sacchi & Rastelli and a taxonomic consideration. *Journal of Conchology. London* **30:** 341–350.

**Reimchen, T. E. 1982.** Shell size divergence in *Littorina mariae* and *L. obtusata* and predation by crabs. *Canadian Journal of Zoology* **60:** 687–695.

**Reinke, E. E. 1912.** A preliminary account of the development of the apyrene spermatozoa in *Strombus* and of the nurse cells in *Littorina. Biological Bulletin. Marine Biological Laboratory Woods Hole* **22:** 319–327.

**Remmert, H. 1969.** Die *Littorina*-arten: Kein Modell für die Entstehung der Landschnecken. *Oecologia. Berlin* **2:** 1–6.

**Risbec, J. 1942.** Recherches anatomiques sur les prosobranches de Nouvelle-Calédonie, pt. 3. *Annales des Sciences Naturelles. Paris. Zoologie et Biologie Animale* **4:** 57–64.

**Robertson, R. 1974.** Marine prosobranch gastropods: larval studies and systematics. *Thalassia Jugoslavica* **10:** 213–238.

**Rosen, B. R. 1981.** The tropical high diversity enigma—the corals'-eye view. *In* P. L. Forey, *The Evolving Biosphere*: 103–129. Cambridge.

**Rosewater, J. 1963.** Problems of species analogues in world Littorinidae. *Report. American Malacological Union and A.M.U. Pacific Division* **30:** 5–6.

**Rosewater, J. 1966.** Reinstatement of *Melarhaphe* Menke, 1828. *Nautilus* **80:** 37–38.

**Rosewater, J. 1970.** The family Littorinidae in the Indo-Pacific. Part I. The subfamily Littorininae. *Indo-Pacific Mollusca* **2:** 417–506.

**Rosewater, J. 1972.** The family Littorinidae in the Indo-Pacific. Part II. The subfamilies Tectariinae and Echinininae. *Indo-Pacific Mollusca* **2:** 507–533.

**Rosewater, J. 1980*a*.** A close look at *Littorina* radulae. *Bulletin of the American Malacological Union* 1979: 5–8.

**Rosewater, J. 1980*b*.** Subspecies of the gastropod *Littorina scabra*. *Nautilus* **94:** 158–162.

**Rosewater, J. 1981.** The family Littorinidae in tropical West Africa. *Atlantide Report. Copenhagen* **13:** 7–48.

**Rosewater, J. 1982.** A new species of the genus *Echininus* (Mollusca: Littorinidae: Echinininae) with a review of the subfamily. *Proceedings of the Biological Society of Washington* **95:** 67–80.

**Rosewater, J. & Vermeij, G.J. 1972.** The amphi-Atlantic distribution of *Littorina meleagris*. *Nautilus* **86:** 67–69.

**Rumphius, G.E. 1705.** *D'Amboinische Rariteitkamer*. Amsterdam.

**Sacchi, C.F. 1968.** Sur le dimorphisme sexuel de *Littorina mariae* Sacchi et Rast. (Gastr., Prosobranchia). *Compte Rendu Hebdomadaire des Séances de l'Académie des Sciences. Paris. Sér. D* **266:** 2483–2485.

**Sacchi, C.F. 1975.** *Littorina nigrolineata* Gray (Gastropoda, Prosobranchia). *Cahiers de Biologie Marine* **16:** 111–120.

**Sacchi, C.F. & Rastelli, M. 1967.** *Littorina mariae*, nov. sp.: les differences morphologiques et écologiques, entre 'nains' et 'normaux' chez l' "espèce" *L. obtusata* (L.) (Gastr, Prosobr.) et leur signification adaptive et évolutive. *Atti della Società Italiana di Scienze Naturali e del Museo Civico di Storia Naturale. Milano* **105:** 351–370.

**Salvat, B. & Rives, C. 1975.** *Coquillages de Polynésie*. Papeete.

**Sasekumar, A. 1974.** Distribution of macrofauna on a Malayan mangrove shore. *Journal of Animal Ecology* **43:** 51–69.

**Scheltema, R. 1971.** Larval dispersal as a means of genetic exchange between geographically separated populations of shallow-water benthic marine gastropods. *Biological Bulletin. Marine Biological Laboratory Woods Hole* **140:** 284–322.

**Schepman, M.M. 1909.** The Prosobranchia of the Siboga Expedition. Part 2. Taenioglossa and Ptenoglossa. *Siboga-Expeditie* **49.** Leiden.

**Schmitt, R.J. 1979.** Mechanics and timing of egg capsule release by the littoral fringe periwinkle *Littorina planaxis* (Gastropoda: Prosobranchia). *Marine Biology. Berlin* **50:** 359–366.

**Schumacher, C.F. 1838.** *Essai d'un Nouveau Système des Habitations des Vers Testacés*. Copenhagen.

**Schuto, T. 1974.** Larval ecology of prosobranch gastropods and its bearing on biogeography and paleontology. *Lethaia* **7:** 239–256.

**Semeniuk, V., Kenneally, K.F. & Wilson, P.G. 1978.** *Mangroves of Western Australia. Western Australian Naturalists' Club Handbook* **12.** Perth.

**Sewell, R.B.S. 1924.** Observations on growth in certain molluscs and on changes correlated with growth in the radula of *Pyrazus palustris*. *Records of the Indian Museum* **26:** 529–548.

**Shikama, T. & Horikoshi, M. 1963.** *Selected Shells of the World Illustrated in Colours* **1.** Hokuryu-Kan.

**Smith, D.A.S. 1975.** Polymorphism and selective predation in *Donax faba* Gmelin (Bivalvia: Tellinacea). *Journal of Experimental Marine Biology and Ecology* **17:** 205–219.

**Smith, D.A.S. 1976.** Disruptive selection and morph-ratio clines in the polymorphic snail *Littorina obtusata* (L.) (Gastropoda: Prosobranchia). *Journal of Molluscan Studies* **42:** 114–135.

**Smith, E.A. 1906.** On South African marine Mollusca, with descriptions of new species. *Annals of the Natal Museum* **1:** 19–71.

**Smith, J.E. 1981.** The natural history and taxonomy of shell variation in the periwinkles *Littorina saxatilis* and *Littorina rudis*. *Journal of the Marine Biological Association of the United Kingdom* **61:** 215–241.

**Smith, S.M. 1982.** A review of the genus *Littorina* in British and Atlantic waters (Gastropoda: Prosobranchia). *Malacologia* **22:** 535–539.

**Solem, A. 1976.** *Endodontoid Land Snails from Pacific Islands (Mollusca: Pulmonata: Sigmurethra). Part I. Family Endodontidae*. Chicago.

**Sowerby, G.B. 1832.** *The Genera of Recent and Fossil Shells*. Part 37.

**Sowerby, G.B. 1892.** *Marine Shells of South Africa*. London.

**Springer, V.G. 1982.** Pacific plate biogeography, with special reference to shorefishes. *Smithsonian Contributions to Zoology* **367:** 1–182.

**Stehli, F. G., MacAlester, A. L. & Helsley, C. E. 1967.** Taxonomic diversity of Recent bivalves and some implications for geology. *Bulletin of the Geological Society of America* **78:** 455–465.

**Stehli, F. G. & Wells, J. W. 1971.** Diversity and age patterns in hermatypic corals. *Systematic Zoology* **20:** 115–126.

**Struhsaker, J. W. 1966.** Breeding, spawning, spawning periodicity and early development in the Hawaiian *Littorina: L. pintado* (Wood); *L. picta* Philippi and *L. scabra* (Linné). *Proceedings of the Malacological Society. London* **37:** 137–166.

**Struhsaker, J. W. 1968.** Selection mechanisms associated with intraspecific shell variation in *Littorina picta* (Prosobranchia: Mesogastropoda). *Evolution. Lancaster, Pa.* **22:** 459–480.

**Struhsaker, J. W. & Costlow, J. D. 1968.** Larval development of *Littorina picta* (Prosobranchia, Mesogastropoda) reared in the laboratory. *Proceedings of the Malacological Society. London* **38:** 153–160.

**Tadjalli-Pour, M. 1974.** *Contribution à l'Etude de la Systématique et de la Répartition des Mollusques des Côtes Iraniennes du Golfe Persique.* Doctoral Thesis: Université des Sciences et Techniques du Languedoc.

**Tapparone-Canefri, C. 1874.** Zoologia del Viaggio Intorno al Globo della Regia Fregata Magenta. Malacologia. *Memorie della Reale Accademia delle Scienze di Torino. Ser.2* **28:** 5–161.

**Tattersall, W. M. 1920.** Notes on breeding habits and life history of the periwinkle. *Scientific Investigations. Fisheries Branch, Department of Agriculture for Ireland* **1:** 1–11.

**Taylor, J. D. 1971*a*.** Reef associated molluscan assemblages in the western Indian Ocean. *Symposia of the Zoological Society of London* **28:** 501–534.

**Taylor, J. D. 1971*b*.** Intertidal zonation at Aldabra Atoll. *Philosophical Transactions of the Royal Society. London. Ser. B* **260:** 173–213.

**Thiele, J. 1929.** *Handbuch der Systematischen Weichtierkunde* **1.** Jena.

**Thiriot-Quiévreux, C. 1972.** Microstructures de coquilles larvaires de prosobranches au microscope électronique à balayage. *Archives de Zoologie Expérimentale et Générale. Paris* **113:** 553–564.

**Thiriot-Quiévreux, C. 1980.** Identification of some planktonic prosobranch larvae present off Beaufort, North Carolina. *Veliger* **23:** 1–9.

**Thiriot-Quiévreux, C. & Babio, C. R. 1975.** Etude des protoconques de quelques prosobranches de la région de Roscoff. *Cahiers de Biologie Marine* **16:** 135–148.

**Troschel, F. H. 1858.** *Das Gebiss der Schnecken* **1.** Berlin.

**Tryon, G. W. 1883.** *Structural and Systematic Conchology* **2.** Philadelphia.

**Tryon, G. W. 1887.** *Manual of Conchology* **9.** Philadelphia.

**Turton, W. H. 1932.** *The Marine Shells of Port Alfred, South Africa.* Oxford.

**Underwood, A. J. & McFadyen, K. E. 1983.** Ecology of the intertidal snail *Littorina acutispira* Smith. *Journal of Experimental Marine Biology and Ecology* **66:** 169–197.

**Vermeij, G. J. 1971.** Gastropod evolution and morphological diversity in relation to shell geometry. *Journal of Zoology. London* **163:** 15–23.

**Vermeij, G. J. 1972*a*.** Endemism and environment: some shore molluscs of the Tropical Atlantic. *American Naturalist* **106:** 89–101.

**Vermeij, G. J. 1972*b*.** Intraspecific shore-level size gradients in intertidal molluscs. *Ecology* **53:** 693–700.

**Vermeij, G. J. 1973*a*.** Morphological patterns in high intertidal gastropods: adaptive strategies and their limitations. *Marine Biology. Berlin* **20:** 319–346.

**Vermeij, G. J. 1973*b*.** Molluscs in mangrove swamps: physiognomy, diversity and regional differences. *Systematic Zoology* **22:** 609–624.

**Vermeij, G. J. 1978.** *Biogeography and Adaptation: Patterns of Marine Life.* Cambridge, Mass.

**Vermeij, G. J. 1980.** Gastropod shell growth rate, allometry, and adult size: environmental implications. *In* D. C. Rhoads & R. A. Lutz, *Skeletal Growth of Aquatic Organisms: Biological Records of Environmental Change:* 379–394. New York.

**Vermeij, G. J. & Covich, A. P. 1978.** Co-evolution of freshwater gastropods and their predators. *American Naturalist* **112:** 833–843.

**Viader, R. 1937.** Revised catalogue of the testaceous Mollusca of Mauritius and its dependencies. *Bulletin of the Mauritius Institute* **1:** i–xii, 1–111.

**Ward, R. D. & Warwick, T. 1980.** Genetic differentiation in the molluscan species *Littorina rudis* and *Littorina arcana* (Prosobranchia: Littorinidae). *Biological Journal of the Linnean Society of London* **14:** 417–428.

**Weinkauff, H. C. 1878, 1882.** Die Gattung *Litorina*. *In* H. C. Küster, *Systematisches Conchylien-Cabinet von Martini und Chemnitz* **2** (abt. 9, parts 269, 315, 318): 25–114. Nürnberg.
**Wenz, W. 1938.** Gastropoda. Teil 1: Allgemeiner Teil und Prosobranchia. *In* O. H. Schindewolf, *Handbuch der Paläozoologie* **6.** Berlin.
**Werner, H. J. 1950.** Observations on the external anatomy and morphology of the alimentary tract of *Littorina irrorata* (Say). *Journal of the Tennessee Academy of Science* **25:** 187–194.
**Werner, H. J. 1951.** Observations on the histology of the alimentary tract of *Littorina irrorata* (Say). *Journal of the Tennessee Academy of Science* **26:** 85–88.
**Wester, L. 1981.** Introduction and spread of mangroves in the Hawaiian Islands. *Association of Pacific Coast Geographers Yearbook* **43:** 125–137.
**Whipple, J. A. 1965.** Systematics of the Hawaiian *Littorina* Férussac (Mollusca: Gastropoda). *Veliger* **7:** 155–166.
**Whitelegge, T. 1889.** List of the marine and fresh-water invertebrate fauna of Port Jackson and the neighbourhood. *Transactions and Proceedings of the Royal Society of New South Wales* **23:** 163–323.
**Wiley, E. O. 1981.** *The Theory and Practice of Phylogenetic Systematics*. New York.
**Wilkins, N. P. & O'Regan, D. 1980.** Genetic variation in sympatric sibling species of *Littorina*. *Veliger* **22:** 355–359.
**Williams, E. E. 1964.** The growth and distribution of *Littorina littorea* (L.) on a rocky shore in Wales. *Journal of Animal Ecology* **33:** 413–432.
**Wilson, B. R. & Gillett, K. 1979.** *A Field Guide to Australian Shells. Prosobranch Gastropods*. Sydney.
**Woodard, T. M. 1942*a*.** Development of the nurse-cells of *Littorina irrorata* (Say). *Transactions of the American Microscopical Society* **61:** 361–372.
**Woodard, T. M. 1942*b*.** Behaviour of the nurse-cells of *Littorina irrorata* (Say). *Biological Bulletin. Marine Biological Laboratory Woods Hole* **82:** 461–466.
**Wright, C. A. 1966.** The pathogenesis of helminths in the Mollusca. *Helminthological Abstracts* **35:** 207–224.
**Yen, T.-C. 1933.** The molluskan fauna of Amoy and its vicinal regions. *Report. Marine Biological Association of China* **2:** 1–120.
**Yen, T.-C. 1942.** A review of Chinese gastropods in the British Museum. *Proceedings of the Malacological Society. London* **24:** 170–289.
**Yoo, J.-S. 1976.** *Korean Shells in Colour*. Seoul.
**Zar, J. H. 1974.** *Biostatistical Analysis*. Englewood Cliffs, N. J.
**Zipser, E. & Vermeij, G. J. 1978.** Crushing behaviour in tropical and temperate crabs. *Journal of Experimental Marine Biology and Ecology* **31:** 155–172.

# INDEX TO NAMES OF *LITTORARIA* SPECIES

Valid species appear in bold type.

Linnean Society
Burlington House
London WC
21.11.94
Sale £1.50